MODULES OVER VALUATION DOMAINS

PURE AND APPLIED MATHEMATICS

A Program of Monographs, Textbooks, and Lecture Notes

EXECUTIVE EDITORS

Earl J. Taft
Rutgers University
New Brunswick, New Jersey

Zuhair Nashed
University of Delaware
Newark, Delaware

CHAIRMEN OF THE EDITORIAL BOARD

S. Kobayashi
University of California, Berkeley
Berkeley, California

Edwin Hewitt
University of Washington
Seattle, Washington

EDITORIAL BOARD

M. S. Baouendi
Purdue University

Donald Passman
University of Wisconsin

Jack K. Hale
Brown University

Fred S. Roberts
Rutgers University

Marvin Marcus
University of California, Santa Barbara

Gian-Carlo Rota
Massachusetts Institute of Technology

W. S. Massey
Yale University

David Russell
University of Wisconsin-Madison

Leopoldo Nachbin
Centro Brasileiro de Pesquisas Físicas
and University of Rochester

Jane Cronin Scanlon
Rutgers University

Anil Nerode
Cornell University

Walter Schempp
Universität Siegen

Mark Teply
University of Florida

LECTURE NOTES

IN PURE AND APPLIED MATHEMATICS

1. *N. Jacobson*, Exceptional Lie Algebras
2. *L. -Å. Lindahl and F. Poulsen*, Thin Sets in Harmonic Analysis
3. *I. Satake*, Classification Theory of Semi-Simple Algebraic Groups
4. *F. Hirzebruch, W. D. Newmann, and S. S. Koh*, Differentiable Manifolds and Quadratic Forms (out of print)
5. *I. Chavel*, Riemannian Symmetric Spaces of Rank One (out of print)
6. *R. B. Burckel*, Characterization of C(X) Among Its Subalgebras
7. *B. R. McDonald, A. R. Magid, and K. C. Smith*, Ring Theory: Proceedings of the Oklahoma Conference
8. *Y.-T. Siu*, Techniques of Extension on Analytic Objects
9. *S. R. Caradus, W. E. Pfaffenberger, and B. Yood*, Calkin Algebras and Algebras of Operators on Banach Spaces
10. *E. O. Roxin, P.-T. Liu, and R. L. Sternberg*, Differential Games and Control Theory
11. *M. Orzech and C. Small*, The Brauer Group of Commutative Rings
12. *S. Thomeier*, Topology and Its Applications
13. *J. M. Lopez and K. A. Ross*, Sidon Sets
14. *W. W. Comfort and S. Negrepontis*, Continuous Pseudometrics
15. *K. McKennon and J. M. Robertson*, Locally Convex Spaces
16. *M. Carmeli and S. Malin*, Representations of the Rotation and Lorentz Groups: An Introduction
17. *G. B. Seligman*, Rational Methods in Lie Algebras
18. *D. G. de Figueiredo*, Functional Analysis: Proceedings of the Brazilian Mathematical Society Symposium
19. *L. Cesari, R. Kannan, and J. D. Schuur*, Nonlinear Functional Analysis and Differential Equations: Proceedings of the Michigan State University Conference
20. *J. J. Schäffer*, Geometry of Spheres in Normed Spaces
21. *K. Yano and M. Kon*, Anti-Invariant Submanifolds
22. *W. V. Vasconcelos*, The Rings of Dimension Two
23. *R. E. Chandler*, Hausdorff Compactifications
24. *S. P. Franklin and B. V. S. Thomas*, Topology: Proceedings of the Memphis State University Conference
25. *S. K. Jain*, Ring Theory: Proceedings of the Ohio University Conference
26. *B. R. McDonald and R. A. Morris*, Ring Theory II: Proceedings of the Second Oklahoma Conference
27. *R. B. Mura and A. Rhemtulla*, Orderable Groups
28. *J. R. Graef*, Stability of Dynamical Systems: Theory and Applications
29. *H.-C. Wang*, Homogeneous Branch Algebras
30. *E. O. Roxin, P.-T. Liu, and R. L. Sternberg*, Differential Games and Control Theory II
31. *R. D. Porter*, Introduction to Fibre Bundles
32. *M. Altman*, Contractors and Contractor Directions Theory and Applications
33. *J. S. Golan*, Decomposition and Dimension in Module Categories
34. *G. Fairweather*, Finite Element Galerkin Methods for Differential Equations
35. *J. D. Sally*, Numbers of Generators of Ideals in Local Rings
36. *S. S. Miller*, Complex Analysis: Proceedings of the S.U.N.Y. Brockport Conference
37. *R. Gordon*, Representation Theory of Algebras: Proceedings of the Philadelphia Conference
38. *M. Goto and F. D. Grosshans*, Semisimple Lie Algebras
39. *A. I. Arruda, N. C. A. da Costa, and R. Chuaqui*, Mathematical Logic: Proceedings of the First Brazilian Conference

40. F. Van Oystaeyen, Ring Theory: Proceedings of the 1977 Antwerp Conference
41. F. Van Oystaeyen and A. Verschoren, Reflectors and Localization: Application to Sheaf Theory
42. M. Satyanarayana, Positively Ordered Semigroups
43. D. L. Russell, Mathematics of Finite-Dimensional Control Systems
44. P.-T. Liu and E. Roxin, Differential Games and Control Theory III: Proceedings of the Third Kingston Conference, Part A
45. A. Geramita and J. Seberry, Orthogonal Designs: Quadratic Forms and Hadamard Matrices
46. J. Cigler, V. Losert, and P. Michor, Banach Modules and Functors on Categories of Banach Spaces
47. P.-T. Liu and J. G. Sutinen, Control Theory in Mathematical Economics: Proceedings of the Third Kingston Conference, Part B
48. C. Byrnes, Partial Differential Equations and Geometry
49. G. Klambauer, Problems and Propositions in Analysis
50. J. Knopfmacher, Analytic Arithmetic of Algebraic Function Fields
51. F. Van Oystaeyen, Ring Theory: Proceedings of the 1978 Antwerp Conference
52. B. Kedem, Binary Time Series
53. J. Barros-Neto and R. A. Artino, Hypoelliptic Boundary-Value Problems
54. R. L. Sternberg, A. J. Kalinowski, and J. S. Papadakis, Nonlinear Partial Differential Equations in Engineering and Applied Science
55. B. R. McDonald, Ring Theory and Algebra III: Proceedings of the Third Oklahoma Conference
56. J. S. Golan, Structure Sheaves over a Noncommutative Ring
57. T. V. Narayana, J. G. Williams, and R. M. Mathsen, Combinatorics, Representation Theory and Statistical Methods in Groups: YOUNG DAY Proceedings
58. T. A. Burton, Modeling and Differential Equations in Biology
59. K. H. Kim and F. W. Roush, Introduction to Mathematical Consensus Theory
60. J. Banas and K. Goebel, Measures of Noncompactness in Banach Spaces
61. O. A. Nielson, Direct Integral Theory
62. J. E. Smith, G. O. Kenny, and R. N. Ball, Ordered Groups: Proceedings of the Boise State Conference
63. J. Cronin, Mathematics of Cell Electrophysiology
64. J. W. Brewer, Power Series Over Commutative Rings
65. P. K. Kamthan and M. Gupta, Sequence Spaces and Series
66. T. G. McLaughlin, Regressive Sets and the Theory of Isols
67. T. L. Herdman, S. M. Rankin, III, and H. W. Stech, Integral and Functional Differential Equations
68. R. Draper, Commutative Algebra: Analytic Methods
69. W. G. McKay and J. Patera, Tables of Dimensions, Indices, and Branching Rules for Representations of Simple Lie Algebras
70. R. L. Devaney and Z. H. Nitecki, Classical Mechanics and Dynamical Systems
71. J. Van Geel, Places and Valuations in Noncommutative Ring Theory
72. C. Faith, Injective Modules and Injective Quotient Rings
73. A. Fiacco, Mathematical Programming with Data Perturbations I
74. P. Schultz, C. Praeger, and R. Sullivan, Algebraic Structures and Applications Proceedings of the First Western Australian Conference on Algebra
75. L. Bican, T. Kepka, and P. Nemec, Rings, Modules, and Preradicals
76. D. C. Kay and M. Breen, Convexity and Related Combinatorial Geometry: Proceedings of the Second University of Oklahoma Conference
77. P. Fletcher and W. F. Lindgren, Quasi-Uniform Spaces
78. C.-C. Yang, Factorization Theory of Meromorphic Functions
79. O. Taussky, Ternary Quadratic Forms and Norms
80. S. P. Singh and J. H. Burry, Nonlinear Analysis and Applications
81. K. B. Hannsgen, T. L. Herdman, H. W. Stech, and R. L. Wheeler, Volterra and Functional Differential Equations

82. *N. L. Johnson, M. J. Kallaher, and C. T. Long*, Finite Geometries: Proceedings of a Conference in Honor of T. G. Ostrom
83. *G. I. Zapata*, Functional Analysis, Holomorphy, and Approximation Theory
84. *S. Greco and G. Valla*, Commutative Algebra: Proceedings of the Trento Conference
85. *A. V. Fiacco*, Mathematical Programming with Data Perturbations II
86. *J.-B. Hiriart-Urruty, W. Oettli, and J. Stoer*, Optimization: Theory and Algorithms
87. *A. Figa Talamanca and M. A. Picardello*, Harmonic Analysis on Free Groups
88. *M. Harada*, Factor Categories with Applications to Direct Decomposition of Modules
89. *V. I. Istrătescu*, Strict Convexity and Complex Strict Convexity: Theory and Applications
90. *V. Lakshmikantham*, Trends in Theory and Practice of Nonlinear Differential Equations
91. *H. L. Manocha and J. B. Srivastava*, Algebra and Its Applications
92. *D. V. Chudnovsky and G. V. Chudnovsky*, Classical and Quantum Models and Arithmetic Problems
93. *J. W. Longley*, Least Squares Computations Using Orthogonalization Methods
94. *L. P. de Alcantara*, Mathematical Logic and Formal Systems
95. *C. E. Aull*, Rings of Continuous Functions
96. *L. Fuchs and L. Salce*, Modules Over Valuation Domains

Other Volumes in Preparation

MODULES OVER VALUATION DOMAINS

Laszlo Fuchs
Department of Mathematics
Tulane University
New Orleans, Louisiana

Luigi Salce
Università di Padova
Padova, Italy

MARCEL DEKKER, INC.   New York and Basel

Library of Congress Cataloging in Publication Data
Main entry under title:

Modules over valuation domains.

   (Lecture notes in pure and applied
mathematics ; v. 96)
   Bibliography: p.
   Includes indexes.
   1. Rings (Algebra)   2. Modules (Algebra)   3. Valuation
theory.   I. Fuchs, Laszlo. II. Salce, Luigi. III. Series.
IV. Title: Valuation rings.
QA251.3.M64 1985   512'.4   84-23763
ISBN 0-8247-7326-8

COPYRIGHT © 1985 by MARCEL DEKKER, INC.   ALL RIGHTS RESERVED

Neither this book nor any part may be reproduced or transmitted in
any form or by any means, electronic or mechanical, including photo-
copying, microfilming, and recording, or by any information storage
and retrieval system, without permission in writing from the pub-
lisher.

MARCEL DEKKER, INC.
270 Madison Avenue, New York, New York   10016

Current printing (last digit):
10 9 8 7 6 5 4 3 2 1

PRINTED IN THE UNITED STATES OF AMERICA

Preface

A great deal of research has been done recently on module theory, and there is now a great body of material available on modules of various kinds. However, it seems that one important aspect, namely, the structure theory of modules over a fixed ring, has not received due attention; it is something of a stepchild in today's algebraic investigations. This is in sharp contrast to abelian group theory (that is, the theory of modules over principal ideal domains) where the structure theory plays a dominating role. It is no secret that the classification of abelian groups has made remarkable advances in the past quarter of a century and reached a stage of logical clarity and exceptional effectiveness.

It is therefore reasonable to expect that the theory of abelian groups can provide us with models in order to venture in the as yet unexplored areas of module structures. It is this expectation which formed the starting point for our study of modules over commutative domains. We were hopeful that such an undertaking would be feasible, though the field appears frightfully forbidding, not only through the absence of any general theory, but also through the immense complications occurring even in restricted cases.

Bearing commutative domains in general as a distant goal in mind, we decided to concentrate on valuation domains (i.e. commutative domains whose ideals form a chain under inclusion). Firstly, non-Noetherian domains looked more challenging. Secondly, the advantage of working with valuation domains lies in their well-balanced position in the hierarchy of commutative rings: they are close enough to (localized) Dedekind domains to retain some of their pleasant features, but at the same time sufficiently general to display new phenomena and to raise hard problems. Thirdly, there is a fairly large supply of valuation domains, depending on the particular properties of their value groups as well as on completeness properties. We are now more convinced than ever before that this was a right choice: modules over valuation domains represent a unique blending of features of abelian groups and general modules.

In our study, the immediate goals have been to transplant some of the methods from abelian group theory, to scrutinize the basic concepts, to adapt them to the more general situation, and to combine them with powerful techniques developed in module theory. In order not to be carried away by our enthusiasm for abelian groups and at the same time to ensure that right generalizations will be introduced (which are often results of a compromise between various alternatives), we used as testing grounds a number of problems we were working on simultaneously: some originating from abelian groups, some inspired by recent publications on modules, and some arising naturally from our own investigations.

At a certain stage of our study, we felt we had accumulated enough material to develop these in a volume. We realized that this might lack the polish and the depth of a monograph. Since a more polished and in-depth treatment can be given only after time has prepared the way, we settled for a less exhaustive volume on the subject to present our results in an accessible form and to invite our readers to join in.

PREFACE

The theory of modules over commutative non-Noetherian domains and, in particular, over valuation domains goes back thirty years or so. Results by several authors, including I. Kaplansky, B. Osofsky, R. B. Warfield, and above all E. Matlis, were the leading forces in the development. Our debt to their works is enormous.

This volume has evolved as an outgrowth of our search for a systematic treatment of modules over valuation domains. Important results (by other authors) which are not intimately related to ours will not be discussed here. On the other hand, some results which are so far quite meagre will be included wherever it seems worthwhile at least to set out the problems. In fact, there are many exciting problems in this subject. In our presentation, we rely heavily both on material available in the literature and on our own research, including joint papers with other authors and works of doctoral students. As we prefer the general framework of arbitrary domains, wherever not too much extra apparatus is required, we state the results in a general form, though it has not been our endeavor to phrase the theorems under most general hypotheses. We are fully aware of the fact that a full-fledged theory over arbitrary domains would be a formidable task, beyond the scope of the present volume and certainly far beyond the authors' knowledge.

Numerous papers on modules have been inspired by abelian group theory which has been a continuous source of ideas for new research. Our treatment owes much more to abelian group theory than is apparent from the text. Throughout our work, enormous stimulus has been given by ideas and methods from abelian groups. A reader familiar with abelian groups will frequently discover their catalytic effects on our subject, even at places which have apparently nothing to do with them. Though our choice of topics occasionally reflects a close kinship to abelian groups, we have not compromised with the guiding principle that a viable theory ought to be developed on its own foundation and to satisfy its own needs.

We have tried to write this volume with both experts and students in mind. We start from no more than a reasonable acquaintance with elementary facts concerning commutative rings, modules and homological algebra. Chapter I collects most of the background material on valuation rings, Chapter II on modules and Chapter III on homological algebra. The remaining chapters can be read, in general, consecutively, without referring to the rest of the volume, though occasionally it was unavoidable not to invoke results developed in later chapters. The chapters end with comments and open problems.

Our theorems, propositions, lemmas, corollaries, and examples are numbered by pairs (a.b) of positive integers where a stands for the section number. Cross-references within a chapter are done by giving the appropriate pair, while the chapter numbers (Roman numerals) are also listed when referring to a different chapter. Numerous exercises are inserted at the end of the sections, partly to cover additional material and partly to assist students in testing their skills.

In order to avoid boring repetitions of the hypotheses, but to make these clear, we insert symbols

$$D, \quad PD, \quad VD, \quad VR$$

in square brackets before formulating theorems, etc. to indicate that our assertion will be proved for domains in general, for Prüfer domains, for valuation domains and valuation rings, respectively. No such symbol will be used if the rings can be arbitrary commutative rings or if the precise hypotheses are spelled out in the statement.

*

The authors have given series of talks on the subject of this volume at several universities, including Tulane University, Università di Padova, Udine, Universities of Calgary, Florida, Orange Free State and Pretoria, Universität Essen, Bar Ilan University and Charles University of Prague. We wish to extend our gratitude to these universities.

PREFACE

We express our appreciation to the Italian Ministero della Pubblica Istruzione for partial support as well as to the U.S. National Science Foundation for partially supporting an early phase of this project.

The manuscript was typed with a great deal of care by Mrs. Meredith R. Mickel of Tulane University. For her patience and cooperation we owe a considerable debt. We would like to thank Marcel Dekker, Inc. and its staff for their assistance and cooperation in producing this book.

Laszlo Fuchs
Luigi Salce

Contents

Preface   iii

I.   VALUATION RINGS   1

     1. Valuation rings   1
     2. Totally ordered abelian groups   7
     3. Valuations   10
     4. Ideals   13
     5. Maximal and almost maximal valuation rings   20
     6. Prüfer domains   26
        Notes (Problem 1)   30

II.   PRELIMINARIES ON MODULES   31

     1. Modules   31
     2. Divisibility   35
     3. Relative divisibility (RD)   39
     4. Pure submodules   44
     5. Lemmas on pure submodules   48
     6. Cyclic purity   51
     7. Modules with local endomorphism rings   53
        Notes   55

III.   HOMOLOGICAL PRELIMINARIES   57

     1. Homological background   57
     2. Lemmas on Hom and Ext   61
     3. Lemmas on tensor and torsion products   65

IV. PROJECTIVITY AND PROJECTIVE DIMENSION — 70

1. Projective and flat modules — 70
2. Projective dimension — 72
3. Projective dimensions of torsion-free modules — 77
4. Projective dimension one — 82
5. Tight systems — 87
6. Quasi-projectivity — 90
   Notes (Problems 2, 3) — 93

V. TOPOLOGY AND FILTRATIONS — 94

1. The R-topology — 94
2. R-complete modules — 102
3. Filtration and ultracompleteness — 107
4. The annihilator filtration — 110
5. R-ultracomplete modules — 112
   Notes (Problem 4) — 115

VI. DIVISIBILITY AND INJECTIVITY — 116

1. Divisible modules — 116
2. h-divisible modules — 119
3. Divisible modules of projective dimension one — 123
4. Injective modules — 130
5. The injective dimension — 134
6. Quasi-injectivity — 136
   Notes (Problems 5, 6) — 139

VII. UNISERIAL MODULES — 140

1. Uniserial modules — 140
2. Endomorphism rings of uniserial modules — 144
3. Non-standard uniserial modules — 147
4. Direct sums of uniserial modules — 152
   Notes (Problem 7) — 155

VIII. HEIGHTS AND INDICATORS — 156

1. Heights — 156
2. Equiheight submodules — 161
3. Indicators — 162
4. Irregularities of indicators — 164
5. Smoothness — 168
   Notes (Problem 8) — 171

IX. FINITELY GENERATED AND POLYSERIAL MODULES — 173

1. Finitely generated modules — 174
2. The Goldie dimension — 178
3. Indecomposable finitely generated modules — 181
4. Decompositions of finitely generated modules — 185
5. Polyserial modules — 189
   Notes (Problems 9-17) — 193

CONTENTS

X. INVARIANTS AND BASIC SUBMODULES — 195

1. α-Invariants — 196
2. α-Invariants of equiheight submodules — 199
3. α-Basic submodules — 201
4. Modules with trivial α-invariants — 205
   Notes (Problems 18, 19) — 208

XI. RD-INJECTIVITY AND PURE-INJECTIVITY — 209

1. RD-injective modules — 210
2. Pure-injective modules — 214
3. Pure-injective modules over Prüfer domains — 220
4. Pure-injectivity over valuation domains — 223
5. Pure-injective hulls of polyserial modules — 228
   Notes (Problems 20-23) — 231

XII. TORSION-COMPLETE AND COTORSION MODULES — 233

1. Torsion-complete modules — 233
2. Torsion-ultracomplete modules — 238
3. Cotorsion modules — 242
4. The cotorsion hull — 245
   Notes (Problem 24) — 248

XIII. TORSION MODULES — 250

1. Embedding in pure polyserial submodules — 251
2. Separable modules — 253
3. Submodules of separable modules — 256
4. Direct sums of cyclic modules — 259
5. Torsion modules of projective dimension one — 262
6. Modules with zero α-invariants — 264
   Notes (Problem 25) — 268

XIV. TORSION-FREE MODULES — 269

1. Preliminaries — 269
2. Completely decomposable modules — 273
3. Finite rank modules over almost maximal valuation domains — 277
4. Rank one dense basic submodules — 281
5. Chains of pure submodules — 286
6. Pure submodules of free modules — 292
7. Slender modules — 294
   Notes (Problem 26) — 301

References — 303
Notation — 309
Author Index — 312
Subject Index — 314

I. Valuation Rings

This chapter is of introductory nature. Mostly, known results are recorded here for the benefit of a reader who does not want to search for these in the literature. In addition, our basic terminology will be introduced.

§1. VALUATION RINGS

By a __valuation ring__ R will be meant a commutative ring with 1 whose ideals are totally ordered by inclusion; equivalently, for all a, b ∈ R, either a ∈ bR or b ∈ aR. If, in addition, R is an integral domain, it is said to be a __valuation domain__. It is tacitly assumed that our rings are not fields.

It is obvious that factor rings of valuation rings are again valuation rings.

The easiest examples of valuation rings are the following.

EXAMPLE 1.1. The localization $\mathbb{Z}_p$ of the ring $\mathbb{Z}$ of integers at a prime ideal $p\mathbb{Z} \neq 0$. The non-zero ideals of $\mathbb{Z}_p$ are of the form

$p^n \mathbb{Z}_p$ ($n \in \omega$).

EXAMPLE 1.2. The ring J_p of the p-adic integers is the completion of $\mathbb{Z}_p$ in the p-adic topology. Its non-zero ideals are of the form $p^n J_p$ ($n \in \omega$).

EXAMPLE 1.3. The ideals of the ring $\mathbb{Z}/p^n\mathbb{Z}$ ($n \geq 1$) are $p^k\mathbb{Z}/p^n\mathbb{Z}$ ($k = 0, 1, \ldots, n$). This is not a domain if $n > 1$; its non-invertible elements are nilpotent.

A valuation ring R contains a unique maximal ideal $\neq R$ that will consistently be denoted by P. It consists of the non-invertible elements (i.e. non-units) of R. A valuation ring is a local ring (not necessarily Noetherian). The field R/P is called the <u>residue class field</u> of R.

An element $r \in R$ is <u>regular</u> if its annihilator $\text{Ann } r = \{x \in R \mid rx = 0\}$ is 0 and a <u>zero-divisor</u> if $\text{Ann } r \neq 0$.

LEMMA 1.4. The set $Z = Z(R)$ of zero-divisors of a valuation ring R is a prime ideal.

<u>Proof</u>. If $a, b \in Z$, then $ac = 0 = bd$ for some non-zero $c, d \in R$. If $Rc \leq Rd$, then c annihilates $a - b$, so $a-b \in Z$. Also $ra \in Z$ for all $r \in R$. The primeness of Z is immediate. □

In any commutative ring R, the intersection of all prime ideals is the set of nilpotent elements of R, called the <u>nilradical</u> $N = N(R)$ of R.

LEMMA 1.5. The nilradical of a valuation ring R is the minimal prime ideal of R.

<u>Proof</u>. It suffices to show that if ab ($a, b \in R$) is nilpotent, then either a or b is nilpotent. If e.g $b \in aR$, then $(ab)^n = 0$ implies $b^{2n} = 0$, indeed. □

The next result collects some basic facts on the ideals of a valuation ring.

PROPOSITION 1.6. For a valuation ring R, we have:
    (a) every finitely generated ideal of R is principal;
    (b) P can be the only non-zero principal prime ideal of R;

1. VALUATION RINGS

(c) an idempotent ideal $< R$ is either 0 or not principal;

(d) for a proper ideal I of R, $\cap\{I^n \mid n \in \omega\}$ is either 0 or a prime ideal.

Proof. (a) is trivial from the definition.

(b) Let $S = pR$ ($p \in R$) be a prime ideal, different from P. Then $S < P$ implies that, for any $a \in P \setminus S$, $p = ab$ holds for some $b \in R$. Hence $b \in S$ and $b = pc$ for some $c \in R$. Thus $p = ba = pca$, $p(1-ca) = 0$. But $ca \in P$ implies $1-ca$ is a unit whence $S = 0$.

(c) Let $I = pR$ ($p \in P$) satisfy $I^2 = I$. Then $p = rp^2$ for some $r \in R$ and we derive $(1-rp)p = 0$. Here $1-rp$ is a unit, thus $I = 0$.

(d) Suppose $x, y \notin J = \cap\{I^n \mid n < \omega\} \neq 0$. Then there exist $m, n \in \omega$ such that $xR > I^m$ and $yR > I^n$. Hence $xyR \geqq I^m y \geqq I^{m+n}$, and $xy \notin J$ follows if we can show that $I^{m+n+1} = xyR$ is impossible. But such an equality would imply that I^{m+n} would be an idempotent principal ideal, thus 0 by (c), contradicting $J \neq 0$. □

The Artinian and Noetherian valuation rings are easy to describe.

PROPOSITION 1.7. (a) A valuation ring R is Artinian if and only if it has but a finite number of ideals: $R > Rp > \ldots > Rp^n = 0$ for some $p \in R$ and $n \in \omega$.

(b) A valuation ring R is Noetherian, but not Artinian, if and only if its non-zero ideals are: $R > Rp > \ldots > Rp^n > \ldots$ for $n \in \omega$. In this case, R is said to be a discrete rank 1 valuation domain.

The proof is easy and left to the reader. □

For a commutative ring R, $Q(R)$, or simply Q, will denote the classical ring of quotients (in which all non-divisors of zero of R have inverses). If R is a valuation ring, then the elements of Q not in R are of the form r^{-1} for regular elements $r \in R$. The next lemma collects information about Q.

LEMMA 1.8. Let Q be the ring of quotients of the valuation ring R. Then

(i) Q is a field or a valuation ring in which every regular element is a unit;

(ii) every proper R-submodule of Q is isomorphic to an ideal of R;

(iii) S is a subring of Q containing R if and only if it is the localization R_T of R at a prime ideal T containing Z; S itself is a valuation ring.

Proof. (i) is obvious.

(ii) Let M be a proper submodule of Q and t a regular element in R with $t^{-1} \notin M$. Then $M < Rt^{-1}$, i.e. $Mt \leq R$. Multiplication by t being a monomorphism, $M \cong Mt$, as asserted.

(iii) For a subring S, the set $A = \{t \in R \mid t$ regular and $t^{-1} \in S\}$ is a multiplicatively closed subset of R such that $I = R \setminus A$ is a prime ideal of R. This, along with the rest of the claim, is straightforward to verify. □

For a while, the only known examples of valuation rings were valuation domains modulo ideals. Examples of different kinds were discovered recently by using the following construction (see (VII.3.5)).

EXAMPLE 1.9. Let R be a valuation domain and U a torsion R-module satisfying $rU = U$ for every $0 \neq r \in R$. Let R_U be the ring whose additive group is $R \oplus U$ and in which multiplication is defined via

$$(r,u) \cdot (s,v) = (rs, rv + su)$$

($r,s \in R$, $u,v \in U$). It follows easily that R_U is again a valuation ring with nilradical $N = \{(0,u) \mid u \in U\}$ satisfying $N^2 = 0$.

The notion of immediate extension is of utmost importance; we formulate it for valuation rings in general.

Let R, S be valuation rings and $\phi: R \to S$ an injective ring homomorphism (viewed, if convenient, as an embedding). S is said to be an <u>immediate extension</u> of R if

(a) the maps $I \longmapsto (\phi I)S$ and $J \longmapsto \phi^{-1}(J \cap \phi R)$ are inverse to each other and establish a bijection between the sets of ideals I of R and J of S;

(b) ϕ induces an isomorphism between R/P and $S/(\phi P)S$.

Identifying R with its image under ϕ, we will consider R as a subring of S. Then condition (a) becomes

$$I = IS \cap R \quad \text{and} \quad J = (J \cap R)S \qquad (1)$$

for all ideals I of R and ideals J of S. Condition (b) is equivalent to

$$S = R + PS \qquad (2)$$

which is clear from the canonical isomorphism $(R + PS)/PS \cong R/(R \cap PS)$.

LEMMA 1.10. *Let S be an immediate extension of the valuation ring R. Then every $s \in S$ is of the form $s = r\varepsilon$ where $r \in R$ and ε is a unit of S.*

Proof. Given $s \in S$, because of (1) we can write $J = sS$ in the form $J = (J \cap R)S$. Hence for some $r \in J \cap R$, $s \in rS$ holds. Therefore $sS = rS$, and we conclude that s and r differ by a unit factor. □

Note that (2) implies that every unit in S is congruent to a unit in R modulo the maximal ideal PS of S.

A valuation ring is called <u>maximally complete</u> if it does not admit any proper immediate extension. The following result is of great importance:

THEOREM 1.11. *Every valuation domain R can be embedded in a maximally complete valuation domain S which is an immediate extension of R.*

For the proof we refer to Schilling [1]. It should be pointed out that the ring S is not unique up to isomorphism. (However, it will follow from (XI.5.9) that S is unique up to isomorphism as an R-module.)

EXERCISES

1. Let R be a valuation domain with quotient field Q, and Q' a subfield of Q. Then $R' = Q' \cap R$ is a valuation domain with quotient field Q'.

2. In a valuation ring R, every non-trivial idempotent ideal is prime.

3. Find the maximal ideal P and the ideal Z of zero-divisors in the ring of (1.9).

4. Show that J_p is an immediate extension of $\mathbb{Z}_p$.

5. Let S be an immediate extension of the valuation ring R. Show that an ideal I of R is principal exactly if the corresponding ideal SI of S is principal. Extend this result to κ-generated ideals where κ is an infinite cardinal number.

6. A domain R with quotient field Q is said to be integrally closed if every $a \in Q$ which is a root of a monic polynomial $f(x) = x^n + r_{n-1}x^{n-1} + \ldots + r_1 x + r_0$ ($r_i \in R$) necessarily belongs to R. Prove that valuation domains are integrally closed.

7. Prove that a domain R with quotient field Q is integrally closed if and only if it coincides with the intersection of all valuation domains in Q which contain R.

8. A domain R with quotient field Q is completely integrally closed if every $a \in Q$ such that, for some $b \in R$, $ba^n \in R$ for all $n \in \omega$, necessarily belongs to R. Show that a valuation domain R is completely integrally closed exactly if P is the only prime ideal $\neq 0$ in R.

9. Let J be a prime ideal $\neq 0$ of the domain R. There exists a valuation ring V with $R \leq V < Q$ such that $J = R \cap P$ where P denotes the maximal ideal of V.

10. (Gill [1]) Let R be a valuation ring. Show that $P = Z$ if and only if every $a \in R$ satisfies $\text{Ann Ann } Ra = Ra$.

11. Let R be a valuation ring with that $Z^2 = 0$. Verify the R/Z-isomorphism $ZR_Z \cong Q(R/Z)$.

§2. TOTALLY ORDERED ABELIAN GROUPS

An (additively written) abelian group Γ which is at the same time a totally ordered set under a binary relation $\leq$ is called a _totally ordered abelian group_ if it satisfies:

$$\alpha \leq \beta \text{ implies } \alpha + \gamma \leq \beta + \gamma \text{ for all } \alpha, \beta, \gamma \in \Gamma.$$

This condition is equivalent to the following one: $\alpha \leq \beta$ and $\gamma \leq \delta$ imply $\alpha + \gamma \leq \beta + \delta$ ($\alpha, \beta, \gamma, \delta \in \Gamma$).

Needless to say, $\beta \geq \alpha$ means $\alpha \leq \beta$, and $\alpha < \beta$ ($\beta > \alpha$) means $\alpha \leq \beta$ but $\alpha \neq \beta$.

An element $\alpha \in \Gamma$ is _positive_ (_strictly positive_) if $\alpha \geq 0$ ($\alpha > 0$). The _positivity domain_ Γ^+ of Γ is defined via

$$\Gamma^+ = \{\alpha \in \Gamma \mid \alpha \geq 0\}.$$

The positivity domain Γ^+ completely determines the order on Γ, namely, $\alpha \leq \beta$ if and only if $\beta - \alpha \in \Gamma^+$. It is well-known (and easy to show) that the following four conditions characterize positivity domains:

(i) $0 \in \Gamma^+$;
(ii) $\Gamma^+ + \Gamma^+ \subseteq \Gamma^+$;
(iii) $\Gamma^+ \cap (-\Gamma^+) = 0$;
(iv) $\Gamma^+ \cup (-\Gamma^+) = \Gamma$.

A totally ordered abelian group has to be torsion-free. Conversely, every torsion-free abelian group admits a total order making it into a totally ordered group.

A group homomorphism $\phi : \Gamma \to \Gamma'$ between the totally ordered groups Γ and Γ' is called an _order-homomorphism_ if $\alpha \leq \beta$ in Γ implies $\phi(\alpha) \leq \phi(\beta)$ in Γ'. An _order-isomorphism_ ϕ is a group isomorphism such that both ϕ and ϕ^{-1} are order-homomorphisms.

A subgroup Δ of Γ is said to be _convex_ if $0 \leq \gamma \leq \alpha$ with $\gamma \in \Gamma$, $\alpha \in \Delta$ implies $\gamma \in \Delta$. The kernels of order-homomorphisms are precisely the convex subgroups. The factor group Γ/Δ (with Δ convex in Γ) carries a natural order structure where $(\Gamma/\Delta)^+$ is the image of Γ^+ under the natural homomorphism.

By a <u>filter</u> in Γ^+ is meant a non-empty proper subset F of Γ^+ such that

$$\alpha \in F \text{ and } \alpha \leq \beta \in \Gamma^+ \text{ imply } \beta \in F.$$

F is a <u>prime filter</u> if $\beta, \gamma \in \Gamma^+ \backslash F$ implies $\beta + \gamma \notin F$, and a <u>principal filter</u> if, for some $\alpha \in \Gamma^+$, $F = \{\beta \in \Gamma^+ \mid \beta \geq \alpha\}$.

LEMMA 2.1. A subset F of Γ^+ is a prime filter if and only if $\Gamma^+ \backslash F$ is the positivity domain of a convex subgroup Δ of Γ. □

The next result is a famous theorem by O. Hölder.

PROPOSITION 2.2. For a totally ordered group Γ, the following are equivalent.

(a) Γ has no non-trivial convex subroups;

(b) Γ is archimedean, i.e. for every pair α, β of strictly positive elements in Γ there exists an $n \in \omega$ such that $\beta \leq n\alpha$;

(c) Γ is order-isomorphic to a subgroup of the additive group $\mathbb{R}$ of real numbers.

For the proof we refer e.g. to Fuchs [1]. □

An ordered subgroup of $\mathbb{R}$ is called <u>discrete</u> if it is order-isomorphic to $\mathbb{Z}$ ($\mathbb{Z}$ is equipped with the natural order). Otherwise, it is called <u>dense</u>; in this case, the subgroup contains, with any two distinct elements, a third one in between. $\mathbb{Q}$ is, for instance, a dense subgroup of $\mathbb{R}$.

Given a family $\{\Gamma_\lambda; \lambda \in \Lambda\}$ of totally ordered abelian groups, a new group can be constructed as follows. Assume the index set Λ is totally ordered. In the cartesian product $\Pi \Gamma_\lambda$ consider the vectors $\alpha = (\alpha_\lambda)_{\lambda \in \Lambda}$ (where $\alpha_\lambda \in \Gamma_\lambda$ for each λ) such that $\operatorname{supp} \alpha = \{\lambda \in \Lambda \mid \alpha_\lambda \neq 0\}$ is well-ordered in the ordering of Λ. These α's form a subgroup of $\Pi\Gamma_\lambda$, called the <u>Hahn product</u> $H\Gamma_\lambda$ of the family $\{\Gamma_\lambda; \lambda \in \Lambda\}$. The Hahn product becomes a totally ordered abelian group by declaring $\alpha = (\alpha_\lambda)_{\lambda \in \Lambda}$ positive

2. ORDERED ABELIAN GROUPS

if $\alpha = 0$ or if $\alpha_{\lambda_0} > 0$ where λ_0 is the first element of supp α in its well-ordering.

PROPOSITION 2.3. (H. Hahn) Every totally ordered abelian group is order-isomorphic to a subgroup of a Hahn product of copies of $\mathbb{R}$ (over an appropriate totally ordered set Λ).

For a proof, see e.g. Fuchs [1]. □

The totally ordered abelian group Γ is called <u>discrete</u> if it is order-isomorphic to a subgroup of the Hahn product of copies of $\mathbb{Z}$. This is the case exactly if $\Delta'/\Delta \cong \mathbb{Z}$ for any two convex subgroups $\Delta < \Delta'$ where Δ' covers Δ.

If Λ is a well-ordered set, the Hahn product of the family $\{\Gamma_\lambda; \lambda \in \Lambda\}$ is also called the <u>lexicographic product</u>.

EXAMPLE 2.4. Let $\Gamma = \prod_{i=1}^{n} \Gamma_i$ where each Γ_i is order-isomorphic to $\mathbb{Z}$ ($i = 1, \ldots, n$). Thus Γ is the additive group of the n-tuples of integers $(\alpha_1, \ldots, \alpha_n)$ such that $(\alpha_1, \ldots, \alpha_n) > 0$ if and only if the first non-zero coordinate is > 0. For every i ($1 \leq i \leq n$), the vectors of the form $(0, \ldots, 0, \alpha_i, \ldots, \alpha_n)$ form a convex subgroup Δ_i, and these are the only non-zero convex subgroups in Γ.

EXAMPLE 2.5. Let Γ be the lexicographic product of $\Gamma_1 \cong \mathbb{Q}$ and $\Gamma_2 \cong \mathbb{Z}$. The only non-trivial convex subgroup Δ consists of all pairs $(0, \alpha_2)$, thus $\Delta \cong \mathbb{Z}$ and $\Gamma/\Delta \cong \mathbb{Q}$. Manifestly, $(0,1)$ is the smallest strictly positive element in Γ. (Observe that if $\Gamma_1 \cong \mathbb{Z}$ and $\Gamma_2 \cong \mathbb{Q}$, then Γ has no such element.)

EXERCISES

1. In a totally ordered group Γ, an element α is positive if for some positive integer n, $n\alpha$ is positive. Conclude that a totally ordered group has to be torsion-free.

2. (a) Using Hahn products, prove that every divisible torsion-free abelian group can be made into a totally ordered group.

(b) Conclude that every torsion-free abelian group is group-isomorphic to a totally ordered group.

3. Let Γ be the Hahn product of the totally ordered groups Γ_λ with λ running over a totally ordered index set Λ. For every $\mu \in \Lambda$, the sets

$$\Delta_\mu = \{(\alpha_\lambda)_{\lambda \in \Lambda} \in \Gamma \mid \alpha_\lambda = 0 \text{ for all } \lambda < \mu\}$$

and

$$\Delta_\mu^* = \{(\alpha_\lambda) \in \Gamma \mid \alpha_\lambda = 0 \text{ for all } \lambda \leq \mu\}$$

are convex subgroups of Γ. Verify the order-isomorphism $\Delta_\mu / \Delta_\mu^* \cong \Gamma_\mu$.

4. Construct a totally ordered abelian group Γ which has neither a largest nor a smallest non-trivial convex subgroup.

5. If Γ contains a smallest strictly positive element, then it has a smallest convex subgroup $\neq 0$ which has to be order-isomorphic to $\mathbb{Z}$.

§3. VALUATIONS

Valuation domains were discovered by W. Krull who defined them with the aid of general valuations he introduced. The interrelation between valuation domains and valuations of fields is one of the most important features in our study.

Let Γ be a totally ordered abelian group and ∞ a symbol regarded as larger than any element of Γ. We set $\alpha + \infty = \infty$ for all $\alpha \in \Gamma$. A map

$$v : Q \to \Gamma \cup \{\infty\}$$

of a field Q is said to be a <u>valuation</u> if it satisfies:

V1. $v(x) = \infty$ if and only if $x = 0$;
V2. $v(xy) = v(x) + v(y)$ for all $x, y \in Q$;
V3. $v(x+y) \geq \min\{v(x), v(y)\}$ for all $x, y \in Q$.

There is no loss of generality in assuming that v is onto, since the image of $Q^\times$ is necessarily a subgroup of Γ.

The subset

3. VALUATIONS

$$R_v = \{x \in Q \mid v(x) \geq 0\}$$

is a domain called the <u>valuation domain of</u> v. It is in fact a valuation ring in the sense of §1. Now $P = \{x \in Q \mid v(x) > 0\}$ will be the maximal ideal of R.

Actually, every valuation domain arises in this way. In fact, let R be a valuation domain and Q its field of quotients. The units of R form a subgroup U in the multiplicative group $Q^\times$, and define Γ as an additive group isomorphic to $Q^\times/U$. Setting $aU \leq bU$ for $a, b \in Q^\times$ if and only if $ba^{-1} \in R$, Γ becomes a totally ordered group, as a straightforward computation shows. It is readily checked that the map

$$v : Q \to \Gamma \cup \{\infty\}$$

acting as $v(a) = aU$ for $a \in Q^\times$ and $v(0) = \infty$ satisfies the conditions V1-V3. Since the positive elements of Γ correspond to cosets of U represented by elements of R, it follows at once that the valuation domain of v coincides with the ring R we started with. Γ is called the <u>value group</u> $\Gamma(R)$ of R.

PROPOSITION 3.1. (W. Krull) Every valuation domain is the valuation domain R_v of a valuation v of its quotient field. □

There is a canonical correspondence between the non-zero ideals of a valuation domain R and the filters in Γ^+ where Γ is the value group of R. In order to describe this, we set, for a filter F in Γ^+,

$$I(F) = \{x \in R \mid v(x) \in F\}$$

which is a non-zero ideal of R.

PROPOSITION 3.2. Let R be the valuation domain of the valuation v of its quotient field. The correspondences

$$I \longmapsto v(I) \quad \text{and} \quad F \longmapsto I(F)$$

define a bijection between the set of non-zero ideals of R and the set of filters in Γ^+. In particular, prime (principal) ideals correspond to prime (principal) filters.

Proof. This is readily checked from the definitions. □

From (3.2) it is evident that an immediate extension S of a valuation domain R must have the same value group as R (more precisely, the canonical map from the value group of R into that of S is an isomorphism). If this happens, then condition (1) in §2 is satisfied. Hence we conclude:

PROPOSITION 3.3. Let R and S be valuation domains such that $R \leq S$. S is an immediate extension of R if and only if

(i) R and S have the same value group Γ;

(ii) the canonical map $R/P \to S/SP$ is an isomorphism between the residue class fields. □

The major question which arises now is whether or not every totally ordered abelian group Γ is the value group of some valuation domain. The next theorem answers this question in the affirmative.

THEOREM 3.4. (W. Krull) Given a field K and a totally ordered abelian group Γ, there exists a valuation domain R with residue class field $\cong K$ and value group order-isomorphic to Γ.

Proof. Consider the polynomial ring over K with indeterminates $\{x_g \mid g \in \Gamma\}$, and let Q be its quotient field. We define a valuation $v: Q \to \Gamma \cup \{\infty\}$ by first setting $v(x_g) = g$ and then by extending it multiplicatively to monomials by the rule
$$v(ax_{g_1}^{k_1} \cdots x_{g_n}^{k_n}) = k_1 g_1 + \cdots + k_n g_n \quad (0 \neq a \in K).$$
Here k_i denote positive integers. Next we set for polynomials
$$v(\Sigma a x_{g_1}^{k_1} \cdots x_{g_n}^{k_n}) = \min v(ax_{g_1}^{k_1} \cdots x_{g_n}^{k_n}).$$
This defines v as a homomorphism of $Q^\times$. The valuation properties of v are readily verified. The value group is evidently Γ, while the residue class field turns out to be isomorphic to K.

For an alternative proof, see Ex. 6. □

This result is a major tool in constructing valuation domains with certain properties.

EXERCISES

1. Show that in V3, $v(x+y) = \min(v(x), v(y))$ holds if $v(x) \neq v(y)$.

2. Let $v : Q \to \Gamma \cup \{\infty\}$ be a valuation and R_v the valuation domain of v. Show that
   (a) $v(x) = 0$ if and only if x is a unit in R_v;
   (b) $v(x) = v(y)$ exactly if $R_v x = R_v y$;
   (c) Q is the field of quotients of R_v.

3. Let $v : Q \to \Gamma \cup \{\infty\}$ be an onto valuation and R_v its valuation domain. Prove that the value group of R_v is order-isomorphic to Γ.

4. Let R be a valuation domain with value group Γ. Show that the maximal ideal P of R is principal exactly if Γ contains a minimal convex subgroup $\neq 0$ which is order-isomorphic to $\mathbb{Z}$.

5. Show that for non-zero ideals I, J of R, $J = rI$ holds if and only if $v(J) = v(r) + v(I)$ in the correspondence of (3.2).

6. Prove (3.4) as follows.
   (a) First show that the semigroup algebra of Γ^+ over K is an integral domain $K[\Gamma^+]$.
   (b) Define $v : K[\Gamma^+] \to \Gamma^+$ via $v(f) = \gamma$ if γ is the minimal element in supp f ($f \in K[\Gamma^+]$), and verify that v extends uniquely to a valuation of the quotient field Q of $K[\Gamma^+]$.
   (c) Finally, show that R_v contains $K[\Gamma^+]$ and is as desired.

7. Give an example for an uncountably generated ideal of a valuation domain, by using (3.4).

8. For every cardinal number $\kappa \geq 1$, there exists a valuation domain with κ prime ideals $\neq 0$.

§4. IDEALS

Some results on ideals have already been discussed in §1; here we continue developing their properties.

Let R be any commutative ring and Q its ring of quotients. For submodules I and J of Q, we set

$$J : I = \{x \in Q \mid xI \leq J\}.$$

This <u>residual</u> is an R-submodule of Q. For $r \in R$, we write simply $J:r$ rather than $J:Rr$.

The following two results are obvious for domains R.

LEMMA 4.1. Let I and J be proper non-zero ideals of a valuation ring R. For $0 \neq r \in R$, $I = J:r$ if and only if $J = rI$.

<u>Proof</u>. Let $J = rI \neq 0$. Then $I \leq J:r$. If $a \in J:r$, $ra \in rI$, then $ra = rb$ for some $b \in I$, and $a-b \in \mathrm{Ann}\ r$. As $I \nleq \mathrm{Ann}\ r$, we have $\mathrm{Ann}\ r < I$. Thus $a-b \in I$ and $a \in I$. Conversely, if $I = J:r$, then $rI \leq J$. Suppose $a \in J\setminus rI$; either $r = ab$ or $a = rb$ for some $b \in R$. In the first case, $J:r \geq J:a \geq R > I$, a contradiction. In the second case, $b \in J:r = I$, thus $a \in rI$, again a contradiction. □

An immediate consequence of (4.1) is:

COROLLARY 4.2. If I and J are proper ideals of a valuation ring R and $r \in R$, then $rI = rJ \neq 0$ implies $I = J$.

<u>Proof</u>. In fact, $I = rI:r = rJ:r = J$. □

Two non-trivial ideals, I and J, of a valuation ring R are said to be <u>equivalent</u> (Nishi [1]), in notation: $I \sim J$, if for a suitable $r \in R$, either $I = rJ$ or $J = rI$; equivalently, if $J = I:r$ or $I = J:r$. It is readily seen that this is in fact an equivalence relation on the set of non-trivial ideals of R.

If R happens to be a valuation domain, then $I \sim J$ means that $I \cong J$. In particular, all principal ideals $\neq 0$ are equivalent.

EXAMPLE 4.3. Let R be a valuation domain with value group $\Gamma \cong \mathbb{R}$. In this case, all non-principal ideals belong to the same equivalence class.

4. IDEALS

EXAMPLE 4.4. Let R be a valuation domain with value group $\Gamma \cong \mathbb{Q}$. There are continuously many different equivalence classes of ideals. (Cf. Ex. 5 in §3.)

With a non-zero ideal L of a valuation ring R, we associate a prime ideal as follows. First we define

$$U_L = \{r \in R \mid rL = L\} ;$$

this is a subsemigroup of the multiplicative semigroup of R. We then set $L^\# = R \setminus U_L$, i.e.

$$L^\# = \{r \in R \mid rL < L\} ;$$

this is an ideal of R.

LEMMA 4.5. [VR] Let L be a non-trivial ideal of R.
 (i) $L^\#$ is a prime ideal containing L;
 (ii) $K \sim L$ implies $K^\# = L^\#$;
 (iii) $L^\# = L$ if L is a prime.

Proof. (i) Suppose $a \in L$ satisfies $aL = L$. Thus, for some $b \in L$, $ab = a$ and $a(1-b) = 0$. As $L < R$, b is a non-unit; so $1-b$ is a unit and $a = 0$. Hence $L \leq L^\#$. Assume $a, b \notin L^\#$, i.e. $aL = L = bL$. Therefore, $abL = L$, i.e. $ab \notin L^\#$ and $L^\#$ is prime.

(ii) Let $L = rK$ ($r \in R$). Clearly, $tK = K$ ($t \in R$) implies $tL = L$. On the other hand, $tL = L$ ($t \in R$) implies $trK = rK$ whence by (4.2) $tK = K$.

(iii) Let L be a prime and $a \in R \setminus L$. As $Ra > L$, every $b \in L$ is of the form $b = ac$ ($c \in R$). Here necessarily $c \in L$ whence $aL = L$ and $L^\# = L$. □

An ideal $L \neq 0$ of a valuation ring R is called <u>archimedean</u> (Matlis [1]) if $L^\# = P$. In other words, $rL = L$ ($r \in R$) implies r is a unit of R. Clearly, all principal ideals are archimedean, and P is the only archimedean prime ideal.

PROPOSITION 4.6. Let L be a regular ideal of the valuation ring R. Then:

(i) $L^{\#}$ contains the ideal Z of zero-divisors of R;

(ii) L carries an $R_{L^{\#}}$-module structure in a natural way;

(iii) $\text{End}_R L \cong R_{L^{\#}}$ and $\text{Aut}_R L \cong U_L \cup U_L^{-1}$.

Proof. (i) is evident.

(ii) Let $b \in R \setminus L^{\#}$, i.e. $b \in U_L$. Then b is a regular element, hence multiplication by b is an automorphism of L. This leads immediately to (ii).

(iii) Clearly, L is a faithful $R_{L^{\#}}$-module, and all R-endomorphisms of L are $R_{L^{\#}}$-endomorphisms. Let η be an $R_{L^{\#}}$-endomorphism of L, and $a \in L$ a regular element. For $u = \eta a$ we have either $u = ta$ or $a = tu$ for some $t \in R$. In the first alternative, $rb = a$ ($b \in L$, $r \in R$) implies $\eta rb = trb$, thus $\eta b = tb$, r being regular; this shows η acts as multiplication by $t \in R$. In the second alternative, both t and u are regular, and it follows similarly that η acts as multiplication by t^{-1}. It is readily seen that $t \notin L^{\#}$. Consequently, (iii) holds. □

It is often convenient to identify the endomorphism ring of L with the subring $R_{L^{\#}}$ of Q, in the obvious way. If this is done, we can claim:

COROLLARY 4.7. For a regular ideal L of a valuation ring R, the following are equivalent:

(a) L is archimedean;

(b) $\text{End}_R L = R$;

(c) $\text{Aut}_R L = U(R)$. □

We will find the following observation useful.

LEMMA 4.8. Let R be a valuation domain and L an ideal of R. Then $L^{\#}L < L$ exactly if $L \cong R_{L^{\#}}$.

Proof. First suppose $a \in L \setminus L^{\#}L$. Define $\phi : R_{L^{\#}} \to L$ via $\phi(rs^{-1}) = rs^{-1}a$ ($r \in R$, $s \notin L^{\#}$); as $sL = L$, $rs^{-1}a$ is a well-defined element of L. Evidently, ϕ is monic. To show it is

4. IDEALS

epic, let $b \in L$. Only the case $b \notin Ra$ needs to be considered. Then $a = tb$ for some $t \in R$ which cannot belong to $L^\#$ by the choice of a. Hence $t^{-1} \in R_{L^\#}$ and $\phi(t^{-1}) = b$.

Conversely, let $L \cong R_{L^\#}$; thus we can set $L = R_{L^\#}a$ for some $a \in L$. We claim that $a \notin L^\# L$. Otherwise $a = rb$ for some $r \in L^\#$, $b \in L$. There is a $t \notin L^\#$ such that $b = t^{-1}a$. Hence we arrive at the contradiction $r = t$. □

We wish to record two cases where the product of two ideals is equivalent to one of them.

PROPOSITION 4.9. [VD] (Soileau [1]) If J, L are non-zero ideals of R, then $JL \cong J$ in either of the following two cases:
1) $J^\# < L^\#$;
2) $J^\# = L^\#$ and $L^\# L < L$.

Proof. Set $JL = \{a_i b_i \mid a_i \in J, b_i \in L\}$. In the first case, choose $r \in L^\# \setminus J^\#$ and let $b \in L \setminus rL$. Then $a_i b_i = (ra_i')b_i = a_i'(rb_i) \in JRb$ for some $a_i' \in J$. We obtain $JL = Jb \cong J$.

In the second case, by (4.8) we have $L^\# \cong R_{L^\#}$. Thus $JL \cong R_{L^\#} \cdot J = R_{J^\#} \cdot J = J$, indeed. □

In order to characterize the endomorphism rings of non-regular ideals L in valuation rings R, it is worth introducing another ideal associated with L. This is defined as

$$L_\# = \{r \in R \mid rx = 0 \text{ for some } 0 \neq x \in L\},$$

i.e. as the set union of $\{\text{Ann } x \mid 0 \neq x \in L\}$. This is indeed an ideal of R. It is readily seen that it is necessarily a prime ideal.

LEMMA 4.10. Let L be a faithful, non-regular ideal of the valuation ring R. Then we have
(i) $L_\#$ contains both $L^\#$ and Z;
(ii) L is in a natural way an $R_{L_\#}$-module;
(iii) the R-endomorphisms of L are $R_{L_\#}$-endomorphisms.

Proof. (i) Let $t \in R$ and $a \in L$ satisfy $tL < Ra < L$. As L is non-regular, some $0 \neq b \in R$ satisfies $ab = 0$, thus $btL = 0$. Because L is faithful, we have $bL \neq 0$; thus $t \in L_{\#}$. The inclusion $Z \leq L_{\#}$ is easy.

(ii) and (iii) follow from (i) in the same way as in (4.6). □

For the R-topology we refer to V.§1.

PROPOSITION 4.11. Let L be a faithful, non-regular ideal of the valuation ring R. Then $End_R L$ is isomorphic to the completion of $R_{L_{\#}}$ in its $R_{L_{\#}}$-topology.

Proof. By (4.10), $End_R L$ is isomorphic to the ring S of $R_{L_{\#}}$-endomorphisms of L. By faithfulness, $R_{L_{\#}}$ is embedded in S. As an endomorphism ring, S is complete in the finite topology. First we show that $R_{L_{\#}}$ is dense in S. In the present case, a base of neighborhoods of 0 is given by the S-ideals

$$V_x = \{\eta \in S \mid \eta x = 0\}$$

with x running over all elements of L. Let $\alpha + V_x$ be a neighborhood of $\alpha \in S$. If $\alpha x = 0$, $0 \in \alpha + V_x$. If $\alpha x \neq 0$, then either $\alpha x = rx$ or $x = r\alpha x$ for some $r \in R$. In the first case $r \in \alpha + V_x$, while in the second case we must have $r \notin L_{\#}$. Otherwise, $ry = 0$ for some $0 \neq y \in L$, thus $y = t\alpha x$ for some $t \in R$, and $0 = ry = rt\alpha x = tx$ leads to the contradiction $y = 0$. Thus for $s = r^{-1} \in R_{L_{\#}}$, we have $\alpha x = sx$ and $s \in \alpha + V_x$. It remains to show that the finite topology on S induces the $R_{L_{\#}}$-topology on $R_{L_{\#}}$. But this is an immediate consequence of $\cap \{V_x \cap R_{L_{\#}} \mid 0 \neq x \in L\} = 0$ where $V_x \cap R_{L_{\#}} \neq 0$. □

It should be pointed out that (4.6) and (4.11) do not describe completely the endomorphism rings of ideals L of valuation rings R. In fact, if L is not faithful, then we can pass to $R/Ann\, L = R'$, and consider L as an R'-module. L will be a uniserial R'-module, but not necessarily isomorphic to an ideal of R' (see Ex. 7). For $End_R L$, see VII.§2.

4. IDEALS

EXERCISES

1. Let R be a valuation domain with value group Γ which is the lexicographic product of copies of $\mathbb{Z}$ over an index set of type ω_1 (the first uncountable ordinal). Prove that P is the only prime ideal of R that is uncountably generated.

2. [VR] Show that if L is an ideal and $r \in R$ such that $rL \neq 0$ is a principal ideal, then L itself is a principal ideal.

3. [VD] All non-zero ideals of R are archimedean if and only if the value group Γ of R is an archimedean ordered group.

4. (Nishi [1]) [VR] For any ideal $L \neq 0$, $L^\#$ is the union of all proper ideals of R equivalent to L.

5. (Nishi [1]) [VR] For an ideal $L \neq 0$ of R, set $L_0 = \cap \{K < R \mid K \neq 0 \text{ and } K \sim L\}$. Show that
   (a) $L_0 \leq Z$;
   (b) L_0 is the annihilator of the injective hull of R/L.

6. [VR] Let I, J be non-trivial ideals of R, and S an immediate extension of R. Show that
   (a) $I \sim J$ if and only if $IS \sim JS$;
   (b) $I^\# S = (IS)^\#$.

7. Let R be a valuation domain with value group $\mathbb{Z} \oplus \mathbb{Z}$ (lexicographic ordering), and I the ideal of R with $v(I) = \{(m,n) \in \Gamma \mid m \geq 2\}$. Show that
   (a) the nilradical N of $S = R/I$ is its unique non-maximal prime ideal;
   (b) Ann $N = N$;
   (c) N is a faithful S/N-module not isomorphic to any ideal of S/N.

8. (Shores and Lewis [1]) Let L be a faithful non-regular ideal of a valuation ring R and let $S = \text{End}_R L$. Show that $\alpha \in S$ is regular exactly if $\alpha L = L$. If α is regular, then αS is comparable with βS for every $\beta \in S$.

9. Let I, J, K be ideals of the ring R. Show that
   (a) $I \leq J$ implies $I : K \leq J : K$ and $K : I \geq K : J$;
   (b) $IJ : J \geq I$;
   (c) $I : (I : J) \geq J$ and $I : [I : (I : J)] = I : J$;
   (d) $(I \cap J) : K = I : K \cap J : K$ and $K : (I \cup J) = K : I \cap K : J$;
   (e) $(I : J) : K = I : JK = (I : K) : J$.

§5. MAXIMAL AND ALMOST MAXIMAL VALUATION RINGS

There are two conditions on valuation rings R which induce far-reaching consequences for R-modules. We devote this section to the discussion of these two conditions.

A valuation ring R is called <u>maximal</u> if every system of finitely solvable congruences of the form

$$x \equiv a_i \mod L_i \quad (i \in I) \qquad (1)$$

where $a_i \in R$, L_i are ideals of R and I is an arbitrary index set, has a simultaneous solution in R. Equivalently, every family of cosets $\{a_i + L_i \mid i \in I\}$ with the finite intersection property has a non-void intersection; in topological terms: R is linearly compact in the discrete topology.

A valuation ring R is said to be <u>almost maximal</u> if the above condition holds whenever $\cap L_i \neq 0$; in other words, R/L is maximal for every ideal $L \neq 0$.

EXAMPLE 5.1. An Artinian valuation ring is always maximal. A Noetherian valuation ring is almost maximal.

The following result reveals the basic connection between maximality and almost maximality. (R-completeness will be discussed in Ch. V.)

LEMMA 5.2. An almost maximal valuation ring R is maximal exactly if it is R-complete.

<u>Proof</u>. As linear compactness implies completeness in any topology, it suffices to show that an almost maximal, R-complete valuation ring R is maximal. Let L_i $(i \in I)$ be ideals of R,

5. MAXIMAL VALUATION RINGS

and let the family of cosets $\{a_i + L_i \mid i \in I\}$ have the finite intersection property. If $\cap L_i \neq 0$, then by almost maximality, $\cap \{a_i + L_i \mid i \in I\}$ is not empty. If $\cap L_i = 0$, then for each $i \in I$, there is a $b_i \in L_i$ such that $b_i \notin L_{i'}$ for some $L_{i'} < L_i$. Hence

$$a_{i'} + L_{i'} \subseteq a_{i'} + Rb_i = a_i + (a_{i'} - a_i) + Rb_i \subseteq a_i + L_i,$$

which shows that the cosets $\{a_{i'} + Rb_i \mid i \in I\}$ have the finite intersection property. Consequently, $\{a_{i'} \mid i \in I\}$ is a Cauchy net which must have, by completeness, a limit $a \in R$. This a obviously belongs to all of $a_i + L_i$. □

EXAMPLE 5.3. $\mathbb{Z}_p$ is an almost maximal valuation domain, while its completion, the ring J_p of p-adic integers, is maximal.

The proof of (5.2) applies to verify:

REMARK 5.4. In checking the maximality of a valuation ring R, it suffices to consider only principal ideals L_i in (1).

It is somewhat surprising that for non-domains, the concepts of maximality and almost maximality coincide. This is a consequence of the following lemma.

LEMMA 5.5. (Brandal [1]) Let R be a valuation ring and L an ideal of R. If R/L is linearly compact in the discrete topology, then so is R/L^2.

Proof. Let $\{a_i + L_i \mid i \in I\}$ be a family of cosets of ideals L_i of R with the finite intersection property such that $L_i \geq L^2$ for all $i \in I$. We must show that this family has a non-empty intersection. If $L_i \geq L$ for all $i \in I$, the result follows from the hypothesis. Thus assume that $L > L_{i_0}$ for some $i_0 \in I$, and let $y \in L \setminus L_{i_0}$. Define $\phi : R/L \to R/L^2$ by $\phi(r+L) = ry + L^2$; hence we deduce that $\text{Im } \phi = Ry/L^2$ is linearly compact. Since $\{\overline{a_i} - \overline{a_{i_0}} + \overline{L_i} \mid i \in I, L_i \leq L_{i_0}\}$ is a family of cosets of submodules of Ry/L^2 ($\overline{a_i} = a_i + L^2$, $\overline{L_i} = L_i/L^2$) with the finite intersection property, there exists an $\overline{a} = a + L^2$ in the intersection of this family. It follows that $a + a_{i_0}$ belongs to $\cap \{a_i + L_i \mid i \in I\}$. □

As an immediate consequence we derive:

THEOREM 5.6. (Gill [1]) Let R be a valuation ring which is not a domain. If R is almost maximal, then it is maximal.

Proof. There exists a $0 \neq t \in R$ such that $t^2 = 0$. R/tR is linearly compact in the discrete topology, hence, by (5.5), $R = R/t^2R$ is likewise linearly compact. □

Consider the following family of ideals of a valuation ring R:

$$\mathcal{F} = \{I \text{ ideal of } R \mid R/I \text{ is not maximal}\}.$$

Obviously, $\mathcal{F} = \emptyset$ exactly if R is maximal, and, if $I \leq J$ and $J \in \mathcal{F}$, then also $I \in \mathcal{F}$. If R is not maximal, we consider the following ideal of R:

$$I_\mathcal{F} = \cup\{I \mid I \in \mathcal{F}\}.$$

LEMMA 5.7. Let R be a valuation ring which is not maximal. Then $I_\mathcal{F}$ is a prime ideal of R, and R/I is an almost maximal valuation ring if and only if $I \geq I_\mathcal{F}$.

Proof. From (5.5) it follows that $I_\mathcal{F}$ is contained in $\cap_{n \in \omega} L^n$ for all $L \notin \mathcal{F}$. Conversely, it is obvious that $I_\mathcal{F}$ contains $\cap\{L \mid L \notin \mathcal{F}\}$. Therefore, we have

$$I_\mathcal{F} = \cap\{L^n \mid L \notin \mathcal{F};\ n \in \omega\},$$

so $I_\mathcal{F}$ is prime by (1.6). The second claim is clear. □

The ideal $I_\mathcal{F}$ itself may or may not belong to $\mathcal{F}$. It does exactly if $R/I_\mathcal{F}$ is R-complete (see V. §1). For instance, in a discrete rank one valuation ring which is not complete, $I_\mathcal{F} = 0 \in \mathcal{F}$. For the second possibility we give the following

EXAMPLE 5.8. Let R be a valuation domain with $\Gamma(R) = \mathbb{R}$. If R is not almost maximal, then $I_\mathcal{F}$ necessarily coincides with P. Here obviously $P \notin \mathcal{F}$, R/P being trivially linearly compact.

Maximality and almost maximality are preserved under localizations:

THEOREM 5.9. (Gill [1]) Let R be a maximal (almost maximal)

5. MAXIMAL VALUATION RINGS

valuation ring and S a prime ideal of R containing Z. Then the localization R_S is likewise maximal (almost maximal).

Proof. Hypothesis implies that $R \leq R_S \leq Q$. Let $\{a_i + \overline{L}_i \mid i \in I\}$ be a family of cosets modulo ideals $\overline{L}_i$ of R_S with the finite intersection property. We can clearly assume that $L_i \neq Q$, i.e. they are proper R_S-submodules of Q. Hence choosing a fixed $i_0 \in I$, there is a regular $t_0 \in R$ such that $t_0 \overline{L}_{i_0} < R$. Moreover, $t_0 a_{i_0} \in R$ can also be assumed. If we ignore those i's for which $t_0 (a_i + L_i)$ is not contained in R, then we obtain a family of cosets $\{t_0 a_0 + t_0 \overline{L}_i \mid i \in I'\}$ in R where $\cap t_0 \overline{L}_i \neq 0$ whenever $\cap \overline{L}_i \neq 0$. We appeal to the (almost) maximality of R to conclude that some $a \in R$ satisfies $a \in t_0 \overline{a}_i + t_0 \overline{L}_i$ for all i. Therefore, $t_0^{-1} a \in \overline{a}_i + \overline{L}_i$ for all i. □

The following theorem is fundamental. Its proof can be found in several of the available treatises; we refer e.g. to Brandal [2] for a lucid proof.

THEOREM 5.10. For a valuation domain R, these are equivalent:

(1) R is maximal;

(2) R is maximally complete. □

The rest of this section is devoted to the discussion how to construct a maximal valuation domain R with prescribed residue class field K and value group Γ. This method is more explicit than the one given in §3 for constructing valuation domains from the same data, followed by the embedding of (1.11).

Let f be a function from the totally ordered abelian group Γ into the field K. By its <u>support</u> is meant

$$\text{supp } f = \{\alpha \in \Gamma \mid f\alpha \neq 0\} \subseteq \Gamma.$$

We set

$$Q = \{f \in K^\Gamma \mid \text{supp } f \text{ is well-ordered}\},$$

and define sum and product of two elements $f, g \in Q$ as follows:

$$(f + g)\alpha = f\alpha + g\alpha, \tag{2}$$

$$(f \cdot g)\alpha = \sum_{\beta+\gamma=\alpha} f\beta \cdot g\gamma \qquad (3)$$

where $\alpha, \beta, \gamma \in \Gamma$. One can check without difficulty that both $\mathrm{supp}(f+g)$ and $\mathrm{supp}(f \cdot g)$ are well-ordered, and that in (3) the sum is finite. Hence it follows easily that Q becomes a commutative ring whose identity is the function f_0 defined as $f_0 \alpha = 1$ or 0 according as $\alpha = 0$ or $\alpha \neq 0$.

More generally, every $\gamma \in \Gamma$ defines an element $f_\gamma \in Q$ via $f_\gamma \alpha = \delta_{\gamma\alpha}$ (the Kronecker delta). Obviously, f_γ is a unit of Q with $f_{-\gamma}$ as inverse such that the map $\gamma \longmapsto f_\gamma$ is a monomorphism of the additive group Γ into the semigroup $Q^\times = Q \setminus 0$. With the aid of these "characteristic functions" f_γ, we can write every $f \in Q$ as a formal infinite sum

$$f = \sum_{\alpha \in \Gamma} (f\alpha) f_\alpha \qquad (4)$$

with coefficients $f\alpha \in K$.

Often it is convenient to replace f_α by symbols x^α behaving as indeterminates, subject to the rules

$$x^\alpha x^\beta = x^{\alpha+\beta} \qquad (\alpha, \beta \in \Gamma).$$

Then f assumes the form of a formal power series

$$f = \sum_{\alpha \in \Gamma} (f\alpha) x^\alpha \qquad (5)$$

over K, and (3) transforms into the "convolution product". (It should be emphasized that the support of a formal power series is by definition a well-ordered subset of Γ.)

Define for $f \in Q^\times$

$$v(f) = \inf \mathrm{supp}\, f.$$

From (2) and (3) it follows at once that this is a valuation of the ring Q with values in Γ. Moreover, we have:

THEOREM 5.11. *Let K be a field and Γ a totally ordered abelian group. The formal power series (5) form a field with valuation v. The valuation domain of v, denoted by $K[[\Gamma]]$ and called the* formal power series ring of Γ over K, *consists of all formal*

5. MAXIMAL VALUATION RINGS

power series (5) with support in the positivity domain Γ^+ of Γ. This $K[[\Gamma]]$ is a maximal valuation domain with value group Γ and residue class field K.

<u>Proof.</u> We only sketch the proof, for details we refer to Schilling [1].

The first step is to establish, by transfinite induction, an inverse g of each $f \in K[[\Gamma]]$ with $v(f) = 0$. To find g with $fg = 1$, set $g0 = (f0)^{-1}$. Let $0 < \gamma \in S = \text{supp } f$ and suppose $g\alpha$ is defined for every $\alpha < \gamma$ ($\alpha \in \Gamma^+$) such that $(fg)\alpha = f_0\alpha$. To define $g\gamma$, observe that

$$0 = f_0\gamma = (fg)\gamma = f0 \cdot g\gamma + \sum_{0 \neq \alpha \in S} f\alpha \cdot g(\gamma - \alpha)$$

where the $g(\gamma-\alpha)$ have already been defined. Hence $g\gamma$ can be computed.

As every formal power series f can be written as $f = f_{v(f)} f'$ with $v(f') = 0$, it is evident that Q is the quotient field of $K[[\Gamma]]$. It is also clear that Γ is the value group of v.

In order to establish maximality, consider a family of cosets with the finite intersection property: $\{g_i + L_i \mid i \in I\}$ where L_i are ideals and g_i elements of $K[[\Gamma]]$. In view of (3.2), we consider the filters $v(L_i)$ in Γ^+, and note that, for $L_i \leq L_j$, we have $g_i - g_j \in L_j$ which amounts to $g_i\alpha = g_j\alpha$ for all $\alpha \notin v(L_i)$. Thus the function g is well-defined by setting $g\alpha = g_i\alpha$ if $\alpha \notin v(L_i)$ for some i and $g\alpha = 0$ otherwise. It is straightforward to check that supp g is well-ordered and $g \in g_i + L_i$ for all $i \in I$. □

EXERCISES

1. Show that a valuation domain R is maximal if and only if Q is a linearly compact R-module in the discrete topology.

2. A valuation domain R is almost maximal exactly if Q/L is linearly compact in the discrete topology, for every ideal $L \neq 0$ of R.

3. Verify (5.9) for primes S contained in $\mathbb{Z}$ (in this case, R_S is no longer a subring in Q).

4. Let R_v be the valuation domain containing $K[\Gamma^+]$ as defined in Ex.6 in §3. Show that $K[[\Gamma^+]]$ is an immediate extension of R_v.

5. Let R_v have the same meaning as in Ex.4. It is almost maximal exactly if $\Gamma \cong \mathbb{Z}$.

6. (Klatt and Levy [1]) Show that a maximal valuation ring R with $P = \mathbb{Z}$ is self-injective (i.e. injective as an R-module). (The converse holds true as well.)

7. Using the preceding exercise, show that $R_{\mathbb{Z}}$ is the injective hull of R.

8. Let S be a maximal immediate extension of the valuation domain R. Show that

$$I_F = \{a \in R \mid aS + R < S\}.$$

§6. PRÜFER DOMAINS

From time to time, we shall have occasion to state and prove theorems for more general domains than valuation domains. Of these, the most important ones are the Prüfer domains.

Throughout this section, R will denote an integral domain and Q its field of quotients.

By a <u>fractional ideal</u> I of R is meant an R-submodule of Q such that $rI \leq R$ for some $0 \neq r \in R$. A fractional ideal I is <u>invertible</u> if $IJ = R$ for a suitable fractional ideal J of R. Principal fractional ideals $\neq 0$ are always invertible. It is easy to see that invertible ideals are finitely generated; in fact, if $\sum_{i=1}^{n} a_i b_i = 1$ ($a_i \in I$, $b_i \in J$), then $I = \sum_{i=1}^{n} Ra_i$ follows easily.

The next result is a most important characterization of invertible ideals.

THEOREM 6.1. [D] (Cartan-Eilenberg [1]) A fractional ideal $\neq 0$ of

6. PRÜFER DOMAINS

R is invertible if and only if it is a projective R-module.

Proof. Assume I invertible and $IJ = R$. Setting $a_1 b_1 + \ldots + a_n b_n = 1$ with $a_i \in I$, $b_i \in J$, the map $\psi : I \to \bigoplus_i^n R$ defined by $\psi : x \longmapsto (b_1 x, \ldots, b_n x)$ $(x \in I)$ is a monomorphism. The homomorphism $\phi : \bigoplus_1^n R \to I$ defined by $\phi : (r_1, \ldots, r_n) \longmapsto a_1 r_1 + \ldots + a_n r_n$ satisfies $\phi\psi = 1_I$. Hence I is a summand of $\bigoplus_1^n R$ and I is projective.

Conversely, let I be a projective ideal of R, and $\phi : F \to I$, $\psi : I \to F$ homomorphisms such that F is a free R-module and $\phi\psi = 1_I$. As R is a domain, $\text{Im}\,\psi$ is contained in a finitely generated summand of F; hence without loss of generality we can assume that F (and hence I) is finitely generated. Write $F = \bigoplus_1^n R x_i$ where let ε_i denote the ith projection. As ϕ has to be epic, we have $I = \sum_{i=1}^n R a_i$ with $a_i = \phi x_i$. The map $\varepsilon_i \psi : I \to R x_i$ can be viewed as a homomorphism between two submodules of Q, therefore it is a multiplication by an element $b_i \in Q$, i.e. $\varepsilon_i \psi a = b_i a x_i$ for all $a \in I$. Setting $J = \sum_1^n R b_i$, we have $IJ = R$, as is clear from

$$a = \phi\psi a = \phi(a \Sigma b_i x_i) = a \Sigma b_i a_i \,. \quad \square$$

By a <u>Prüfer domain</u> is mean a domain in which all finitely generated ideals $\neq 0$ are invertible. In view of (6.1), this means that it is a semihereditary domain. A Noetherian Prüfer domain is called a <u>Dedekind domain</u>; here all non-zero ideals are invertible. A <u>Bézout domain</u> is a domain in which the finitely generated ideals are principal.

LEMMA 6.2. A domain is a valuation domain if and only if it is a local Bézout domain.

Proof. It suffices to show that a local Bézout domain R has to be a valuation ring. Given $a, b \in R$, $Ra + Rb = Rc$ for some $c \in R$. If we set $a = a'c$, $b = b'c$ for $a', b' \in R$, then $Ra' + Rb' = R$. As R is local, either $Ra' = R$ or $Rb' = R$. Thus either $Rb \leqq Ra$ or $Ra \leqq Rb$. $\quad \square$

The fundamental relation between Prüfer domains and valuation domains is formulated in the next theorem. First, a preliminary lemma where R_M denotes R localized at M.

LEMMA 6.3. [D] Let I be a finitely generated ideal $\neq 0$ of R such that $I_M = IR_M$ is a principal ideal for every maximal ideal M of R. Then I is invertible.

Proof. Manifestly, $J = \{q \in Q \mid qI \leq R\}$ is a fractional ideal of R. If IJ were a proper ideal of R, say contained in the maximal ideal M of R, then writing $I_M = R_M a$ with $a \in I$ and choosing $s \in R \backslash M$ with $sI \leq Ra$, we would have $a^{-1}s \in J$. This would lead to the contradiction $s = a \cdot a^{-1}s \in IJ \leq M$. □

THEOREM 6.4. For a domain R, the following are equivalent:
  (a) R is a Prüfer domain;
  (b) R_T is a valuation domain for every prime $T \neq 0$;
  (c) R_M is a valuation domain for every maximal ideal M.

Proof. (a) $\implies$ (b) Let T be a prime ideal $\neq 0$ of the Prüfer domain R, and $J \neq 0$ a finitely generated ideal of R_T, say $J = \sum_{1}^{n} R_T a_i s_i^{-1}$ ($a_i \in R$, $s_i \in R \backslash T$). Then $J = I_T$ where $I = \sum_{1}^{n} Ra_i$ is by hypothesis invertible. It is readily checked (Ex.1) that an invertible ideal of a local domain is principal. Hence R_T is a local Bézout domain; thus (6.2) implies (b).

  (b) $\implies$ (c) There is nothing to prove.

  (c) $\implies$ (a) Let R_M be a valuation domain for every maximal ideal M of R, and assume I is a finitely generated ideal $\neq 0$ of R. Evidently, I_M is a finitely generated R_M-module, hence singly generated. From (6.3) we infer that I is invertible. □

A domain R is said to be <u>h-local</u> if it satisfies the following conditions:
  (i) every non-zero prime ideal of R is contained in exactly one maximal ideal of R;
  (ii) every non-zero element of R is contained in but a finite number of maximal ideals of R.

h-local domains admit various characterizations (see Matlis [3]); the one given in the following proposition will suffice for our purposes.

6. PRÜFER DOMAINS

LEMMA 6.5. A Prüfer domain R is h-local if and only if every torsion R-module T decomposes as $T \cong \oplus_P T_P$, where P ranges over all the maximal ideals of R and $T_P = T \otimes_R R_P$.

Proof. In order to prove sufficiency, let I be a non-zero ideal of R; notice that $(R/I)_P \neq 0$ if and only if $I \leq P$. From $R/I \cong \oplus_P (R/I)_P$ and from the obvious fact that R/I cannot be an infinite direct sum of submodules, it follows that I is contained but in a finite number of maximal ideals. In particular, (ii) holds. If I is prime, then it cannot be a finite intersection of properly larger ideals. Therefore, in the above decomposition of R/I only one summand is not zero; thus I is contained only in one maximal ideal.

Conversely, assume R to be h-local and T to be a torsion R-module. Using the notation $K = Q/R$, we have $T \cong \text{Tor}_1(K,T)$ and $T_P \cong \text{Tor}_1(K_P, T)$ for maximal ideals P of R; indeed the first isomorphism is in (III.3.1), while the second follows similarly. R h-local implies that there is a canonical isomorphism $K \cong \oplus_P K_P$ (see Ex.5), therefore we deduce:

$$T \cong \text{Tor}_1(K,T) \cong \text{Tor}_1(\oplus_P K_P, T) \cong \oplus_P \text{Tor}_1(K_P, T) \cong \oplus_P T_P . \quad \square$$

EXERCISES

1. Either use Kaplansky's theorem that projective modules over local rings are free or prove directly that projective ideals in a local domain are principal.

2. Let R be a Prüfer domain, S a multiplicatively closed subset of R and T a prime ideal of R. Show that R_S and R/T are Prüfer domains.

3. If V is a valuation domain between a Prüfer domain R and its quotient field, then V is a localization of R at a prime ideal.

4. The intersection of two finitely generated ideals in a Prüfer domain R is likewise finitely generated.

5. Show that if R is an h-local Prüfer domain and $K = Q/R$, then $K \cong \oplus_P K_P$, where P ranges over the maximal ideals of R. (Hint: let $R^P = \cap_{M \neq P} R_M$; show that $K_P \cong R^P/R$ and $Q = \sum_P R^P$.)

6. The ring of all algebraic integers is a Prüfer domain.

NOTES

Valuation domains were discovered by W. Krull in his fundamental paper on the general (non-archimedean) valuation of fields, published in 1932. He also proved that the integrally closed domains are precisely the intersections of valuation domains. All Prüfer domains are integrally closed. Integrally closed domains D satisfying the following condition are Prüfer domains: there is an integer $n > 1$ such that, for all $a, b \in D$, $(a,b)^n = (a^n, b^n)$ holds.

Valuation rings are natural generalizations of valuation domains by admitting zero-divisors. Shores [1] has proved that they can be obtained in the same fashion as valuation domains via valuations of rings with values in certain totally ordered semigroups rather than groups. Surprisingly, some of the relevant properties of fields with valuations carry over to this more general case. A well-known open question attributed to I. Kaplansky has been solved recently: Not every valuation ring is a factor ring of a valuation domain modulo an ideal; see (VII.3.5).

PROBLEM 1. Find conditions on a valuation ring to be a quotient of a valuation domain.

For a partial solution, see Ohm-Vicknair [1].

Most recently, an unexpected application of valuation domains has surfaced. A real-closed domain is defined as a totally ordered domain, not a field, which satisfies the Intermediate Value Theorem for polynomials (see Cherlin-Dickmann [1], Becker [1]). Real-closed domains turn out to be Henselian valuation domains.

II. Preliminaries on Modules

This chapter provides the setting for these notes. The notion of divisibility and its relative version occupy a central position in this chapter. Their roles will become increasingly important as our subject develops.

We refer to Anderson-Fuller [1] for basic definitions and fundamental results on modules.

§1. MODULES

Throughout we shall consider modules either over domains or over valuation rings, unless stated otherwise. All modules are unital. As our rings are commutative, we need not specify the side on which ring elements operate on modules.

The rudiments of module theory will be taken for granted. Our convention is to write $N \leq M$ ($N < M$) if N is a submodule (proper submodule) of M. The submodule generated by a subset $X = \{\ldots, x_i, \ldots\}$ will be denoted by $\langle X \rangle$ or $\langle \ldots, x_i, \ldots \rangle$ or

$\Sigma R x_i$. The symbols $\oplus$ and Π stand for direct sums and products, respectively.

Let R denote any commutative ring.

If M is an R-module and $r \in R$, we write

$$rM = \{ra \mid a \in M\}.$$

This is always a submodule of M. If I is an ideal of R, then IM will denote the submodule of M consisting of all finite sums $\Sigma r_k a_k$ with $r_k \in I$ and $a_k \in M$. In a valuation ring R, one of these r_k divides all the rest, so we have

$$IM = \{ra \mid r \in I, a \in M\}.$$

It is routine to check that for an ideal I of R and a submodule N of M, we have always

$$I(M/N) = (IM + N)/N. \tag{1}$$

By the <u>annihilator</u> of an element a of a module M is meant the ideal

$$\text{Ann}_R a = \{r \in R \mid ra = 0\}.$$

Frequently, we write simply $\text{Ann } a$ if there is no danger of confusion. For an R-module M, for $r \in R$ and an ideal I of R, we set

$$M[r] = \{a \in M \mid ra = 0\}$$

and

$$M[I] = \{a \in M \mid \text{Ann } a \geq I\}.$$

These are submodules of M, and so is

$$M[I^+] = \{a \in M \mid \text{Ann } a > I\}$$

provided that R is a valuation ring.

The <u>annihilator</u> $\text{Ann } M$ of an R-module M will be defined to distinguish between the cases when M contains an element of minimal annihilator or does not. Setting $I = \cap \{\text{Ann } a \mid a \in M\}$, we define

$$\text{Ann } M = I \text{ or } I^+$$

according as $\text{Ann } a_0 = I$ for some $a_0 \in M$, or $\text{Ann } a > I$ for all

1. MODULES

$a \in M$. We agree to regard I^+ as a larger annihilator than I, but smaller than J if J is an ideal of R properly containing I.

The following is a useful rule for annihilators over any commutative ring.

LEMMA 1.1. *If* $ra \neq 0$ *and* $rM \neq 0$, *then*

$$\text{Ann } ra = (\text{Ann } a):r \quad \text{and} \quad \text{Ann } rM = (\text{Ann } M):r.$$

Proof. This is readily checked. □

If R is a domain, then it makes sense to talk of torsion and torsion-free R-modules. The definitions are obvious: an R-module M is **torsion** if Ann $a \neq 0$ for all $a \in M$, and **torsion-free** if Ann $a = 0$ for all $0 \neq a \in M$. A **mixed** module is one which is neither torsion nor torsion-free. In a mixed module M, the elements with non-zero annihilators form a submodule of M, the **torsion submodule** tM. We say M is **splitting** if tM is a summand in M.

M is **bounded** if $rM = 0$ for some nonzero $r \in R$.

A **cyclic** R-module is a module of the form Ra (i.e. it can be generated by a single element). Obviously, $Ra \cong R/\text{Ann } a$. If Ann a is a principal ideal Rx ($x \in R$), then $Ra \cong R/Rx$ is **cyclically presented**. As usual, we say that the module M is **finitely generated** if it has a finite set $\{a_1,\ldots,a_k\}$ of generators, i.e. $M = \sum_{i=1}^{k} Ra_i$. It is, moreover, **finitely presented** if there exist a finitely generated free R-module F and a finitely generated submodule G of F such that $M \cong F/G$. It is well known that then in every presentation, $M \cong F'/G'$ where F' is finitely generated free, G' is necessarily finitely generated.

A decisive role is played in our discussions by the uniserial modules. An R-module U is called **uniserial** if its submodules form a chain under inclusion. Equivalently, for all $u,v \in U$, either $u \in Rv$ or $v \in Ru$. Submodules and quotients of uniserial modules are likewise uniserial. The most frequently used uniserial modules are the valuation ring R itself, the ideals of R and the ring Q of quotients of R, as well as $K = Q/R$, and last but not least the cyclic R-modules.

For a module M, let us set
$$A(M) = \{\text{Ann } a \mid 0 \neq a \in M\}.$$

PROPOSITION 1.2. [VR] Let U be a uniserial R-module. Then $A(U)$ consists of all proper ideals in an equivalence class of ideals which are $\geq \text{Ann } U$.

<u>Proof.</u> All what we have to note is that $\text{Ann } ru = \text{Ann } u:r$, and so $\text{Ann } u$ and $\text{Ann } ru$ are always equivalent ideals for all $u \in U$, $ru \neq 0$. □

Clearly, $\text{Ann } U \in A(U)$ or $\notin A(U)$ according as $\text{Ann } U = I$ or I^+.

EXERCISES

1. [VD] Show that $M[I]$ and $M[I^+]$ are fully invariant submodules of M and every homomorphism $M \to N$ carries $M[I]$ into $N[I]$, and $M[I^+]$ into $N[I^+]$.

2. The <u>socle</u> $S(M)$ of a module M is defined to be the submodule generated by all simple submodules of M. Prove that if R is a valuation ring, then all simple R-modules are isomorphic and $S(M) = M[P]$.

3. Show that for a valuation ring R and an ideal I of R, $S(R/I) \cong R/P$ or 0 according as I is equivalent to P or not.

4. [VD] Describe $A(Q/J)$ for an ideal J of R.

5. [VD] For an R-module M, define $M^\# = \{r \in R \mid rM < M\}$ and $M_\# = \{r \in R \mid ra = 0 \text{ for some } 0 \neq a \in M\}$. Show that both are prime ideals of R.

6. [D] Prove that direct limits of torsion-free modules are torsion-free, and those of torsion modules are torsion.

7. [D] Inverse limits of torsion-free modules are torsion-free, but those of torsion modules may even be torsion-free.

§2. DIVISIBILITY

The notion of divisibility is absolutely fundamental for modules over domains R. We depart from the standard conventions and use the same definition for divisibility if R is any valuation ring. However, mostly divisibility by regular elements of R will be considered, even if this is not always stated explicitly.

Let M be an R-module and $a \in M$, $r \in R$. We say that a is <u>divisible</u> by r in M, or r <u>divides</u> a (in notation: $r|a$) if the equation

$$rx = a \qquad (1)$$

is solvable in M, i.e. some $b \in M$ satisfies $rb = a$. Clearly, $r|a$ if and only if $a \in rM$.

Notice that along with $b \in M$, every element of $b + M[r]$ is a solution to (1). Unique solvability holds in torsion-free modules.

Evidently, $r|a$ and $r|b$ imply $r|a + b$. Moreover, if $M = A \oplus B$ and $a \in A$, $b \in B$, then $r|a + b$ only if $r|a$ and $r|b$.

The submodule

$$M^1 = \cap\{rM \mid 0 \neq r \in R\}$$

of the R-module M consists of all $a \in M$ such that $r|a$ for every $0 \neq r \in R$. It is called the <u>(first) Ulm submodule</u> of M.

Next we formulate and prove a most relevant and frequently used result; it is elementary, but perhaps not as trivial as it sounds.

LEMMA 2.1. [VR] Let U be a uniserial R-module, R a valuation ring, and $a \in U$, $r,s \in R$. If $ra \neq 0$ is divisible in U by rs, then s divides a. In other words, $0 \neq ra \in rsU$ implies $a \in sU$.

<u>Proof</u>. Let $b \in U$ satisfy $rsb = ra$; then $r(sb-a) = 0$. As $ra \neq 0$, we must have $R(sb-a) < Ra$, i.e. $sb - a = pa$ for some $p \in P$. We obtain $sb = (1+p)a$. Here $1+p$ is a unit of R, so s divides a. □

From now on, assume that R is a domain.

An R-module D is said to be <u>divisible</u> if $rD = D$ for all $0 \neq r \in R$. In view of the unique divisibility in torsion-free modules, every element of a torsion-free divisible module can be embedded in a submodule isomorphic to Q. Therefore a torsion-free divisible module is a Q-vectorspace, and hence a direct sum of copies of Q. (As we shall see in (VI.4.1), it is moreover injective.) In any R-module M, the union of all divisible submodules is again one which we shall denote by dM. Extensions of divisible modules by divisible modules ought to be again divisible; hence it is clear that

$$d(M/dM) = 0.$$

A module M satisfying $dM = 0$ may be called <u>reduced</u>.

It is readily verified that an R-module D is divisible if and only if every homomorphism $Rr \to D$ (r any element of R) can be extended to a homomorphism $R \to D$. Using the exact sequence $0 \to Rr \to R \to R/Rr \to 0$ we derive the exact sequence

$$\text{Hom}_R(R,D) \to \text{Hom}_R(Rr,D) \to \text{Ext}^1_R(R/Rr,D) \to \text{Ext}^1_R(R,D) = 0.$$

From our remark we conclude that the divisibility of D is equivalent to the surjectivity of the first map in the last exact sequence which is obviously equivalent to $\text{Ext}^1_R(R/Rr,D) = 0$. This leads to the first part of the following characterization of divisibility.

PROPOSITION 2.2. [D]   An R-module D is divisible if and only if

$$\text{Ext}^1_R(R/Rr,D) = 0 \qquad \text{for every } r \in R.$$

This holds exactly if

$$\text{Ext}^1_R(R/L,D) = 0$$

for each projective ideal L of R.

<u>Proof</u>. As above we can argue that what we have to verify amounts to the extensibility of every homomorphism $\phi : L \to D$ from a projective ideal L into a divisible module D to a map

2. DIVISIBILITY

$\psi: R \to D$. Projective ideals are by (I.6.1) invertible, thus there exist $r_1,\ldots,r_n \in L$, $q_1,\ldots,q_n \in Q$ such that $q_i L \leq R$ and $\sum_{i=1}^{n} q_i r_i = 1$. By divisibility, $\phi r_i = r_i d_i$ for suitable $d_i \in D$ ($i = 1,\ldots,n$). Setting $d = \sum_{i=1}^{n} (q_i r_i) d_i \in D$, we have

$$\phi r = \phi(\sum_i q_i r_i r) = \sum_i (q_i r) \phi r_i = \sum_i q_i r r_i d_i = rd$$

for each $r \in L$. Hence $1 \longmapsto d$ induces a desired extension $\psi: R \to D$. □

A concept that is slightly stronger than divisibility is the so-called h-divisibility (Matlis [2]). An R-module H is h-<u>divisible</u> if it is an epimorphic image of an injective R-module. Manifestly, h-divisibility (just as divisibility) is inherited under epimorphisms. Since every injective R-module is an epic image of a direct sum of copies of Q, it is evident that a module over a domain is h-divisible exactly if it is an epic image of some $\oplus Q$ (i.e. a torsion-free injective R-module).

It is straightforward to check that the last characterization of h-divisibility can be rephrased as follows: H is h-divisible exactly if every homomorphism $R \to H$ can be extended to a homomorphism $Q \to H$.

It is pretty obvious that the union hM of all h-divisible submodules of an R-module M is again h-divisible. Evidently, $hM \leq dM$. However, in general $h(M/hM) \neq 0$ (see (VI.1.4)). It will be convenient to call an R-module M h-<u>reduced</u> whenever $hM = 0$.

It is natural to ask for a characterization of those domains over which divisible modules are necessarily h-divisible. In the next lemma we offer a sufficient condition (which turns out to be necessary for valuation domains, cf. (VI.1.3)).

LEMMA 2.3. [D] If Q is a countably generated R-module, then every divisible R-module is h-divisible.

<u>Proof</u>. Let $t_1,\ldots,t_n,\ldots$ be a sequence in R such that the

elements $t_1^{-1} t_2^{-1} \ldots t_n^{-1}$ (for all n) generate Q. Assume a belongs to a divisible R-module D. Then there is a sequence $a_0 = a, a_1, \ldots, a_n, \ldots$ of elements in D such that $t_n a_n = a_{n-1}$ ($n \geq 1$). Evidently, the correspondence $1 \longmapsto a$, $t_1^{-1} \ldots t_n^{-1} \longmapsto a_n$ ($n \geq 1$) gives rise to a homomorphism $\eta : Q \to D$. Hence D is h-divisible. □

The following example shows that hM need not be a summand of M.

EXAMPLE 2.4. Let H be the direct sum of countably many copies of Q/R; thus H is h-divisible. If R is not Noetherian, then H is not injective. We choose for M a submodule of the injective hull of H, generated by H and an element $a \notin H$. Then hM = H (as M/H is cyclic), but H is essential in M, so it is not a summand.

We will resume the study of divisible and h-divisible modules in Chapter VI.

EXERCISES

1. [VD] Prove that an R-module D is divisible if and only if, for every countably generated ideal J of R and for every $a \in J$, each homomorphism $Ra \to D$ extends to a homomorphism $J \to D$.

2. [D] Let M be a torsion-free R-module and $\alpha : M \to H$ where H is h-divisible. Then α extends to a homomorphism $M \otimes_R Q \to H$.

3. [D] A direct product of modules is divisible (h-divisible) exactly if all components are.

4. [D] Divisibility and h-divisibility are preserved under direct limits (but not under inverse limits).

5. [D] Show that every injective R-module is the epic image of a direct sum of copies of Q, and every injective torsion R-module is the epic image of a direct sum of copies of Q/R.

§3. RELATIVE DIVISIBILITY (RD)

A submodule N of the module M is called <u>relatively divisible</u>, or briefly an <u>RD-submodule</u> (Warfield [1]), if

$$rN = N \cap rM \qquad \text{for all } r \in R.$$

As the inclusion $rN \leq N \cap rM$ holds for all submodules N of M, it is clear that the relative divisibility of N amounts to the condition that if $a \in N$ and the equation $rx = a$ has a solution in M, then it is solvable in N too.

Our discussion will culminate in showing that for modules over Prüfer domains, relative divisibility is equivalent to purity (see (4.5)).

To start with, we list a few elementary properties of relative divisibility over arbitrary domains. The proofs are left to the reader.

(A) Direct summands are RD-submodules.

(B) If, for a submodule N of M, M/N is torsion-free, then N is RD in M.

(C) If M is torsion-free, then a submodule N of M is RD in M exactly if M/N is torsion-free. Therefore intersections of RD-submodules of a torsion-free module M are again RD-submodules, and it makes sense to talk of the RD-submodule $\langle X \rangle_*$ generated by a subset X of M. This consists of all $y \in M$ such that, for some $r \neq 0$ in R (depending on y), ry belongs to the submodule generated by X.

(D) The RD-property is inductive: the union of a chain of RD-submodules of M is again an RD-submodule of M.

(E) The RD-property is transitive: if $H \leq N \leq M$ where H is RD in N, and N is RD in M, then H is an RD-submodule of M.

(F) If $H \leq N \leq M$, then N RD in M implies that N/H is RD in M/H.

(G) If $H \leq N \leq M$ such that H is RD in M and N/H is RD in M/H, then N is RD in M.

From these properties it is evident that relative divisibility gives rise to a relative homological theory.

We continue with various characterizations of relative divisibility. Let us agree to call an exact sequence $0 \to N \to M \to H \to 0$ <u>RD-exact</u> if the map $N \to M$ embeds N in M as an RD-submodule.

LEMMA 3.1. For an exact sequence $0 \to N \to M \to H \to 0$ of R-modules, the following are equivalent:

(i) the sequence is RD-exact;

(ii) the induced map $\operatorname{Hom}_R(R/Rr, M) \to \operatorname{Hom}_R(R/Rr, H)$ is surjective for all $r \in R$;

(iii) the induced map $R/Rr \otimes_R N \to R/Rr \otimes_R M$ is injective for all $r \in R$.

<u>Proof.</u> Notice that (i) and (ii) are equivalent, since (ii) holds exactly if, for all $x \in M$ with $rx \in N$ there is a $y \in N$ such that $rx = ry$. In order to verify the equivalence of (i) and (iii), consider the commutative diagram

$$\begin{array}{ccccccc} R \otimes_R N \cong N & \xrightarrow{r} & R \otimes_R N \cong N & \longrightarrow & R/Rr \otimes_R N & \longrightarrow & 0 \\ \downarrow & & \downarrow & & \downarrow \delta & & \\ R \otimes_R M \cong M & \xrightarrow{r} & R \otimes_R M \cong M & \longrightarrow & R/Rr \otimes_R M & \longrightarrow & 0 \end{array}$$

where the rows are induced by the exact sequence $0 \to R \xrightarrow{r} R \to R/Rr \to 0$ (here r denotes multiplication by r). Easy diagram chasing shows that $\operatorname{Ker} \delta = 0$ if and only if $rN = N \cap rM$. □

Observe that (ii) amounts to the projective property of cyclically presented modules R/Rr relative to RD-exact sequences. This leads inevitably to the question of enough relative projectives.

Before entering into the discussion of this question let us point out that the proof of (2.2) applies to conclude from the

3. RELATIVE DIVISIBILITY

projectivity of R/Rr relative to RD-exact sequences that the following holds true over domains:

LEMMA 3.2. [D] The cyclic R-modules of projective dimension 1 have the projective property relative to RD-exact sequences.

Proof. All what we have to note that R/L has projective dimension 1 exactly if L is projective; cf. Kaplansky's lemma in IV.§2. □

For a generalization of (3.2), see (4.4).

LEMMA 3.3. Every R-module M can be embedded in an RD-exact sequence

$$0 \longrightarrow N \longrightarrow C \overset{\phi}{\longrightarrow} M \longrightarrow 0$$

where C is a direct sum of cyclically presented R-modules.

Proof. For every $a \in M$ and every $r \in \text{Ann } a$, consider maps

$$\phi_{a,r} : R/Rr \to Ra$$

defined via $1 + Rr \longmapsto a$, and set $C = \oplus R/Rr$ for all these $\phi_{a,r}$. These maps induce an epimorphism $\phi : C \to M$ whose kernel N has to be RD in C as condition (ii) of (3.1) is satisfied. □

Call an R-module <u>RD-projective</u> if it has the projective property relative to RD-exact sequences. We can easily derive:

THEOREM 3.4. A module over a ring R is RD-projective if and only if it is a summand of a direct sum of cyclically presented R-modules. Such a summand is itself a direct sum of cyclically presented modules whenever R is a valuation ring.

Proof. We have already noticed that cyclically presented modules are RD-projective; hence so are the summands of their direct sums. Standard arguments apply to derive from (3.3) that these are the only RD-projectives.

As the R-endomorphism ring of R/Ra is evidently isomorphic to R/Ra, the second claim follows from (7.3). □

In the balance of this section, we consider valuation rings. The following lemma is quite easy.

LEMMA 3.5. Let R be a valuation ring. A cyclic submodule Ra in an R-module M is RD exactly if

$$\text{Ann } a \geq \text{Ann } x \quad \text{for all } x \in a + PM.$$

Proof. Suppose Ra is RD in M, and $a+y$ ($y \in PM$) satisfies $r(a+y) = 0$, $ra \neq 0$ for some $r \in R$. Writing $y = px$ for some $p \in P$, $x \in M$, we obtain $0 \neq ra = -rpx$ which contradicts the RD-property of Ra in M (cf. (2.1)).

Conversely, let a have a maximal annihilator in $a + PM$. If there exist $r,s \in R$ with $Rs < Rr$ and $0 \neq ra = sx$ for some $x \in M$, then setting $s = pr$ ($p \in P$), we see that $a - px \in a + PM$ is annihilated by r. By the maximality of Ann a, we obtain $r \in \text{Ann } a$, thus $ra = 0$, again a contradiction. □

We now have the important

THEOREM 3.6. (Fuchs-Salce [1]) For every finitely generated R-module M over a valuation ring R, there is a finite chain

$$0 = M_0 < M_1 < \ldots < M_n = M \tag{1}$$

of submodules such that
  (i) each M_i is an RD-submodule of M;
  (ii) M_{i+1}/M_i is cyclic, for every i.
Moreover, (1) can be chosen such that

$$\text{Ann } M = \text{Ann } M_1/M_0 \leq \ldots \leq \text{Ann } M_n/M_{n-1}. \tag{2}$$

Proof. As M is finitely generated, the R/P-vectorspace M/PM has a finite basis; let $\{a_j + PM\}_{j=1,\ldots,m}$ be a basis for M/PM. By Nakayama's lemma, the elements $a_1,\ldots,a_m$ generate M. Manifestly, Ann $M = \cap \text{Ann } a_j$, so Ann $M = \text{Ann } a_j$ for at least one index j, and at least one of these a_j, say a_m, cannot be replaced by any element in its coset mod PM with a larger annihilator. By (3.5), Ra_m is an RD-submodule of M; and we set

3. RELATIVE DIVISIBILITY

$M_1 = Ra_m$. We now induct on the number of generators, applying the induction hypothesis to M/M_1 and noting that RD-submodules of M/M_1 have RD preimages in M (see (G)). The way $M_1, \ldots$ are chosen guarantees that (2) can be satisfied. □

In the special case when M is finitely presented, a more powerful result can be established:

THEOREM 3.7. (Warfield [3]) A finitely presented module over a valuation ring is a finite direct sum of cyclically presented modules.

Proof. The quotient of a finitely presented module M mod a finitely generated submodule is again finitely presented. Therefore in (1), M/M_{n-1} is cyclic and finitely presented, so it is cyclically presented. Hence it is RD-projective, and we conclude that M_{n-1} is a summand of M, $M \cong M_{n-1} \oplus M/M_{n-1}$. Here M_{n-1} is likewise finitely presented and has a smaller number of generators than M. An obvious induction completes the proof. □

There is an additional aspect of (3.6) which we shall explore later on [cf. (IX.1.1)].

EXERCISES

1. [D] If N is an RD-submodule of M, then $N \cap rN:s = N \cap (rM:s)$ for all $r, s \in R$.

2. [D] If N is RD in M, then rN is RD in rM for every $r \in R$.

3. [D] An exact sequence $0 \to N \to M \to H \to 0$ is RD-exact if and only if the induced sequence $0 \to rN \to rM \to rH \to 0$ is exact for every $r \in R$.

4. [D] Show that N is RD in M exactly if
$$(M/N)[r] = (M[r] + N)/N \quad \text{for each } r \in R.$$

5. [D] The direct limit of RD-exact sequences is again RD-exact.

6. [VR] Let $a \in M$ be such that $\text{Ann } a = \text{Ann } M$. If $\text{Ann } M$ is an archimedean ideal, then Ra is RD in M.

7. [VR] Let I be a non-archimedean ideal of R and $r \in P$ such that $rI = I$. Then R/I is not RD in $r^{-1}R/I$. Conclude that in Ex.6, "archimedean" is a relevant condition.

8. (Stacked Basis Theorem for Valuation Domains) If F is a finitely generated free R-module and H is a finitely generated submodule of F, then there is a decomposition $F = Rx_1 \oplus \ldots \oplus Rx_n$ such that $H = Rr_1 x_1 \oplus \ldots \oplus Rr_n x_n$ for suitable $r_1, \ldots, r_n \in R$.

§4. PURE SUBMODULES

Our intention is to consider briefly the notion of purity, in the sense generally accepted in module theory, and to compare it with the RD-property discussed in the preceding section. The main goal is to show that for valuation rings and Prüfer domains the two concepts coincide.

Recall the customary definition of purity: a submodule A of an R-module B is <u>pure</u> if every finite system of equations over A:

$$\sum_{j=1}^{m} r_{ij} x_j = a_i \in A \quad (i = 1, \ldots, n) \tag{1}$$

with $r_{ij} \in R$ and unknowns $x_1, \ldots, x_m$, has a solution in A whenever it is solvable in B. Evidently, purity implies relative divisibility.

If F is the free R-module on $\{x_1, \ldots, x_m\}$ and H is the submodule generated by the left members of (1), then the correspondence $\Sigma r_{ij} x_j \longmapsto a_i$ ($i = 1, \ldots, n$) gives rise to a homomorphism ϕ of H into A, and a solution $x_j = b_j \in B$ ($j = 1, \ldots, m$) of (1) defines an extension $\psi : F \to B$ (via $x_j \longmapsto b_j$) of ϕ. Hence the purity of A in B can be rephrased as follows: for a commutative diagram

4. PURE SUBMODULES

$$\begin{array}{ccccccc}
0 & \longrightarrow & H & \xrightarrow{\gamma} & F & & \\
& & \phi \downarrow & \swarrow \sigma & \downarrow \psi & & \\
0 & \longrightarrow & A & \longrightarrow & B & \longrightarrow & B/A & \longrightarrow & 0
\end{array}$$

with exact rows where F is finitely generated free and H finitely generated, there exists a map $\sigma : F \to A$ such that $\sigma\gamma = \phi$.

A technical lemma (valid over arbitrary rings) will be required.

LEMMA 4.1. Let

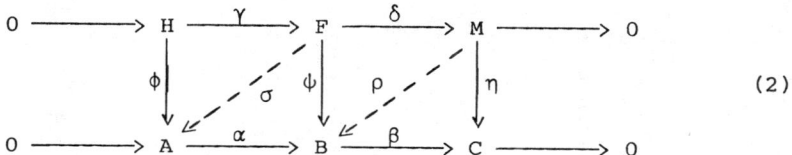 (2)

be a diagram with exact rows and commutative squares. There exists a map $\sigma : F \to A$ such that $\sigma\gamma = \phi$ if and only if there exists a map $\rho : M \to B$ such that $\beta\rho = \eta$.

Proof. Suppose $\rho : M \to B$ exists such that $\beta\rho = \eta$. Then $\sigma' = \psi - \rho\delta$ satisfies $\beta\sigma' = \beta\psi - \beta\rho\delta = \eta\delta - \eta\delta = 0$, so there is a map $\sigma : F \to A$ with $\alpha\sigma = \sigma'$. It satisfies $\alpha\sigma\gamma = \psi\gamma - \rho\delta\gamma = \alpha\phi$, i.e. $\sigma\gamma = \phi$ as desired.

Conversely, if $\sigma : F \to A$ satisfies $\sigma\gamma = \phi$, then $(\psi - \alpha\sigma)\gamma = \alpha\phi - \alpha\phi = 0$ implies the existence of a $\rho : M \to B$ such that $\rho\delta = \psi - \alpha\sigma$. Hence $\beta\rho\delta = \beta\psi - \beta\alpha\sigma = \eta\delta$ implies $\beta\rho = \eta$. □

It is now easy to establish:

COROLLARY 4.2. An exact sequence is pure-exact if and only if all finitely presented R-modules have the projective property relative to it.

Proof. It suffices to note that, given an $\eta : M \to C$ and the bottom row in (2), the rest of (2) can be filled in provided that F is projective. Furthermore, if M is finitely presented and F

finitely generated, then H is again finitely generated. □

By virtue of (3.7), a finitely presented module over a valuation ring is a direct sum of cyclically presented modules. Hence the comparison of (3.1) with (4.2) leads at once to

PROPOSITION 4.3. (Warfield [1]) For modules over valuation rings, RD-property and purity are equivalent. □

We wish to show that this generalizes to Prüfer domains. In the following lemma, by a <u>special</u> finitely presented module M is meant one which has a chain

$$0 = M_0 < M_1 < \ldots < M_n = M \qquad (3)$$

of submodules such that each M_i/M_{i-1} is cyclic of projective dimension 1. An example of such a module is the module generated by $\{x, y_1, \ldots, y_{n-1}\}$ subject to the defining relations

$$rx = 0, \quad t_1 y_1 = s_1 x, \ldots, t_{n-1} y_{n-1} = s_{n-1} x$$

for $r, t_i, s_i \in R$. In fact, the chain

$$0 \leq Rx \leq Rx + Ry_1 \leq Rx + Ry_1 + Ry_2 \leq \ldots \leq M$$

has cyclically presented factors.

LEMMA 4.4. [D] All special finitely presented R-modules have the projective property relative to all RD-exact sequences.

<u>Proof</u>. Let $0 \longrightarrow A \xrightarrow{\alpha} B \xrightarrow{\beta} C \longrightarrow 0$ be an RD-exact sequence. Owing to (3.2), the cyclic R-modules of projective dimension 1 have the projective property relative to it. Now if M has a sequence (3) with cyclic factors of projective dimension 1, then pick a projective resolution $0 \longrightarrow H \longrightarrow F \xrightarrow{\delta} M \longrightarrow 0$ of M with F finitely generated projective. Setting $H_i = \delta^{-1} M_i$ ($i = 0, 1, \ldots, n$), we obtain a chain

$$H = H_0 < H_1 < \ldots < H_n = F \qquad (4)$$

such that $H_i/H_{i-1} \cong M_i/M_{i-1}$ for $i = 1, \ldots, n$. From Kaplansky's Lemma in IV. §2 we obtain successively that $H_{n-1}, \ldots, H_0$ are necessarily projective.

4. PURE SUBMODULES

By making use of our two exact sequences and a map $\eta : M \to C$, we can construct a commutative diagram (2) with the solid arrows. This will induce a commutative diagram

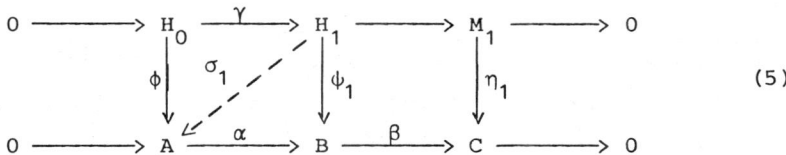

(5)

where ϕ_1, η_1 denote the obvious restrictions of ϕ, η. As M_1 is cyclic of projective dimension 1, (3.2), (4.1) and what has been said in the preceding paragraph imply the existence of a map $\sigma_1 : H_1 \to A$ with $\sigma_1 \gamma = \phi$. Using $\sigma_1 : H_1 \to A$ and $\psi_2 = \psi|H_2$, in the next step we extend σ_1 to a map $\sigma_2 : H_2 \to A$, and keep going until we arrive at a map $\sigma : F \to A$ satisfying $\sigma \gamma = \phi$. A simple appeal to (4.1) concludes the proof. □

We are now able to verify:

THEOREM 4.5. (Warfield [1]) Over Prüfer domains, the RD-property and purity are equivalent.

Proof. In view of (4.4), it suffices to show that over Prüfer domains, all finitely presented modules M are special. Manifestly, M has a chain (3) with cyclic factors. Then the quotients M/M_i are all finitely presented which implies that all H_i in (4) are finitely generated. But over Prüfer domains, finitely generated submodules of projective modules are likewise projective. Hence $\text{p.d.} M_i/M_{i-1} \leq 1$ becomes evident. □

It is the coincidence of the RD-property and purity over valuation rings and Prüfer domains that makes them so useful, as will be seen in the sequel.

For the sake of reference we record here:

PROPOSITION 4.6. An exact sequence $0 \to A \to B \to C \to 0$ is pure-exact if and only if, for every R-module M, the induced sequence

$$0 \to A \otimes_R M \to B \otimes_R M \to C \otimes_R M \to 0$$

is exact. This sequence will then also be pure-exact.

Proof. The first part is verified as the analogue in (3.1). The second part follows easily from the first part. □

EXERCISES

1. [D] Show that direct limits of pure-exact sequences are again pure-exact.

2. [D] Using the fact that every module is the direct limit of finitely presented modules, prove that every pure-exact sequence is the direct limit of splitting exact sequences.

3. Over a Prüfer domain, a finitely generated submodule of a finitely presented module is likewise finitely presented.

4. (Megibben [1]) Let R be a Prüfer domain. An R-module is absolutely pure (i.e. it is a pure submodule in every R-module containing it) exactly if it is divisible.

§5. LEMMAS ON PURE SUBMODULES

Next we wish to establish a few useful properties of pure submodules some of which we shall use later on. Throughout this section, R will denote a valuation domain.

We start with reiterating §3(C): if the R-module M is torsion-free, then the intersection of all pure submodules of M, containing a subset X of M, is again pure in M; this is the pure submodule $\langle X \rangle_*$ generated by X or the purification of X. In particular, every non-zero element x of M generates a pure submodule $\langle x \rangle_*$ of rank 1, and every submodule A can be embedded in a pure submodule of the same rank as A.

We consider the question of embeddability of submodules in possibly small pure submodules. An R-module is called κ-generated (for an infinite cardinal κ) if it has a generating system of cardinality at most κ. It is a useful convention to mean by $\aleph_{-1}$-

5. LEMMAS ON PURE SUBMODULES

generated simply finitely generated.

Denote by λ the minimal cardinal number such that all rank one torsion-free modules over the (fixed) valuation domain R are λ-generated.

The first two lemmas are preparatory in character.

LEMMA 5.1. [VD] A torsion-free R-module of rank κ is at most $\kappa\lambda$-generated.

Proof. Let M be a torsion-free module. Select a rank one pure submodule M_1, then a rank one pure submodule M_2/M_1 of M/M_1, etc. Continuing in this way and taking unions at limit ordinals, we obtain, for some ordinal τ, a well-ordered ascending chain of pure submodules

$$0 = M_0 < M_1 < M_2 < \ldots < M_\alpha < \ldots < M_\tau = M \quad (\alpha < \tau) \qquad (1)$$

such that, for each α, $M_{\alpha+1}/M_\alpha$ is torsion-free of rank 1, and the chain is <u>continuous</u> in the sense that $M_\beta = \bigcup_{\alpha < \beta} M_\alpha$ holds for limit ordinals $\beta \leq \tau$. Let X_α be a subset of M_α which generates $M_{\alpha+1}$ modulo M_α; thus $|X_\alpha| \leq \lambda$ can be assumed. It is straightforward to show that $\{X_\alpha \mid \alpha < \tau\}$ will be a generating system for M. As $|\tau|$ is just the rank κ of M, the assertion follows. □

LEMMA 5.2. [VD] A submodule of a κ-generated R-module is $\kappa\lambda$-generated.

Proof. First, let F be a free R-module of rank κ and H a submodule of F. As the rank of H is $\leq \kappa$, in view of (5.1), H is $\kappa\lambda$-generated.

Now let M be an arbitrary κ-generated R-module. Then $M \cong F/G$ with F a κ-generated free R-module. A submodule N of M is of the form $N \cong H/G$ with H a submodule of F. By the first part of the proof, H is $\kappa\lambda$-generated, and thus the same holds for N. □

We can now prove:

LEMMA 5.3 [VD] Every κ-generated submodule of an R-module M can be embedded in a $\kappa\mu$-generated pure submodule of M where μ

denotes the cardinality of the set of principal ideals of R (or, equivalently, the cardinality of the value group of R).

Proof. First of all, observe that $\lambda \leq \mu$ holds obviously.

Let N be a κ-generated submodule of the R-module M. For every principal ideal Rr of R, select a generating system $\{a_{ri} \mid i \in I_r\}$ for $N \cap rM$. By (5.2), $|I_r| \leq \kappa\lambda$ can be assumed. Let $b_{ri} \in M$ satisfy $rb_{ri} = a_{ri}$ ($i \in I_r$), and define $N_1 = \langle N, b_{ri} \mid i \in I_r,$ principal ideals $Rr\rangle$. Clearly, N_1 is $\kappa\mu$-generated. Repeat this process with N_1 in the place of N, and continue to obtain an ascending chain of submodules $N \leq N_1 \leq \ldots \leq N_k \leq \ldots$ ($k < \omega$). We claim $N^* = \cup_{k<\omega} N_k$ is a pure submodule in M. In fact, if $r \in R$ divides $a \in N^*$ in M, then r divides $a \in N_k$ in N_{k+1}, so in N^*. As each N_k is $\kappa\mu$-generated, the proof is complete. □

A result of a different nature is as follows.

LEMMA 5.4. [VD] (Fuchs [4]) If a torsion-free R-module M is κ-generated, then the same holds for all pure submodules of M.

Proof. Let N be a rank one pure submodule of M, and $\{a_i \mid i \in I\}$ with $|I| \leq \kappa$ a generating set of M. By way of contradiction, assume that the generation of N requires more than κ elements, i.e. N is the union of a well-ordered ascending chain of cyclic submodules Rb_α ($\alpha < \Omega$) where Ω is an initial ordinal of cardinality > κ and is as small as possible. Evidently, every b_α is an R-linear combination of a finite number of generators a_i, therefore, there must exist a subset of $\{b_\alpha \mid \alpha < \Omega\}$ of cardinality $|\Omega|$ such that every b_α in this subset is an R-linear combination of the same finite subset of the a_i's, say, of $\{a_1, \ldots, a_k\}$. This subset of the b_α's has to generate N, consequently, $N \leq \sum_{i=1}^{k} Ra_i = F$. We infer that N is a pure submodule in a finitely generated torsion-free R-module F. By (IV.1.1), F is free, and so is F/N. Consequently, N is a summand of F, and so $N \cong R$. This contradiction shows that N is likewise κ-generated.

6. CYCLIC PURITY

Now let N be a pure submodule of arbitrary rank. As in (1), we can write N as the union of a well-ordered continuous ascending chain of submodules, $0 = N_0 < N_1 < \ldots < N_\alpha < \ldots < N_\tau = N$, where each $N_{\alpha+1}/N_\alpha$ is torsion-free of rank 1, $\alpha+1 \leq \tau$. If M is κ-generated, then so are all the factor modules M/N_α, and in view of the preceding paragraph, the same holds true for $N_{\alpha+1}/N_\alpha$. Since the rank of N cannot exceed the rank of M which is $\leq \kappa$, we see that $|\tau| \leq \kappa$. We conclude that N is κ-generated as well. □

EXERCISES

1. Extend (5.1) and (5.2) to modules over arbitrary domains.
2. Extend (5.3) to RD-submodules over arbitrary domains.
3. [VD] Show that a κ-generated module is at most $\kappa\lambda$-related.
4. [VD] A κ-generated torsion-free module is κ-presented.

§6. CYCLIC PURITY

There is a somewhat stronger version of purity which is of interest for modules over valuation rings. It has been investigated by Simmons [1]. It is a useful concept in studying direct sums of cyclic modules (cf. XIII.§4).

A standard result in abelian group theory states that a subgroup C of a group A is pure if and only if every coset $a + C$ ($a \in A$) contains an element whose order is the same as the order of $a + C$ in A/C. With this motivation in mind, define a submodule N of a module M <u>cyclically pure</u> if, for every $a \in M$, there is a $c \in N$ such that

$$\text{Ann}(a + c) = \text{Ann}(a + C).$$

It is immediate that cyclic purity implies RD-property. Even for valuation rings the two concepts are in general different as is shown by the following example.

EXAMPLE 6.1. Let R be a valuation domain with value group $\mathbb{Q}$. Suppose α is a rational number > 0, and $r \in R$ has value α. Choose a sequence $\{p_n\}_{n=1,2,\ldots}$ with $p_n \in P$ such that the values $v(p_1 \ldots p_n) \to \alpha$ as $n \to +\infty$. Set

$$M = \bigoplus_{n=0}^{\infty} Ra_n \quad \text{with Ann } a_n = Rr.$$

Consider the submodule

$$N = \bigoplus_{n=0}^{\infty} Rb_n \quad \text{where } b_n = a_n - p_{n+1} a_{n+1}.$$

Here $a_0 \notin N$, since if $a_0 = \sum_{i=0}^{n} r_i b_i = r_0 a_0 + (r_1 - r_0 p_1) a_1 + \ldots + (r_n - r_{n-1} p_n) a_n - r_n p_{n+1} a_{n+1}$, then $r_0 \equiv 1$, $r_1 \equiv p_1$, $r_2 \equiv p_1 p_2, \ldots$, $r_n \equiv p_1 \ldots p_n$ and $0 \equiv r_n p_{n+1} \equiv p_1 \ldots p_{n+1} \mod Rr$, where the last congruence is impossible. Furthermore, for every n, $a_0 = b_0 + p_1 b_1 + p_1 p_2 b_2 + \ldots + p_1 \ldots p_n b_n + p_1 \ldots p_{n+1} a_{n+1}$ holds, whence $\text{Ann}(a_0 + N) = P$. But the annihilators of elements of M are all principal ideals, while P is not, hence $a_0 + N$ can not contain any element whose annihilator is P. It is readily checked that N is RD in M.

All the properties (A)-(G) listed in §3 for RD-submodules hold for cyclic purity with the exception of (D).

The next lemma justifies terminology.

LEMMA 6.2. [D] (Simmons [1]) For an exact sequence $0 \to N \to M \to H \to 0$ of R-modules, the following conditions are equivalent:

    (i) the sequence is cyclically-pure-exact;

    (ii) for all cyclic R-modules C, the induced map $\text{Hom}_R(C,M) \to \text{Hom}_R(C,H)$ is surjective;

    (iii) if a system of equations $r_i x = a_i \in N$ ($r_i \in R$, $i \in I$) with a single unknown x and arbitrary index set I is solvable in M, then it is also solvable in N.

Proof is essentially the same as that for (3.1). □

The analogues of (3.3) and (3.4) hold for cyclic purity:

PROPOSITION 6.3. [D] (Simmons [1]) Every R-module M can be embedded in a cyclically-pure-exact sequence

7. LOCAL ENDOMORPHISM RINGS

$$0 \to N \to C \to M \to 0$$

where C is a direct sum of cyclic R-modules. An R-module is cyclically-pure-projective exactly if it is a summand of a direct sum of cyclic R-modules.

If R is a valuation ring, then an R-module is cyclically-pure-projective if and only if it is a direct sum of cyclic R-modules. □

EXERCISES

1. [D] Give an example of an ascending chain of cyclically pure submodules whose union fails to be cyclically pure.

2. (Simmons [1]) A domain R is Noetherian if and only if purity implies cyclic purity.

§7. MODULES WITH LOCAL ENDOMORPHISM RINGS

In our discussion of direct decompositions, an important role will be played by the so-called exchange property and Azumaya type theorems. Let us pause briefly to discuss these here.

An extremely useful but rather rare property is shared by certain modules which will occur in our discussions. An R-module M is said to have the (<u>finite</u>) <u>exchange property</u> if, for any direct decompositions

$$A = M' \oplus N = \bigoplus_{i \in I} A_i \qquad (1)$$

of an arbitrary R-module A where $M' \cong M$ and A_i, N are any R-modules, I is any (finite) index set, there always exist submodules B_i of A_i such that

$$A = M' \oplus \bigoplus_{i \in I} B_i.$$

It is easy to see that the B_i are necessarily summands of A_i. (For the exchange property, see Crawley-Jónsson [1], Warfield [6].) Injective, quasi-injective, pure-injective, torsion-ultracomplete modules are known to enjoy the exchange property.

The following lemma is straightforward:

LEMMA 7.1. An indecomposable module that has the finite exchange property has also the exchange property. □

Moreover, the finite exchange property is to be checked only for an index set of cardinality 2.

The next result is of great importance for us. It shows that the good behavior of modules is sometimes a consequence of special features of their endomorphism rings.

THEOREM 7.2. A module M whose endomorphism ring is local has the finite exchange property.

Proof. Suppose $A = M \oplus N = B \oplus C$ and let π, ρ, σ denote the projections of A onto M, B and C, respectively. Evidently, $\rho + \sigma = 1$ gives $\pi\rho\pi + \pi\sigma\pi = \pi$ where $\pi\rho\pi, \pi\sigma\pi, \pi$ can be considered as endomorphisms of M, in particular, π as the identity. End M local implies that either $\pi\rho\pi$ or $\pi\sigma\pi$ is an automorphism of M, say, the first is one. Setting $H = \text{Im } \rho\pi$, it is clear that $\pi|H : H \to M$ and $\rho|M : M \to H$ are isomorphisms. Hence $A = H \oplus N$ follows. $H \leq B$ implies $B = H \oplus B'$ for some $B' \leq B$; thus $A = M \oplus N = H \oplus B' \oplus C$. As ρ induces an isomorphism of M with H, $A = M \oplus B' \oplus C$ holds. □

Before stating the next result, we need a definition. An object A in an additive category $\mathcal{A}$ is called (countably) <u>small</u> if for every morphism $\psi : A \to \oplus_{i \in I} A_i$, where A_i are objects in $\mathcal{A}$, there is a finite (countable) subset J of I such that ψ factors through $\oplus_{i \in J} A_i$. Note that finitely (countably) generated objects are always (countably) small in a module category.

The following theorem has several contributors (Crawley-Jónsson [1], Azumaya [1], Warfield [2]); we state it without proof. It holds for modules and for certain additive categories.

THEOREM 7.3. Let $M = \oplus_{i \in I} M_i$ where
(i) each M_i is countably small;
(ii) each M_i has local endomorphism ring.

Then every direct decomposition of M can be refined to a decomposition isomorphic to the given one. In particular, every summand of M is isomorphic to a direct sum $\oplus_{i \in J} M_i$ with J a subset of I. □

EXERCISES

1. Give the details of proof for (7.1).

2. Prove that in the category of torsion-free modules over a domain R, the finitely generated modules are small.

3. In order to check the exchange property for M, it suffices to consider decompositions (1) with A_i isomorphic to submodules of M.

NOTES

The theory of modules over a discrete rank one valuation domain is essentially the same as abelian group theory in the local case (i.e. when the groups are $\mathbb{Z}_p$-modules). Localization simplifies the theory somewhat only in the torsion-free and mixed cases. The theory of abelian p-groups extends easily to torsion modules over Dedekind domains.

Modules over general valuation domains and Prüfer domains were first studied by Kaplansky [1]. Some of the basic concepts were introduced in this paper. Fleischer [1] tried to exploit the new ideas and dealt with the finite rank case. Injectives over Prüfer domains were studied successfully by Matlis [1].

It was Matlis [2] who started to distinguish between divisibility in general and h-divisibility. This distinction became increasingly important. The RD-property was systematically investigated by Warfield [1] who could show its coincidence with purity over Prüfer domains.

Let us note that, over a domain R, purity and RD-property coincide only if R is Prüfer. In fact, if $0 \to H \to F \to G \to 0$ is a short exact sequence with F a free R-module, then H is pure (RD) in F precisely if G is flat (torsion-free). Hattori [1] proved that flatness and torsion-freeness are equivalent if and only if R is Prüfer.

Simmons [1] proved that purity and cyclic purity are equivalent over a domain R exactly if R is Dedekind.

The rather simple (3.5) is based on Kaplansky's proof in [1] that finitely generated modules over almost maximal valuation domains are direct sums of cyclics. (3.6) explores an idea by Warfield [3]. It is a common generalization of Kaplansky's theorem just mentioned and Warfield's (3.7).

III. Homological Preliminaries

Homological machinery is indispensible in the study of modules. For easy reference, we list here some of the basic facts which are of particular importance in our discussions.

The special features that occur in the case of integral domains are particularly noteworthy as they have direct bearing on our results.

§1. HOMOLOGICAL BACKGROUND

In this section, R can be any commutative ring. For the definitions of the functors Hom_R, Ext^n_R, $\otimes_R$ and Tor^R_n we refer, e.g. to Cartan-Eilenberg [1] or any introduction to Homological Algebra. As we are dealing with commutative rings only, one should bear in mind that all of Hom, Ext, $\otimes$ and Tor will carry a natural R-module structure which we shall take for granted.

Let
$$0 \longrightarrow A \xrightarrow{\alpha} B \xrightarrow{\beta} C \longrightarrow 0 \tag{1}$$

be an exact sequence of R-modules (it is implicitly understood that all occurring homomorphisms are R-homomorphisms). Then for any R-module M there are three major induced sequences (with induced maps carrying the same symbols):

(I) $\quad 0 \longrightarrow \text{Hom}_R(M,A) \xrightarrow{\alpha} \text{Hom}_R(M,B) \xrightarrow{\beta} \text{Hom}_R(M,C) \longrightarrow$
$\longrightarrow \text{Ext}_R^1(M,A) \xrightarrow{\alpha} \text{Ext}_R^1(M,B) \xrightarrow{\beta} \text{Ext}_R^1(M,C) \longrightarrow$
$\longrightarrow \text{Ext}_R^2(M,A) \xrightarrow{\alpha} \text{Ext}_R^2(M,B) \xrightarrow{\beta} \text{Ext}_R^2(M,C) \longrightarrow \ldots$

(II) $\quad 0 \longrightarrow \text{Hom}_R(C,M) \xrightarrow{\beta} \text{Hom}_R(B,M) \xrightarrow{\alpha} \text{Hom}_R(A,M) \longrightarrow$
$\longrightarrow \text{Ext}_R^1(C,M) \xrightarrow{\beta} \text{Ext}_R^1(B,M) \xrightarrow{\alpha} \text{Ext}_R^1(A,M) \longrightarrow$
$\longrightarrow \text{Ext}_R^2(C,M) \xrightarrow{\beta} \text{Ext}_R^2(B,M) \xrightarrow{\alpha} \text{Ext}_R^2(A,M) \longrightarrow \ldots$

(III) $\quad \ldots \longrightarrow \text{Tor}_2^R(A,M) \xrightarrow{\alpha} \text{Tor}_2^R(B,M) \xrightarrow{\beta} \text{Tor}_2^R(C,M) \longrightarrow$
$\longrightarrow \text{Tor}_1^R(A,M) \xrightarrow{\alpha} \text{Tor}_1^R(B,M) \xrightarrow{\beta} \text{Tor}_1^R(C,M) \longrightarrow$
$\longrightarrow A \otimes_R M \xrightarrow{\alpha} B \otimes_R M \xrightarrow{\beta} C \otimes_R M \longrightarrow 0$

The following natural isomorphisms are useful to keep in mind:

(IV) $\quad \text{Hom}_R(A, \text{Hom}_R(B,C)) \cong \text{Hom}_R(A \otimes_R B, C)$

(V) $\quad \text{Ext}_R^1(A, \text{Hom}_R(B,E)) \cong \text{Hom}_R(\text{Tor}_1^R(A,B), E)$

where A, B, C are arbitrary R-modules and E is an injective R-module.

Before deriving further properties of these functors, we pause briefly to introduce a functorial submodule of Ext.

Relative divisibility gives rise to a relative homology theory which is of great interest for us. The properties of relative divisibility listed in II.§3 constitute a proof that the RD-exact sequences form a so-called proper class. Consequently, given the

1. HOMOLOGICAL BACKGROUND

R-modules A, C (R any domain), the equivalence classes of RD-extensions represented by RD-exact sequences (1) form a subgroup of $\text{Ext}_R^1(C,A)$ which shall be denoted by

$$\text{RDext}_R^1(C,A).$$

It is straightforward to verify that this is an R-submodule of $\text{Ext}_R^1(C,A)$. From the theory of relative homological algebra we conclude the existence of higher derived functors $\text{RDext}_R^n(C,A)$ ($n \geq 2$). It should be pointed out that starting with an RD-exact sequence (1), one obtains exact sequences like (I) and (II) with Ext_R^n replaced throughout by RDext_R^n.

Let us prove something useful on RDext right away:

LEMMA 1.1. [D] If M is a torsion-free and T a torsion R-module, then

$$\text{RDext}_R^n(T,M) = 0 \quad \text{for} \quad n \geq 1.$$

Proof. Let E be an R-module such that M is an RD-submodule of E and $E/M \cong T$. Let $T^* = tE$. Then $M \oplus T^*$ is essential in E, thus for each $e \in E$, there exists a relation $re = a + u \neq 0$ ($r \in R$, $a \in M$, $u \in T^*$). If $0 \neq s \in S$ satisfies $su = 0$, then $rse = sa \in M$ implies that $rse = rsb$ for some $b \in M$. Hence $e - b \in T^*$ and $e \in M \oplus T^*$. This means that E is necessarily a splitting extension of M by T. This proves the assertion for $n = 1$.

To complete the proof, we induct on n. Let $0 \to H \to F \to T \to 0$ be an RD-projective resolution of T. Consider the induced sequence

$$\text{RDext}^n(H,M) \to \text{RDext}^{n+1}(T,M) \to \text{RDext}^{n+1}(F,M).$$

The end terms vanish by induction hypothesis (H being torsion) and by the RD-projectivity, thus the middle term is 0. □

It is of considerable importance that if $\bar{r} : M \to M$ is the endomorphism acting as multiplication by $r \in R$:

$$\bar{r} : x \longmapsto rx \qquad (x \in M),$$

then, for every additive functor T of R-Mod into itself, the endomorphism $T(\bar{r}) : T(M) \to T(M)$ is likewise multiplication by r.

This simple fact implies, in particular, that if $\bar{r}$ is an automorphism (which happens if M is torsion-free and divisible), then $T(\bar{r})$ is likewise an automorphism. We state this observation as a lemma for future reference.

LEMMA 1.2. [D] (Cartan-Eilenberg [1]) For an additive functor T of R-Mod into itself, $T(M)$ is torsion-free and divisible whenever M is torsion-free and divisible. □

In particular, if R is any domain, the R-module $Q \otimes_R M$ is torsion-free and divisible, for every R-module M. The embedding $R \to Q$ induces a canonical map $\gamma : M \to Q \otimes_R M$ (via $\gamma a = 1 \otimes a$ for $a \in M$), and it is readily checked that $\text{Ker } \gamma = tM$.

Another useful piece of information is the content of the following lemma.

LEMMA 1.3. [D] Let T be an additive functor of R-Mod into itself. If T commutes with direct limits, then $T(M)$ is torsion for all torsion R-modules M.

Proof. If M is a finitely generated torsion module, then $rM = 0$ for some $0 \neq r \in R$, i.e. $\bar{r} : M \to M$ is the zero endomorphism. Therefore $T(\bar{r}) = 0$, i.e. $T(M)$ is r-bounded. Every module is the direct limit of its finitely generated submodules. Hence the assertion is evident. □

We combine the last two lemmas in order to conclude that

$$Q \otimes_R M = 0 \quad \text{and} \quad \text{Tor}_n^R(Q,M) = 0$$

for all torsion R-modules M; indeed, they are both torsion-free and torsion.

It was Matlis [3] who pointed out that the exact sequence arising from the embedding $R \to Q$ is an extremely useful tool in deriving information of various kinds. For instance, (II) implies that $0 \to \text{Hom}_R(Q/R, M) \to \text{Hom}_R(Q,M) \to \text{Hom}_R(R,M) \cong M$ is exact. As the last map sends $\eta \in \text{Hom}_R(Q,M)$ into $\eta(1)$, it is pretty obvious that its image in M is precisely hM. In this way, we obtain the exact sequence

$$0 \to \text{Hom}_R(Q/R, M) \to \text{Hom}_R(Q,M) \to hM \to 0.$$

EXERCISES

1. [D] Show that $\text{RDext}_R^1(M,*) = 0$ for an R-module M if and only if M is RD-projective.

2. [D] $\text{Ext}_R^1(Q,T) = 0$ if T is a bounded R-module.

3. Let Q be the ring of quotients of R and M any R-module. Characterize the kernel of the natural map $M \cong R \otimes_R M \to Q \otimes_R M$.

4. [D] $\text{Tor}_n^R(M,*)$ is torsion for all torsion modules M.

§2. LEMMAS ON HOM AND EXT

In this section, we shall examine in greater detail the behavior of the modules Hom_R and Ext_R. We begin with arbitrary domains R, and then specialize.

The following isomorphism is immediate:

$$\text{Hom}_R(R/L, M) \cong M[L]$$

for any ideal L of R and any R-module M.

LEMMA 2.1. [D] For non-zero ideals $I > L$ and J of R, the following isomorphisms hold:

(i) $\text{Hom}_R(J,Q) \cong Q$;
(ii) $\text{Hom}_R(J,L) \cong L:J$;
(iii) $\text{Hom}_R(R/J, Q/L) \cong (L:J)/L$;
(iv) $\text{Hom}_R(R/J, I/L) \cong [(L:J) \cap I]/L$.

Proof. It is straightforward to see that every homomorphism ϕ of J into a submodule L of Q is completely determined by the image of any non-zero element a of J. It follows that ϕ is multiplication by $q = \phi(a)/a \in Q$, and $\operatorname{Hom}_R(J,L)$ is isomorphic to the submodule of Q consisting of these q's. Evidently, multiplication by q is a homomorphism exactly if $qJ \leq L$, i.e. $q \in L:J$. This proves (ii), while the special case $L = Q$ yields (i).

To verify (iii), notice that every $\phi : R/J \to Q/L$ is completely determined by the image $\phi(1+J) = q+L$ ($q \in Q$). Clearly, $q+L$ can be the image of $1+J$ under some ϕ if and only if $qJ \leq L$, i.e. $q \in L:J$.

(iv) is an immediate consequence of (iii). □

The next two results are concerned with the R-modules Ext.

LEMMA 2.2. [D] For every $r \in R$ and R-module M, $\operatorname{Ext}_R^1(R/Rr,M) \cong M/rM$ and $\operatorname{Ext}_R^n(R/Rr,M) = 0$ ($n \geq 2$).

Proof. Starting off with the exact sequence $0 \longrightarrow R \xrightarrow{r} R \longrightarrow R/Rr \longrightarrow 0$, we form the exact sequence $0 \longrightarrow \operatorname{Hom}_R(R,M) \cong M \xrightarrow{r} \operatorname{Hom}_R(R,M) \cong M \longrightarrow \operatorname{Ext}_R^1(R/Rr,M) \longrightarrow 0$. This establishes the stated isomorphism. The continuation of the last exact sequence proves the second part of our assertion. □

LEMMA 2.3. [D] Suppose that A is a torsion and M a torsion-free R-module. If E is the injective hull of M, then $\operatorname{Ext}_R^1(A,M) \cong \operatorname{Hom}_R(A,E/M)$ and $\operatorname{Ext}_R^n(A,M) \cong \operatorname{Ext}_R^{n-1}(A,E/M)$ ($n \geq 2$).

Proof. These isomorphisms are immediate consequences of the long exact sequence induced by $0 \to M \to E \to E/M \to 0$. □

The following result characterizes almost maximal valuation domains.

THEOREM 2.4. For a valuation domain R, these are equivalent:
 (a) R is almost maximal;
 (b) i.d.$R \leq 1$;
 (c) i.d.$J \leq 1$ for all ideals J of R;

2. LEMMAS ON HOM AND EXT 63

(d) $\operatorname{Ext}_R^1(L,J) = 0$ for all ideals J, L of R.

Proof. Observe that, for a submodule J of Q, i.d.$J \leq 1$ holds exactly if Q/J is injective. Hence the equivalence of (a), (b) and (c) will be an immediate consequence of (VI.4.8).

The exact sequence $0 = \operatorname{Ext}_R^1(R,J) \to \operatorname{Ext}_R^1(L,J) \to \operatorname{Ext}_R^2(R/L,J) \to \operatorname{Ext}_R^2(R,J) = 0$ shows that (d) holds if and only if $\operatorname{Ext}_R^2(R/L,J) = 0$ for all ideals J, L of R. By (VI.5.1), this is equivalent to (c). □

By making use of (2.4), it is now easy to verify:

LEMMA 2.5. If J and $I > L$ are ideals in an almost maximal valuation domain R, then

$$\operatorname{Hom}_R(J, I/L) \cong (I:J)/(L:J).$$

Proof. From the exact sequence $0 \to L \to I \to I/L \to 0$ we derive the exact sequence $0 \to \operatorname{Hom}_R(J,L) \to \operatorname{Hom}_R(J,I) \to \operatorname{Hom}_R(J,I/L) \to \operatorname{Ext}_R^1(J,L) = 0$; the last term vanishes in view of (2.4). The stated isomorphism follows at once from (2.1)(ii). □

We prove one more result on almost maximal valuation domains.

LEMMA 2.6. If R is an almost maximal valuation domain, then for every $0 \neq r \in R$ and R-module M,

$\operatorname{Ext}_R^1(M, R/Rr) \cong \operatorname{Ext}_R^1(M[r], R/Rr)$ and $\operatorname{Ext}_R^n(M, R/Rr) = 0$ $(n \geq 2)$.

Proof. Use the exact sequence $0 \longrightarrow M[r] \longrightarrow M \xrightarrow{r} rM \longrightarrow 0$ to induce the exact sequence $\operatorname{Ext}_R^1(rM,R/Rr) \xrightarrow{r} \operatorname{Ext}_R^1(M,R/Rr) \longrightarrow \operatorname{Ext}_R^1(M[r],R/Rr) \longrightarrow \operatorname{Ext}_R^2(rM,R/Rr)$. As $Q/Rr \cong Q/R$ is injective, i.d.$R/Rr \leq 1$. Thus $\operatorname{Ext}_R^2(*,R/Rr) = 0$ and the assertions follow. □

The corollaries of the next result will find use later on. Our considerations are valid over any domain.

LEMMA 2.7. [D] Let the exact sequence

$$E: \quad 0 \longrightarrow A \xrightarrow{\mu} B \xrightarrow{\nu} C \longrightarrow 0$$

represent an element of $\mathrm{Ext}_R^1(C,A)$. A homomorphism $\alpha: A' \to A$ induces a homomorphism $\alpha_*: \mathrm{Ext}_R^1(C,A') \to \mathrm{Ext}_R^1(C,A)$.

If $E \in \mathrm{Im}\, \alpha_*$, then $\mathrm{Im}\, \alpha/\mathrm{Im}\, \mu\alpha$ is a summand of $B/\mathrm{Im}\, \mu\alpha$. The converse holds whenever $\mathrm{p.d.}\, C \leq 1$.

<u>Proof</u>. Consider the exact sequences

$$0 \longrightarrow \mathrm{Im}\, \alpha \xrightarrow{\phi} A \xrightarrow{\psi} \mathrm{Im}\, \mu/\mathrm{Im}\, \mu\alpha \longrightarrow 0$$

$$0 \longrightarrow \mathrm{Ker}\, \alpha \longrightarrow A' \xrightarrow{\alpha'} \mathrm{Im}\, \alpha \longrightarrow 0.$$

They induce the exact sequences:

$$\ldots \to \mathrm{Ext}_R^1(C, \mathrm{Im}\, \alpha) \xrightarrow{\phi_*} \mathrm{Ext}_R^1(C,A) \xrightarrow{\psi_*} \mathrm{Ext}_R^1(C, \mathrm{Im}\, \mu/\mathrm{Im}\, \mu\alpha) \to \ldots$$

$$\ldots \to \mathrm{Ext}_R^1(C, \mathrm{Ker}\, \alpha) \longrightarrow \mathrm{Ext}_R^1(C,A') \xrightarrow{\alpha'_*} \mathrm{Ext}_R^1(C, \mathrm{Im}\, \alpha) \to \ldots$$

As $\alpha = \phi\alpha'$, we have $\mathrm{Im}\, \alpha_* = \mathrm{Im}\, \phi_*\alpha'_* \leq \mathrm{Im}\, \phi_* = \mathrm{Ker}\, \psi_*$, that is, $\psi_* \alpha_* = 0$. Clearly,

$$\psi_* E: 0 \to \mathrm{Im}\, \mu/\mathrm{Im}\, \mu\alpha \to B/\mathrm{Im}\, \mu\alpha \to C \to 0,$$

which splits if $E \in \mathrm{Im}\, \alpha_*$. This proves the first assertion.

If $\mathrm{p.d.}\, C \leq 1$, then $\mathrm{Ext}_R^2(C,*) = 0$, so α'_* is an epimorphism, and therefore $\mathrm{Im}\, \alpha_* = \mathrm{Ker}\, \psi_*$. Hence the second assertion is evident. □

The following is a noteworthy consequence:

PROPOSITION 2.8. [D] Let $\mathrm{p.d.}\, C \leq 1$. Then the exact sequence $0 \to A \to B \to C \to 0$ represents an element of $r\, \mathrm{Ext}_R^1(C,A)$ if and only if the sequence

$$0 \to A/rA \to B/rA \to C \to 0 \tag{1}$$

is splitting.

<u>Proof</u>. Apply (2.7) to the homomorphism $r: A \to A$. □

In view of the last result, we have a very precise test for determining whether or not an extension of A by C belongs to the first Ulm submodule of $\mathrm{Ext}_R^1(C,A)$:

COROLLARY 2.9. [D] Suppose $\mathrm{p.d.}\, C \leq 1$. The exact sequence

$0 \to A \to B \to C \to 0$ represents an element of the first Ulm submodule of $\text{Ext}_R^1(C,A)$ if and only if (1) splits for every $0 \neq r \in R$. □

EXERCISES

1. [D] Prove that for all R-modules A, M and all ideals L of R, $M[L] = 0$ implies $\text{Hom}_R(A,M)[L] = 0$.

2. [D] $rA = A$ for some $r \in R$ implies $\text{Hom}_R(A,M)[r] = 0$.

3. [D] Show that $\text{Hom}_R(A,M)$ is torsion-free if either A is divisible or M is torsion-free.

4. [D] Prove that for ideals J, L of R,
   (a) $\text{Ext}_R^1(R/J, L) \cong (L:J)/L$.
   (b) $\text{Ext}_R^1(Q/J, J) \cong \text{End}_R Q/J$.

5. If R is a valuation domain, then $\text{Hom}_R(P,R) \cong R$, and $\text{Hom}_R(P,P) \cong R$.

6. Let R be a valuation domain and J an ideal of R.
   (a) $\text{Hom}_R(J, R/P) \cong \text{Ext}_R^1(R/J, R/P)$;
   (b) $\text{Hom}_R(J, R/P) \cong R/P$ or 0 according as J is a principal ideal or not.

7. [VD] For every R-module M,
$$\text{Ext}_R^1(R/P, M) \cong \text{socle of } E/M$$
where E denotes the injective hull of M.

8. [D] Apply (2.7) to the homomorphisms $r : A \to A$ to conclude that $\text{RDext}_R^1(C,A)$ always contains the first Ulm submodule of $\text{Ext}_R^1(C,A)$.

§3. LEMMAS ON TENSOR AND TORSION PRODUCTS

Here we collect information on the tensor and torsion products of modules. We restrict ourselves to results which are specific for domains or for valuation domains.

PROPOSITION 3.1. [D] (Cartan-Eilenberg [1]) For a domain R and R-module M, we have

$$\mathrm{Tor}_n^R(Q,M) = 0 \qquad \text{for all } n \geq 1,$$

and

$$\mathrm{Tor}_1^R(Q/R, M) \cong tM.$$

Proof. Note that $\mathrm{Tor}_n^R(R,M) = 0$ always, $n \geq 1$. For every domain R, $Q = \varinjlim r^{-1}R$ ($0 \neq r \in R$) where $r^{-1}R \cong R$. Since Tor commutes with direct limits, the first assertion is immediate.

Application of (III) in §1 to the exact sequence $0 \to R \to Q \to Q/R \to 0$ yields the exact sequence

$$0 \to \mathrm{Tor}_1^R(Q/R,M) \to R \otimes M \to Q \otimes M \to Q/R \otimes M \to 0.$$

We know that $R \otimes M \cong M$ and the middle map has kernel tM. Hence the second assertion follows. □

LEMMA 3.2. [D] For $r \in R$ and ideal L of R,
 (i) $\mathrm{Tor}_1(R/Rr, M) \cong M[r]$,
 (ii) $\mathrm{Tor}_1(Q/L, M) \cong L \otimes tM$,
 (iii) $(R/L) \otimes M \cong M/LM$,
for every R-module M. All these are natural isomorphisms.

Proof. Using $0 \longrightarrow R \xrightarrow{r} R \longrightarrow R/Rr \longrightarrow 0$ we infer that $\mathrm{Tor}_1(R/Rr, M)$ is naturally isomorphic to the kernel of the homomorphism $Rr \otimes M \to R \otimes M$. This is precisely the set of elements $r \otimes x$ ($x \in M$) such that $1 \otimes rx = 0$ in $R \otimes M$. This holds just in case $rx = 0$.

The exact sequence $0 \to L \to Q \to Q/L \to 0$ induces by (3.1) the exact sequence $0 \to \mathrm{Tor}_1(Q/L, tM) \to L \otimes tM \to Q \otimes tM = 0$.

From $0 \to L \to R \to R/L \to 0$ we derive the exact sequence $L \otimes M \to R \otimes M \to (R/L) \otimes M \to 0$. Here $L \otimes M$ is generated by all $r \otimes x$ (where $r \in L$, $x \in M$) which become $1 \otimes rx$ in $R \otimes M$. This shows that the image of $L \otimes M$ in $R \otimes M$ is naturally isomorphic to LM. Hence (iii) is evident. □

3. TENSOR AND TORSION PRODUCTS

We now move to Prüfer domains; in this case, torsion-freeness and flatness are equivalent; cf. (IV.1.4). This is needed in the following proofs.

LEMMA 3.3. [PD] (Soileau [1]) Let $I > J$ and $K > L$ be ideals of R and M a torsion-free R-module. Then

(i) $I \otimes M \cong IM$;

(ii) $I \otimes K \cong IK$;

(iii) $(I/J) \otimes M \cong IM/JM$;

(iv) $(I/J) \otimes (K/L) \cong IK/(JK + IL)$;

(v) $\mathrm{Tor}_1(I/J, K/L) \cong (JK \cap IL)/JL$.

Proof. (i) M being flat, the induced map $I \otimes M \to R \otimes M$ is monic. As noted in the proof of (3.2), the image of $I \otimes M$ under this map is IM.

(ii) is a special case of (i).

(iii) Again by the flatness of M the sequence $0 \to J \otimes M \to I \otimes M \to (I/J) \otimes M \to 0$ is exact. The claim follows from (i).

(iv) We now use the exact sequence $J \otimes (K/L) \xrightarrow{\alpha} I \otimes (K/L) \longrightarrow (I/J) \otimes (K/L) \longrightarrow 0$. By (iii), the first two modules are isomorphic to JK/JL and IK/IL. The image of JK/JL under the obvious map is easily seen to be $(JK + IL)/IL$ whence (iv) is immediate.

(v) The exact sequence in (iv) can be continued to the left to obtain a longer exact sequence $0 \longrightarrow \mathrm{Tor}_1(I/J, K/L) \longrightarrow JK/JL \longrightarrow IK/IL$. As the cokernel of the first map is $(JK + IL)/IL \cong JK/(JK \cap IL)$, (v) becomes clear. □

We wish to show that $\mathrm{Tor}_1^R(A,C)$ can be defined for valuation domains R easily in terms of generators and defining relations.

Let A and C be R-modules. Define X as the free abelian group on the set

$$\{(a,r,c) \mid a \in A, c \in C, r \in R, ra = 0 = rc\}.$$

Let Y be the subgroup of X generated by all elements of X of the form

(a) $(a_1+a_2, r, c) - (a_1, r, c) - (a_2, r, c)$ $[ra_i = 0 = rc]$,
(b) $(a, r, c_1+c_2) - (a, r, c_1) - (a, r, c_2)$ $[ra = 0 = rc_i]$,
(c) $(a, rs, c) - (ra, s, c)$ $[rsa = 0 = sc]$,
(d) $(a, rs, c) - (a, s, rc)$ $[sa = 0 = rsc]$.

THEOREM 3.4. (Soileau [1]) For a valuation ring R, $T = X/Y$ is an R-module, naturally R-isomorphic to $\operatorname{Tor}_1^R(A,C)$.

Proof. It is straightforward to check that T becomes an R-module by defining $s(a, r, c) = (a, rs, c)$ for $s \in R$. Recall that if $0 \longrightarrow H \xrightarrow{\alpha} F \xrightarrow{\beta} C \longrightarrow 0$ is a free resolution of C, then $\operatorname{Tor}_1^R(A, C)$ is defined as $\operatorname{Ker}(1 \otimes \alpha)$ where $1 \otimes \alpha$ is the induced map $A \otimes H \to A \otimes F$.

Now define the map $\phi : T \to \operatorname{Ker}(1 \otimes \alpha)$ as follows:

$$\phi(a, r, c) = a \otimes y \text{ where } c = \beta b, \; rb = \alpha y \; (b \in F, y \in H).$$

It requires only a routine check to see that ϕ is well-defined and respects the defining relations in T. Hence it is an R-homomorphism; it is natural as is easily seen from its definition.

To prove ϕ is surjective, assume $z = \Sigma(a_j \otimes y_j) \in \operatorname{Ker}(1 \otimes \alpha)$ for $a_j \in A$, $y_j \in H$. The y_j are contained in a finitely generated submodule H' of H whose purification in F is a finitely generated free summand F' of F. In view of Exercise 8 in II, §3, we can write

$$F' = Rx_1 \oplus \ldots \oplus Rx_n \text{ and } H' = Rr_1 x_1 \oplus \ldots \oplus Rr_n x_n$$

with $r_i \in R$, $x_i \in F$. Using this basis of H', z assumes the form $z = \Sigma(b_i \otimes r_i x_i)$ with $b_i \in A$. The inclusion $z \in \operatorname{Ker}(1 \otimes \alpha)$ means that $\Sigma r_i b_i \otimes x_i = 0$ in $A \otimes F'$. i.e. $r_i b_i = 0$ for all i. We conclude that $(b_i, r_i, \beta x_i) \in T$ and $\Sigma(b_i, r_i, \beta x_i)$ is mapped by ϕ upon z.

It remains to show that ϕ is monic. Suppose that $\Sigma(a_i, s_i, c_i) \in \operatorname{Ker} \phi$ (where $a_i \in A$, $c_i \in C$, $s_i \in R$, $s_i a_i = 0 = s_i c_i$), i.e. $\Sigma(a_i \otimes y_i) = 0$ in $A \otimes H$ where $c_i = \beta b_i$, $\alpha y_i = s_i b_i$ ($b_i \in F$, $y_i \in H$). Again, no generality is lost in assuming that everything takes place in F' and H' as above.

3. TENSOR AND TORSION PRODUCTS

Write

$$b_i = \sum_j r_{ij} x_j \quad \text{and} \quad y_i = \sum_j t_{ij}(r_j x_j)$$

with $r_{ij}, t_{ij} \in R$, and note that $\alpha y_i = s_i b_i$ implies $s_i r_{ij} = t_{ij} r_j$. From $0 = \sum_i (a_i \otimes \sum_j t_{ij} r_j x_j) = \sum_j [(\sum_i a_i t_{ij}) \otimes r_j x_j]$ we deduce that $\sum_i a_i t_{ij} = 0$ for each index j. Hence

$$\sum_i (a_i, s_i, c_i) = \sum_i (a_i, s_i, \sum_j r_{ij} \beta x_j) = \sum_i \sum_j (a_i, t_{ij} r_j, \beta x_j)$$

$$= \sum_j (\sum_i a_i t_{ij}, r_j, \beta x_j) = 0. \quad \square$$

EXERCISES

1. [VR] The tensor and torsion products of uniserial modules are again uniserial.

2. [VR] (Soileau [1]) Suppose U, V are uniserial R-modules and $u \otimes v = 0$ ($u \in U$, $v \in V$). Then there exists an $r \in R$ such that $r|u$ and $rv = 0$ or vice-versa.

3. [VR] (Soileau [1]) Let M and N be polyserial R-modules of lengths m and n, respectively. Show that both $M \otimes_R N$ and $\text{Tor}_R(M,N)$ are polyserial of lengths at most mn.

4. [VD] (Soileau [1]) In $\text{Tor}_R(M,N)$, the annihilator of the generator (a,r,c) is precisely $r^{-1} \text{Ann } a \text{ Ann } b$.

IV. Projectivity and Projective Dimension

We have come to the issue which is one of the focal points in the study of module categories, viz. the question of projectives. Fortunately, over valuation domains, it is an easy task to identify the projectives as well as their direct limits: the flat modules.

It is well understood that projective dimension is intimately related to the cardinalities of generators and relations, but nowhere else is this relation so significant as in the case of valuation domains.

We believe that one of the basic ingredients for a structure theory of modules over valuation domains is the device introduced here: 'tight systems'.

§1. PROJECTIVE AND FLAT MODULES

A considerable body of literature is devoted to the study of projectives. We will treat here only those over valuation rings.

We start with the following simple observation:

1. PROJECTIVE AND FLAT MODULES

PROPOSITION 1.1. A finitely generated torsion-free module over a Prüfer domain is projective.

Proof. Let $F = \langle a_1, \ldots, a_n \rangle$ be a torsion-free module over a Prüfer domain R. The correspondence $a_1 \mapsto 1$ induces an isomorphism $Ra_1 \to R$ which extends to a homomorphism $\alpha: F \to Q$. The finitely generated αF is isomorphic to an ideal of R, and so it is projective. We conclude $F \cong \text{Ker } \alpha \oplus \alpha F$. Here Ker α is again finitely generated, so an easy induction (e.g. on the rank of the module) completes the proof. □

The following consequence is obvious.

COROLLARY 1.2. The torsion submodule of a finitely generated module over a Prüfer domain is always a summand. □

Turning our attention to projective modules, we have the important

THEOREM 1.3. (Kaplansky [3]) Projective modules over valuation rings are free.

Proof. (Actually this holds for arbitrary local rings.) A proof can be given by making use of (II.7.3), as projective modules are summands of free R-modules and in the present case R has a local endomorphism ring. □

About flat modules we do not have much to say except for the relevant (1.4). Recall that an R-module F is <u>flat</u> if

$$\text{Tor}_1^R(F, *) = 0.$$

Over any ring, projective modules are flat.

As by (III.3.1), $\text{Tor}_1(Q/R, M) \cong tM$, we can conclude that, over any domain R, flat R-modules ought to be torsion-free. As was shown by Cartan-Eilenberg [1], for Prüfer domains the converse is also true:

THEOREM 1.4. Over a Prüfer domain R, an R-module is flat if and only if it is torsion-free.

Proof. To give a short proof of the "if" part, recall that flatness is preserved under direct limits. Every module is the direct limit of its finitely generated submodules, thus (1.1) implies that a torsion-free R-module is the direct limit of projective (and hence flat) modules, so it is itself flat. □

If $0 \to H \to F \to M \to 0$ is a short free resolution of an R-module M, then H is torsion-free, thus flat. Consequently, $\text{Tor}_2^R(M,*) = \text{Tor}_1^R(H,*) = 0$. We conclude that Tor_2 identically vanishes for Prüfer domains. This is expressed in a different form in

COROLLARY 1.5. The weak global dimension of a Prüfer domain is 1.□

EXERCISES

1. If R is a Prüfer domain, then for all R-modules M and N, there is a natural isomorphism
$$\text{Tor}_1^R(M,N) \cong \text{Tor}_1^R(tM,tN).$$

2. For a Prüfer domain R and for torsion-free R-modules M,N, the R-module $M \otimes_R N$ is likewise torsion-free.

3. [D] A finite rank torsion-free projective module is finitely generated.

4. [D] Suppose R is a domain such that there exists a torsion-free R-module M which is not flat. (As is shown by Hattori [1], this is the case exactly if R is not Prüfer.) Give an example of an RD-submodule that is not pure.

5. [D] Show that the localization of a domain R at any semigroup is a flat R-module.

§2. PROJECTIVE DIMENSION

Turning our attention to the question of projective dimensions, let us first remind the reader that a <u>projective resolution</u> of a module

2. PROJECTIVE DIMENSION

M is an exact sequence

$$\cdots \longrightarrow P_n \xrightarrow{\delta_n} P_{n-1} \longrightarrow \cdots \longrightarrow P_1 \xrightarrow{\delta_1} P_0 \xrightarrow{\delta_0} M \longrightarrow 0$$

of projective modules P_i. The <u>projective dimension</u> of M, in notation p.d.$_R$ M, is equal to n if n is the smallest index with Im δ_n projective, or to ∞ if no such n exists. (It is well known that this definition is independent of the particular choice of projective resolutions.) Equivalently, p.d.$_R$ M = n if n is the smallest index such that $\text{Ext}_R^{n+1}(M,*) = 0$. The <u>injective dimension</u> of M, i.d.$_R$ M, is defined dually, by using an injective resolution of M. The <u>global dimension</u> of the ring R, gl.d.R, is the smallest n for which Ext_R^{n+1} vanishes identically. This is the supremum of all p.d.$_R$ M or, equivalently, that of all i.d.$_R$ M with M ranging over all R-modules.

For many theorems on homological dimensions, a comparison of projective dimensions in an exact sequence is inevitable. In this respect, the following result is decisive (it holds over any ring).

KAPLANSKY'S LEMMA. Let $0 \to A \to B \to C \to 0$ be an exact sequence of R-modules. If two of p.d.A, p.d.B, p.d.C are finite, so is the third, and then only the following cases may occur:

   (i)   p.d.A < p.d.B = p.d.C ;
   (ii)   p.d.B < p.d.A = p.d.C-1;
   (iii) p.d.A = p.d.B $\geq$ p.d.C-1.

A straightforward proof follows by examining the induced long exact sequence (II) in III,§1.

Another major tool in investigating projective dimensions is the next result (also valid over any ring).

AUSLANDER'S LEMMA. Let, for some ordinal τ,

$$0 = M_0 \leq M_1 \leq \cdots \leq M_\alpha \leq \cdots \qquad (\alpha < \tau) \qquad (1)$$

be a well-ordered ascending chain of submodules of a module M such that

   (i) $\bigcup_{\alpha<\tau} M_\alpha = M$;

(ii) $\bigcup_{\alpha<\beta} M_\alpha = M_\beta$ for limit ordinals $\beta < \tau$; i.e. (1) is a continuous chain;

(iii) $p.d._R M_{\alpha+1}/M_\alpha \leq n$ for some fixed integer n and for all $1 \leq \alpha+1 < \tau$.

Then
$$p.d._R M \leq n.$$

The standard proof proceeds by induction on n, the case $n = 0$ being obvious, as then simply $M \cong \bigoplus_{0 \leq \alpha+1 < \tau} M_{\alpha+1}/M_\alpha$.

Kaplansky's and Auslander's Lemmas will be used frequently in the sequel. The following lemma which generalizes Auslander's Lemma for $n = 1$ will be needed later on.

LEMMA 2.1. [D] Let (1) be a well-ordered continuous ascending chain of submodules of an R-module M. Suppose that, for some R-module X,

$$\text{Ext}_R^1(M_{\alpha+1}/M_\alpha, X) = 0 \quad \text{for all } \alpha+1 < \tau.$$

Then
$$\text{Ext}_R^1(\cup M_\alpha, X) = 0.$$

Proof. We can assume $\cup M_\alpha = M$. Let $0 \to X \to E \to M \to 0$ be an extension of X by M; we wish to prove that it splits by constructing a complement to X in E.

Let $0 \to X \to E_\alpha \to M_\alpha \to 0$ be the exact sequence induced by the inclusion $M_\alpha \to M$. Obviously, this splits for $\alpha = 0$. Regard E as the union of the ascending chain $0 = E_0 \leq E_1 \leq \cdots \leq E_\alpha \leq \cdots$ ($\alpha < \tau$), and suppose that we have found R-submodules A_β of E_β for each $\beta < \alpha$ such that $0 = A_0 \leq \cdots \leq A_\beta \leq \cdots$ ($\beta < \alpha$) is a well-ordered continuous ascending chain satisfying $E_\beta = X \oplus A_\beta$ ($\beta < \alpha$). If α is a limit ordinal, then set $A_\alpha = \bigcup_{\beta<\alpha} A_\beta$. This will satisfy $E_\alpha = X \oplus A_\alpha$. If $\alpha-1$ exists, then $E_\alpha/A_{\alpha-1}$ is an extension of $E_{\alpha-1}/A_{\alpha-1} \cong X$ by $E_\alpha/E_{\alpha-1} \cong M_\alpha/M_{\alpha-1}$. By hypothesis, this splits, i.e. $E_\alpha/A_{\alpha-1} = E_{\alpha-1}/A_{\alpha-1} \oplus A_\alpha/A_{\alpha-1}$ for some $A_\alpha \geq A_{\alpha-1}$. Evidently, $E_\alpha = X + A_\alpha$. On the other hand, $X \cap A_\alpha =$

2. PROJECTIVE DIMENSION

$= X \cap E_{\alpha-1} \cap A_\alpha = X \cap A_{\alpha-1} = 0$, thus $E_\alpha = X \oplus A_\alpha$. Manifestly, $A = \cup A_\alpha$ satisfies $X \oplus A$. □

By a κ-<u>filtration</u> of M (where κ is an infinite cardinal) is meant a continuous chain (1) of submodules with union M such that each M_α can be generated by less than κ elements and τ is the first ordinal of cardinality κ.

Let κ be a regular uncountable cardinal, or the first ordinal of this cardinality. A subset C of the set of ordinals $< \kappa$ is called a <u>club</u> (closed and unbounded) if (a) for all $X \subseteq C$, sup $X < \kappa$ implies sup $X \in C$; and (b) sup $C = \kappa$. The intersection of two clubs is again a club. A subset E of the set of ordinals $< \kappa$ is said to be <u>stationary</u> in κ if it intersects every club C in κ.

Recently, Eklof [1] proved a converse to Auslander's Lemma, which we formulate in a weaker form (needed in the sequel). By a <u>coherent</u> module we mean one whose finitely generated submodules are finitely presented. Notice that because of (1.1), all torsion-free modules over Prüfer domains are coherent.

EKLOF'S THEOREM. Let R be a coherent domain and M a coherent R-module. Suppose (1) is a κ-filtration of M (where κ is a regular uncountable cardinal) such that

$$\text{p.d.} M_\alpha \leq n-1 \text{ for every } \alpha < \kappa.$$

Then $\text{p.d.} M \geq n$ if and only if the set

$$E = \{\nu < \kappa \mid \text{there is a } \mu > \nu \text{ with p.d.} M_\mu/M_\nu \geq n\}$$

is stationary in κ.

We can now prove the following results. (For the meaning of $\aleph_{-1}$, see II.§5.)

THEOREM 2.2. (Osofsky [1]) Let R be a valuation domain and J an ideal of R which can be generated by $\aleph_m$, but not by $\aleph_{m-1}$ elements (m an integer). Then

$$\text{p.d.}_R J = m+1.$$

Proof. First we show by induction on m that $\text{p.d.}J \leq m+1$, the finitely generated case being obvious. Thus $m \geq 0$ and J has an $\aleph_m$-filtration $\{J_\nu \mid \nu < \omega_m\}$; as each J_ν is at most $\aleph_{m-1}$-generated, $\text{p.d.}J_\nu \leq m$ by induction hypothesis. From Kaplansky's Lemma we obtain $\text{p.d.}J_{\nu+1}/J_\nu \leq m+1$ whence Auslander's Lemma implies $\text{p.d.}J \leq m+1$.

To prove the converse, we again apply induction. If $m = 0$, J is not finitely generated, so it cannot be a projective ideal, i.e. $\text{p.d.}J \geq 1$. Suppose that $m \geq 1$, and choose, by transfinite induction, a continuous well-ordered chain $\{J_\nu \mid \nu < \omega_m\}$ of submodules such that for every ν, $J_{\nu+1} = Ra_\nu$ for some $a_\nu \in J$. Consider the stationary set:

$$E_0 = \{\nu < \omega_m \mid \nu \text{ is a limit ordinal of cofinality } \omega_{m-1}\}.$$

If $\nu \in E_0$, then J_ν is generated by $\aleph_{m-1}$, but not by fewer elements, thus by induction hypothesis, $\text{p.d.}J_\nu \geq m$. But $J_{\nu+1} = Ra_\nu$ is projective, therefore $\text{p.d.}(J_{\nu+1}/J_\nu) \geq m+1$ by Kaplansky's Lemma. As E_0 is a subset of E in Eklof's Theorem, the claim follows. □

Before proceeding, we draw an easy but important corollary.

COROLLARY 2.3. (Osofsky [1]) Let R be a valuation domain and suppose that there is a smallest cardinal of the form $\aleph_m$ (m an integer) such that every ideal J of R can be generated by at most $\aleph_m$ elements. Then $\text{gl.d.}R = m+2$.

If no such cardinal exists, then $\text{gl.d.}R = \infty$.

Proof. It is a straightforward corollary to Auslander's Lemma that the left $\text{gl.d.}R$ is the supremum of $\text{p.d.}R/J$ with J running over the left ideals of R, plus 1, provided that R is not an Artinian semisimple ring. As $\text{p.d.}R/J = \text{p.d.}J + 1$ ($J \neq 0$), the assertion follows from (2.2). □

A result analogous to (2.2) is as follows.

THEOREM 2.4. (Kaplansky [3], Small [1]) For a valuation domain R, $\text{p.d.}_R Q = m+1$ if and only if Q can be generated by

3. DIMENSIONS OF TORSION-FREE MODULES

$\aleph_m$, but not by $\aleph_{m-1}$ elements.

Proof. Just as in the proof of (2.2), the estimate $p.d.Q \leq m+1$ follows from Auslander's Lemma, and the converse from Eklof's theorem, but in the proof refer to (2.2) rather than to an inductive hypothesis. □

EXERCISES

1. Show that in an exact sequence $0 \to A \to B \to C \to 0$, we always have

$$p.d.B \leq \max\{p.d.A, p.d.C\}.$$

2. Let (1) satisfy (i) and (ii). Prove that $p.d.M \leq n$ if $p.d._R M_\alpha \leq n-1$ for each α.

3. Prove that, over a Prüfer domain R, $p.d._R J \leq n+1$ if J is an ideal of R, generated by $\aleph_n$ elements.

4. Given an ordinal n with $1 \leq n \leq \omega$, construct a valuation domain R of global dimension n.

5. If R is a valuation domain and S an immediate extension of R, then $gl.d.S = gl.d.R$.

6. For any domain R, $p.d.Q/R = p.d.Q$.

7. For any domain R, $p.d.Q = 1$ if Q is countably generated.

§3. PROJECTIVE DIMENSIONS OF TORSION-FREE MODULES

We are in a good position to determine the projective dimension of most torsion-free modules over valuation domains in so far we are able to determine the cardinalities of minimal generating systems (see Fuchs [4]).

In this section, R will mean a valuation domain. First we dispose of the most relevant finite rank case.

THEOREM 3.1. [VD] (Fuchs [4]) Let M be a torsion-free module of finite rank. If a generating system of minimal cardinality of M contains $\aleph_m$ elements, then

$$\text{p.d.} M = m+1.$$

Proof. We induct on the rank k of M. For $k = 1$, this is just (2.2) and (2.4) combined. So assume the assertion true for $k-1$, and let M be torsion-free of rank $k \geq 2$. Let N be a pure submodule of M of rank $k-1$, so that in the arising exact sequence

$$0 \longrightarrow N \longrightarrow M \xrightarrow{\phi} A \longrightarrow 0$$

A is torsion-free of rank 1. Let the cardinalities of minimal generating systems in N and A be denoted by $\aleph_n$ and $\aleph_h$, respectively. In view of (II.5.4), the minimal cardinality of a generating system of M is precisely $\max(\aleph_n, \aleph_h)$. By induction hypothesis, $\text{p.d.} N = n+1$ and $\text{p.d.} A = h+1$, and what we want to show is that $\text{p.d.} M = \max(n,h) + 1$. By Kaplansky's Lemma, $\leq$ holds.

If $n > h$, then by Kaplansky's Lemma, $\text{p.d.} M = \text{p.d.} N$ follows.

If $n = h$, then necessarily $\text{p.d.} M \leq n+1$. If we had here strict inequality, then again by Kaplansky's Lemma, $\text{p.d.} A = \text{p.d.} N+1$ would follow, a contradiction.

Finally, suppose $n < h$. Consider a filtration of A by submodules generated by $\aleph_{h-1}$ elements, $\{A_\nu \mid \nu < \omega_h\}$. Evidently, $\{\phi^{-1} A_\nu \mid \nu < \omega_h\}$ yields a filtration of M by $\aleph_{h-1}$-generated submodules. Then $\text{p.d.} \phi^{-1} A_\nu \leq h$. Since $\phi^{-1} A_\mu / \phi^{-1} A_\nu \cong A_\mu / A_\nu$ for for all $\nu < \mu$, by Eklof's theorem, the set $\{\nu < \omega_h \mid \text{p.d.} \phi^{-1} A_\mu / \phi^{-1} A_\nu \geq h+1 \text{ for some } \mu > \nu\}$ is stationary in ω_h, thus $\text{p.d.} M \geq h+1$. □

Once the finite rank case is settled, it is a relatively easy task to determine the projective dimensions of those infinite rank torsion-free modules whose generating sets are of larger cardinalities. To start with, we establish an upper bound.

3. DIMENSIONS OF TORSION-FREE MODULES

PROPOSITION 3.2. [VD] If M is a torsion-free R-module which can be generated by $\aleph_m$ elements, then

$$p.d._R M \leq m+1.$$

Proof. Represent M as the union of a well-ordered continuous ascending chain of pure submodules M_ν such that all factors $M_{\nu+1}/M_\nu$ are of rank 1. All these factors are, by (II.5.4), generated by at most $\aleph_m$ elements, so by (2.2) their projective dimensions are at most $m+1$. Auslander's Lemma implies the stated inequality on p.d.M. □

The crucial result in our study of projective dimensions is the following.

THEOREM 3.3. [VD] (Fuchs [4]) Let M be a torsion-free R-module of rank $\aleph_n$ which can be generated by $\aleph_m$ but not by fewer elements. If $n < m$, then

$$p.d._R M = m+1.$$

Proof. Let $\{a_i\}_{i \in I}$ be a maximal independent set of elements in M; thus $|I| = \aleph_n$. For each finite subset α of I, let M_α be the pure submodule of M generated by a_i, $i \in \alpha$. Then M is the set union of all these M_α. As α runs over a set of cardinality $\aleph_n$, it is clear that $n < m$ implies the existence of an M_α whose generation requires $\aleph_m$ elements. With such an M_α we form the exact sequence

$$0 \to M_\alpha \to M \to M/M_\alpha \to 0.$$

As M_α is of finite rank, (3.1) implies p.d.$M_\alpha = m+1$, while from (3.2) we infer that p.d.$M/M_\alpha \leq m+1$. From Kaplansky's Lemma, the desired conclusion p.d.$M = m+1$ follows. □

COROLLARY 3.4. [VD] If M is an R-module which is generated by $\aleph_n$ elements subject to $\aleph_m$ ($m > n$) relations and if m is minimal, then

$$p.d._R M = m+2.$$

Proof. Let $0 \to H \to F \to M \to 0$ be an exact sequence where F is free on $\aleph_n$ generators. Then the rank of H is at most $\aleph_n$ and H is generated by $\aleph_m$ but not by fewer elements. In view of (3.3), p.d.$H = m+1$ whence the assertion follows. □

Actually, (3.3) is the best possible result of its kind. In fact, p.d.M can be any integer between 0 and $m+1$ if M is torsion-free of rank $\aleph_m$ and is generated by $\aleph_m$ elements. In this ambiguous case, however, we can offer a useful criterion for p.d.$M \leq m$ which will have interesting consequences. It should be pointed out that, by virtue of (3.6), this criterion can be regarded as a generalization of Pontryagin's criterion for the freeness of countable torsion-free abelian groups.

THEOREM 3.5. [VD] (Fuchs [4]) Suppose M is a torsion-free R-module of rank $\aleph_m$, m a non-negative integer. Then p.d.$_R M \leq m$ exactly if all pure submodules of rank $< \aleph_m$ in M have projective dimensions $\leq m$.

Proof. First assume p.d.$M \leq m$. If a pure submodule N has projective dimension $> m$, then Kaplansky's Lemma implies p.d.M/N = p.d.$N+1 > m+1$. From (3.3) we derive that M, and hence M/N, can be generated by at most $\aleph_m$ elements, thus by (3.2) p.d.$M/N \leq m+1$. This contradiction proves p.d.$N \leq m$.

Conversely, let all pure submodules N of rank $< \aleph_m$ in M have projective dimensions $\leq m$. In view of (3.3), such an N can be generated by $\aleph_{m-1}$ elements. Consider M as the union of a continuous well-ordered ascending chain of pure submodules, say, $0 = M_0 < M_1 < \ldots < M_\nu < \ldots$ ($\nu < \omega_m$); here all ranks are assumed $< \aleph_m$, thus each M_ν is $< \aleph_m$-generated. Since each factor $M_{\nu+1}/M_\nu$ can be generated by less than $\aleph_m$ elements, we clearly have p.d.$M_{\nu+1}/M_\nu \leq m$. From Auslander's Lemma, the inequality p.d.$M \leq m$ follows. □

The special case $m = 0$ yields:

3. DIMENSIONS OF TORSION-FREE MODULES

COROLLARY 3.6. [VD] A torsion-free R-module M of countable rank is free if and only if all finite rank pure submodules of M are free. □

An important consequence of (3.3) and (3.5) is as follows.

COROLLARY 3.7. [VD] (Fuchs [4]) If a torsion-free R-module M of rank $\aleph_m$ ($m \geq 0$) has projective dimension $d \geq m+1$, then M has a finite rank pure submodule of projective dimension d.

Proof. If $d > m+1$, then M requires $\aleph_{d-1}$ generators in view of (3.3), so at least one of its finite rank pure submodules requires this many generators. If $d = m+1$, then M cannot contain any finite rank pure submodules of projective dimension $> d$, because then M would require more than $\aleph_m$ generators, thus p.d.$M > m+1$, a contradiction. Suppose that all finite rank pure submodules of M have projective dimensions $\leq m$, i.e. all of them can be generated by at most $\aleph_{m-1}$ elements. Then all pure submodules of rank $< \aleph_m$ can be generated by at most $\aleph_{m-1}$ elements, i.e. their projective dimensions are $\leq m$. A simple appeal to (3.5) completes the proof. □

Let us return to (3.5) and give a different version which will be needed later on.

THEOREM 3.8. [VD] Let M be a torsion-free R-module and $m \geq 0$. Suppose that

$$0 = M_0 < M_1 < \ldots < M_\alpha < \ldots < M_\lambda = M \qquad (\alpha < \lambda)$$

is a continuous well-ordered ascending chain of submodules such that
  (i)   each M_α is pure in M;
  (ii) each M_α is $\aleph_m$-generated;
  (iii) p.d.$M_\alpha \leq m$ for each $\alpha < \lambda$;
  (iv) cof $\lambda \leq \omega_m$.
Then p.d.$M \leq m$.

Proof. Without loss of generality, $\lambda = \omega_k$ ($k \leq m$) can be assumed. Then M is $\aleph_m$-generated, thus in view of (3.5), it suffices to show that every pure submodule N of M whose rank is

$\leq \aleph_{m-1}$ has projective dimension $\leq m$. If $k = m$, then N is contained in some M_α ($\alpha < \lambda$), and (iii) implies p.d.$N \leq m$. If $k < m$, then (by the same reason) $N \cap M_\alpha$ being at most $\aleph_{m-1}$-generated, N is $\aleph_{m-1}$-generated. Thus p.d.$N \leq m$. □

EXERCISES

1. [VD] Prove that there exists a torsion-free R-module whose projective dimension is equal to the global dimension of R if and only if

$$\text{p.d.}Q = \text{gl.d.}R.$$

2. [VD] A finitely generated R-module M is finitely presented if and only if p.d.$M \leq 1$.

3. [VD] A finitely generated submodule N of a finitely generated R-module M satisfies p.d.$N \leq$ p.d.M.

4. [VD] Let R be any valuation domain of global dimension $> m+1 \geq 2$. Given $0 \leq k \leq m+1$, find a torsion-free R-module of rank $\aleph_m$ and of projective dimension k.

5. [VD] Let M be a torsion R-module of projective dimension $m+1$ such that M can be generated by $\aleph_m$ elements ($m \geq 0$). Then every finitely generated submodule of M has projective dimension $\leq m+1$.

6. [VD] Let I, J be R-submodules of Q such that $0 < J < I \leq Q$. Then

$$\text{p.d.}I/J = \begin{cases} \text{p.d.}J+1 & \text{if p.d.}I \leq \text{p.d.}J, \\ \text{p.d.}I & \text{otherwise.} \end{cases}$$

§4. PROJECTIVE DIMENSION ONE

Of particular interest are the R-modules whose projective dimensions are one. There are indications that these modules are more amenable, though as yet they have not been studied systematically.

4. PROJECTIVE DIMENSION ONE

Our discussion starts with an elementary result.

PROPOSITION 4.1. Let R be a Prüfer domain. A finitely generated R-module M is finitely presented if and only if $\text{p.d.}_R M \leq 1$.

Proof. Let $0 \to H \to F \to M \to 0$ be an exact sequence with F finitely generated free. If R is a Prüfer domain, then H is finitely generated exactly if it is projective (cf. (1.1)). Hence the claim is immediate. □

A less trivial result is contained in the following lemma.

LEMMA 4.2. Let R be a Prüfer domain, F a torsion-free R-module, and H a free submodule of F. If F/H is finitely generated, then F is projective and F/H finitely presented.

Proof. Manifestly, the proof can be restricted to the case in which F/H is cyclic and torsion. Let $\{x_i \mid i \in I\}$ denote a basis of H. As F is torsion-free, we may think of F as being contained in the Q-vectorspace with basis $\{x_i \mid i \in I\}$. Therefore, if F/H is generated, say, by the coset $a + H$ ($a \in F$), then there exist a finite subset $\{1,\ldots,n\}$ of I and non-zero elements $r, r_1, \ldots, r_n \in R$ such that

$$ra = r_1 x_1 + \ldots + r_n x_n. \tag{1}$$

It is obvious that any relation between a and basis elements x_i can be obtained from (1) if we multiply (1) by a suitable $s \in R$ and then divide it by some $t \in R$. Hence $F_0 = Ra + \sum_{i=1}^{n} Rx_i$ is a summand of F. As a finitely generated torsion-free R-module, F_0 is projective in view of (1.1). It follows that F is projective and $F/H \cong F_0/(\sum Rx_i)$ is finitely presented. □

We will find it convenient to refer to a submodule N of an R-module M as a _tight_ submodule (Fuchs [4]) if both

$$\text{p.d.}N \leq \text{p.d.}M \quad \text{and} \quad \text{p.d.}M/N \leq \text{p.d.}M.$$

Observe that the second inequality implies the first.

THEOREM 4.3. Let R be a Prüfer domain. An R-module of projective dimension 1 is coherent and its finitely generated submodules are tight.

Proof. Let N be a finitely generated submodule of an R-module M with $p.d.M = 1$. We can write $M \cong F/H$ with F and H free. N is then of the form $N \cong G/H$ for some G between H and F. From the preceding lemma we conclude that N is finitely presented and G is projective. This completes the proof. □

The following consequence is especially noteworthy.

THEOREM 4.4. (Fuchs [6]) Over a Prüfer domain R, a countably generated module has projective dimension ≤ 1 if and only if it is the union of a countable ascending chain of finitely presented R-modules.

Proof. The necessity is a trivial corollary to (4.3). On the other hand, if M is the union of a chain

$$0 \leq M_1 \leq M_2 \leq \cdots \leq M_n \leq \cdots$$

where each M_n is finitely presented, then all factors M_{n+1}/M_n are likewise finitely presented, and thus of projective dimension ≤ 1. An appeal to Auslander's Lemma concludes the proof. □

Our next objective is to show that in an R-module of projective dimension 1, tight submodules are abundant. The next result holds for arbitrary domains R.

THEOREM 4.5. [D] (Fuchs [6]) Every R-module M of projective dimension 1 has a family $T = \{M_i \mid i \in I\}$ of submodules such that

(i) $0, M$ belong to T;

(ii) T is closed under unions of chains;

(iii) if $M_i < M_j$ in T, then $p.d.M_j/M_i \leq 1$;

(iv) given $M_i \in T$ and a countable subset Δ of M, there is an $M_j \in T$ such that $\langle M_i, \Delta \rangle \leq M_j$ and M_j/M_i is countably generated.

Proof. Start off with a presentation of M:

4. PROJECTIVE DIMENSION ONE

$$0 \longrightarrow H \longrightarrow F \xrightarrow{\phi} M \longrightarrow 0$$

where $F = \oplus\{Rx \mid x \in X\}$ is a free R-module on X and H is projective. By a well-known result of Kaplansky, $H = \oplus\{H_y \mid y \in Y\}$ where the H_y's are countably generated projective R-modules.

Consider all pairs (X_i, Y_i) of subsets $X_i \subseteq X$, $Y_i \subseteq Y$ such that $F_i = \oplus\{Rx \mid x \in X_i\}$ and $H_i = \oplus\{H_y \mid y \in Y_i\}$ satisfy

$$H_i = H \cap F_i.$$

Let i run over an index set I. Note that $H = H_i \oplus H_i^*$ where $H_i^* = \oplus\{H_y \mid y \in Y \setminus Y_i\}$. Consequently, the submodules $F_i + H = F_i \oplus H_i^*$ are projective. Define

$$T = \{M_i \mid i \in I\} \quad \text{where} \quad M_i = (F_i + H)/H.$$

Manifestly, T satisfies (i) and (ii). As for $M_i < M_j$, the isomorphism $M_j/M_i \cong (F_j + H)/(F_i + H)$ holds, (iii) is likewise satisfied by T.

It remains to verify (iv) for T. The proof can be restricted to the case $M_i = 0$; in fact, the general result then follows by applying this special case to the R-module M/M_i and to the family $\{M_j/M_i \mid M_i \leq M_j \in T\}$. Given a countable subset Δ of M, there is a countable subset $X^{(1)}$ of X such that $\phi\langle X^{(1)}\rangle$ contains Δ. The submodule $\langle X^{(1)}\rangle \cap H$ has at most countable rank, thus there is a countable subset $Y^{(1)}$ of Y such that $\langle X^{(1)}\rangle \cap H \leq \langle H_y \mid y \in Y^{(1)}\rangle$. We can select a countable subset $X^{(2)}$ of X that contains $X^{(1)}$ and satisfies $\langle H_y \mid y \in Y^{(1)}\rangle \leq \langle X^{(2)}\rangle$. Repeating this process, we obtain ascending chains of countable subsets

$$X^{(1)} \subseteq X^{(2)} \subseteq \ldots \subseteq X^{(n)} \subseteq \ldots \quad \text{and} \quad Y^{(1)} \subseteq Y^{(2)} \subseteq \ldots \subseteq Y^{(n)} \subseteq \ldots$$

of X and Y, respectively, such that

$$\langle X^{(n)}\rangle \cap H \leq \langle H_y \mid y \in Y^{(n)}\rangle \leq \langle X^{(n+1)}\rangle$$

for each $n \geq 1$. Denoting by X^* and Y^* the unions of these chains, it is clear that they are countable and the pair (X^*, Y^*) is one of the pairs (X_i, Y_i) defined above. Thus $M^* = (F^* + H)/H$ belongs to T and is countably generated; here we used the notation $F^* = \oplus\{Rx \mid x \in X^*\}$. □

Starting from a family T of R-submodules of M as in (4.5), we can construct a well-ordered continuous chain of tight submodules $M_\alpha \in T$,

$$0 = M_0 < M_1 < \ldots < M_\alpha < \ldots < M_\tau = M \qquad (\alpha < \tau) \qquad (2)$$

such that, for each $\alpha < \tau$, $M_{\alpha+1}/M_\alpha$ is countably generated. Thus a comparison of (4.5) with Auslander's Lemma yields:

COROLLARY 4.6. [D] (Fuchs [6]) An R-module M has projective dimension ≤ 1 if and only if it has a continuous well-ordered ascending chain (2) of submodules such that, for each $\alpha < \tau$, $M_{\alpha+1}/M_\alpha$ is countably generated and has projective dimension ≤ 1. □

Combining (4.4) and (4.6), we are led to

COROLLARY 4.7. A module M over a Prüfer domain has projective dimension ≤ 1 exactly if it is the union of a well-ordered continuous chain (2) of submodules such that $M_{\alpha+1}/M_\alpha$ is finitely presented cyclic for each $\alpha < \tau$. □

EXERCISES

1. Verify the following converse of (4.1). Let R be a domain such that a finitely generated R-module M is finitely presented exactly if p.d.$M \leq 1$. Then R is a Prüfer domain.

2. Use Exercise 5 in §3 to prove the following generalization of (4.4) for valuation domains R. A countably generated R-module has projective dimension $\leq m$ $(m \geq 1)$ if and only if it is the union of a countable ascending chain of finitely generated submodules of projective dimensions $\leq m$.

3. A submodule of a projective module is tight exactly if it is a summand.

5. TIGHT SYSTEMS 87

4. [D] If T and T' are two families of submodules of M, p.d.$M \leq 1$, satisfying properties (i)-(iv) in (4.5), then $T \cap T'$ also satisfies the same properties.

5. [VD] Improve on (4.7) for valuation domains R by demanding that, for each $\alpha < \tau$, $M_{\alpha+1}/M_\alpha$ be cyclically presented.

§5. TIGHT SYSTEMS

In view of the results on projective dimensions of modules over valuation domains (see §3), one can expect that modules with small projective dimensions, but with a large number of generators can be built up of modules of the same projective dimensions and with smaller numbers of generators. This is in fact the case, and this is what we intend to prove here.

Motivated by (4.5), we introduce the following terminology (Bazzoni-Fuchs [1]). A **tight system** for M is a family T of submodules M_i ($i \in I$) of M such that

(i) $0, M \in T$;

(ii) T is closed under unions of chains;

(iii) if $M_i, M_j \in T$ and $M_i \leq M_j$, then
$$\text{p.d.} M_j/M_i \leq \text{p.d.} M = n;$$

(iv) if $M_i \in T$ and Δ is a subset of M of cardinality $\leq \aleph_{n-1}$, then there is an $M_j \in T$ satisfying

(a) $\langle M_i, \Delta \rangle \leq M_j$;

(b) M_j/M_i is $\aleph_{n-1}$-generated.

From (i) and (iii) it is clear that all submodules in T are tight in M. It is readily checked that (iv) continues to hold if $\aleph_{n-1}$ is replaced by any cardinal number larger than $\aleph_{n-1}$.

The main result of this section is the following theorem. Observe that (A_1) is a special case of (4.5) for valuation domains.

THEOREM 5.1. (Bazzoni-Fuchs [1]) For a valuation domain R, the following assertions are true for every $n \geq 0$:

(A_n) Every R-module M of projective dimension n has a tight system.

(B_n) Every torsion-free R-module M of projective dimension n has a tight system consisting of pure submodules.

(C_n) In every torsion-free R-module M of projective dimension n, a pure submodule of rank $\leq \aleph_{n-1}$ is $\aleph_{n-1}$-generated.

Proof. We induct on n in the following way. First we prove (A_0), and then verify the implications (A_n) $\Longrightarrow$ (C_n); (A_n) + (C_n) $\Longrightarrow$ (B_n) and (B_n) $\Longrightarrow$ (A_{n+1}).

It is easy to see that (A_0) holds. In fact, let M be a projective R-module. By (1.3), M is free, so we can write $M = \oplus\{Rx \mid x \in X\}$. Let I denote the power set of X and let $M_i = \oplus\{Rx \mid x \in i\}$ for $i \in I$. Then $\mathcal{T} = \{M_i \mid i \in I\}$ is evidently a tight system for M.

(A_n) $\Longrightarrow$ (C_n): Let M be torsion-free of projective dimension n, and N a pure submodule of rank $\leq \aleph_{n-1}$. By way of contradiction, suppose that N requires $\aleph_k (k \geq n)$ generators. By (3.3), p.d.$N = k+1$. From (A_n) it follows that N can be embedded in a submodule $\overline{N}$ of M such that p.d.$\overline{N} \leq n$ and $\overline{N}$ is $\aleph_k$-generated. (II.5.4) shows that $\overline{N}$ cannot be generated by fewer than $\aleph_k$ elements. If the rank of $\overline{N}$ were $< \aleph_k$, then (3.3) would imply p.d.$\overline{N} = k+1 > n$, impossible. Hence $\overline{N}$ has rank $\aleph_k$. Now (3.5) applied to $\overline{N}$ leads us to the conclusion p.d.$N \leq k$ which is a desired contradiction.

(A_n) + (C_n) $\Longrightarrow$ (B_n): Let again M be torsion-free of projective dimension n, and $\mathcal{T}$ a tight system for M as stipulated by (A_n). Let $\mathcal{T}'$ be the subsystem of $\mathcal{T}$ consisting of those $M_i \in \mathcal{T}$ which are pure in M. We claim that $\mathcal{T}'$ is likewise a tight system for M. It is clearly enough to show that (iv) holds for $\mathcal{T}'$, and - as in the proof of (4.5) - it suffices to do this for $M_i = 0$. Any subset Δ of cardinality $\leq \aleph_{n-1}$ is contained in an $\aleph_{n-1}$-generated member $M^{(1)}$ of $\mathcal{T}$. In view of (C_n), the

5. TIGHT SYSTEMS

purification $M_*^{(1)}$ of $M^{(1)}$ is again $\aleph_{n-1}$-generated. There is an $\aleph_{n-1}$-generated $M^{(2)} \in T$ that contains $M_*^{(1)}$ whose purification $M_*^{(2)}$ is again $\aleph_{n-1}$-generated. Thus proceeding, we obtain an ascending chain $M^{(1)} \leq M_*^{(1)} \leq M^{(2)} \leq M_*^{(2)} \leq \ldots$ whose union is in T', as it belongs to T and is pure in M. Manifestly, it is $\aleph_{n-1}$-generated.

$(B_n) \Longrightarrow (A_{n+1})$: Let M be an R-module of projective dimension $n+1$, and choose a presentation of M, $0 \to H \to F \to M \to 0$ where $F = \oplus\{Rx \mid x \in X\}$ is a free R-module and $p.d.H = n$. By (B_n), H has a tight system $\{H_t \mid t \in T\}$ consisting of pure submodules. Consider pairs (X_i, T_i) of subsets $X_i \subseteq X$, $T_i \subseteq T$ satisfying $H_i = H \cap F_i$ where $H_i = \langle H_t \mid t \in T_i \rangle$ and $F_i = \langle X_i \rangle$; here i runs over a suitable index set I. We claim that

$$T = \{M_i \mid i \in I\} \text{ where } M_i = (F_i + H)/H$$

is a tight system for M.

Properties (i) and (ii) are clear for T. Now if $M_i < M_j$ in T, then

$$M_j/M_i \cong (F_j+H)/(F_i+H) \cong F_j/[F_j \cap (F_i+H)] = F_j/(F_i+H_j).$$

Since $(F_i+H_j)/F_i \cong H_j/H_i$ has projective dimension $\leq n$ (H_i being tight in H_j), we have $p.d.(F_i+H_j) \leq n$. Therefore, $p.d.M_j/M_i \leq n+1$, establishing (iii). Property (iv) can be verified by the same back and forth method as in the proof of (4.5), after reducing the proof to the case $M_i = 0$. □

An immediate consequence is the analogue of (4.6):

COROLLARY 5.2. [VD] An R-module M is of projective dimension $\leq m$ exactly if it is the union of a well-ordered ascending chain $\{M_\alpha \ (\alpha < \tau)\}$ of submodules, such that, for each $\alpha < \tau$, $M_{\alpha+1}/M_\alpha$ is $\aleph_{m-1}$-generated and has projective dimension $\leq m$. □

EXERCISES

1. [VD] Show that the intersection of two tight systems for an R-module M is again one.

2. [VD] (Bazzoni-Fuchs [1]) In a torsion-free M with p.d.M = n, a pure submodule of rank $\leq \aleph_{n-1}$ is tight.

3. [VD] (Bazzoni-Fuchs [1]) If G is torsion-free, H is a submodule with p.d.H $\leq$ n and G/H is $\aleph_{n-1}$-generated, then p.d.G $\leq$ n.

§6. QUASI-PROJECTIVITY

We discuss briefly a notion which is a familiar generalization of projectivity.

Let M and A be R-modules. M is called A-<u>projective</u>, if, for every submodule B of A, each diagram

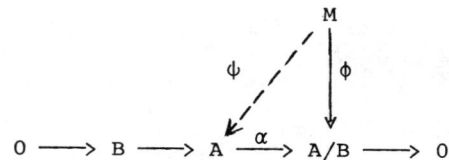

can be completed by a map ψ making the triangle commute. Evidently, M is A-projective if and only if the natural map $\text{Hom}_R(M,A) \to \text{Hom}_R(M,A/B)$ is surjective for every $B \leq A$.

It is readily seen that if M is A-projective, then it is X-projective for every submodule or factor module X of A.

If M is M-projective, it is called <u>quasi-projective</u>.

Our main concern here will be the ideals of valuation domains. We start with a simple observation.

LEMMA 6.1. [VD] (Herrmann [1]) Let I and J be non-zero ideals of R. Then I is J-projective exactly if it is Q-projective.

<u>Proof</u>. In view of our previous remark, it suffices to show that I is Q-projective whenever it is quasi-projective. Let $\alpha : Q \to Q/K$ be an epimorphism and $\phi : I \to Q/K$ any map. As Ker $\phi \cong K$, there is no loss of generality in assuming Ker ϕ = K and Im ϕ = I/K. Thus ϕ factors through the canonical map $\nu : I \to I/K$. By quasi-projectivity, there is a $\mu : I \to I$ such that $\alpha\mu = \phi$.

6. QUASI-PROJECTIVITY

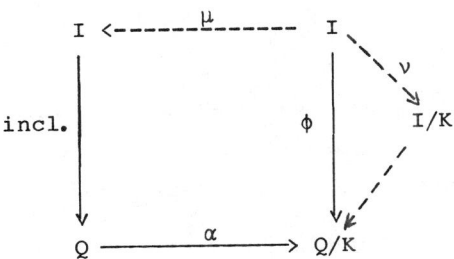

This completes the proof. □

The next result is most relevant.

THEOREM 6.2. [VD] (Herrmann [1]) For an ideal $I \neq 0$ of a valuation domain R, the following are equivalent:

(a) I is quasi-projective;
(b) $\operatorname{Ext}_R^1(I,K) = 0$ for every ideal K of R;
(c) for every ideal K of R, Q/K has the injective property relative to the exact sequence $0 \to I \to R \to R/I \to 0$.

Proof. (a) ⟺ (b). In view of the preceding lemma, (a) is equivalent to the Q-projectivity of I. Consider the exact sequence

$$0 \longrightarrow \operatorname{Hom}_R(I,K) \longrightarrow \operatorname{Hom}_R(I,Q) \xrightarrow{\delta} \operatorname{Hom}_R(I,Q/K) \longrightarrow$$
$$\longrightarrow \operatorname{Ext}_R^1(I,K) \longrightarrow \operatorname{Ext}_R^1(I,Q) = 0 \, . \qquad (1)$$

It is clear that δ is surjective if and only if $\operatorname{Ext}_R^1(I,K) = 0$.

(b) ⟺ (c). The obvious exact sequences

$$0 = \operatorname{Ext}_R^1(R,K) \to \operatorname{Ext}_R^1(I,K) \to \operatorname{Ext}_R^2(R/I,K) \to 0$$

and

$$0 = \operatorname{Ext}_R^1(R/I,Q) \to \operatorname{Ext}_R^1(R/I,Q/K) \to \operatorname{Ext}_R^2(R/I,K) \to 0$$

imply that $\operatorname{Ext}_R^1(I,K) \cong \operatorname{Ext}_R^1(R/I,Q/K)$. Hence the assertion follows. □

We are now able to prove:

THEOREM 6.3. [VD] (Herrmann [1]) Every ideal of a valuation domain R is quasi-projective exactly if R is almost maximal.

Proof. By (VI.4.8), R is almost maximal if and only if Q/K is injective for every ideal K of R. Therefore, in view of Baer's criterion for injectivity, (6.2) implies the claim. □

Maximal valuation domains can similarly be characterized. We require a lemma.

LEMMA 6.4. [VD] (Rangaswamy-Vanaja [1], Herrmann [1]) Q is quasi-projective if and only if R is R-complete.

Proof. From the exact sequence (1), with I replaced by Q, we conclude that for the quasi-projectivity of Q it is necessary and sufficient that $\text{Ext}^1(Q,K) = 0$ for every ideal K of R. By virtue of (V.1.2), this condition means K is R-complete. But if R is R-complete, then so are all ideals K of R, as is shown in (V.1.3). □

We have now at once

THEOREM 6.5. (Herrmann [1]) Let R be a valuation domain. All rank one torsion-free R-modules are quasi-projective exactly if R is a maximal valuation domain.

Proof. This follows from (6.4), (6.5) and from (I.5.2). □

EXERCISES

1. If M is A_i-projective for $i = 1,\ldots,n$, then M is $A_1 \oplus \cdots \oplus A_n$-projective.

2. $\oplus M_i$ is A-projective if and only if each M_i is A-projective.

3. [VD] (Herrmann [1]) If $I_1,\ldots,I_n$ are quasi-projective ideals of R, then $I_1 \oplus \cdots \oplus I_n$ is likewise quasi-projective.

4. Let R be an almost maximal valuation domain which is not maximal. A finite rank torsion-free R-module is quasi-projective exactly if it is isomorphic to a direct sum of ideals of R.

5. Over a maximal valuation domain, all torsion-free modules of finite rank are quasi-projective.

6. [VD] Let F be a projective R-module and I an ideal of R. Then F/IF is a quasi-projective R-module.

NOTES

The rather surprising relation between projective dimension and number of generators has been studied by various authors, and there is an extensive literature on this subject. A most satisfactory, and for us the most significant, result is due to Osofsky; this is our (2.2). Actually, she proved a more general result than our (2.2). Also, (2.4) holds under more general conditions, see Kaplansky [3] and Small [1]. Some results extend to Prüfer domains, cf. Osofsky [4].

Osofsky [3] proved that the global dimension of a valuation ring which is not a domain is necessarily infinity.

The results in §§3-5 are based on three papers; see Fuchs [4], [6] and Bazzoni-Fuchs [1]. In the last paper, the reader can find generalizations of results in §3.

Not much is known of the global pure-projective dimension of valuation domains R. It is easy to see that it must be at least gl.d.R-1 and at least p.d.Q. An upper bound can be given by making use of (II.5.3).

PROBLEM 2. Find the global pure-projective dimension of a valuation domain.

Quasi-projectives over valuation domains have been studied by Herrmann [1], but no satisfactory characterization is known so far.

PROBLEM 3. Characterize quasi-projective modules over valuation domains.

V. Topology and Filtrations

Before launching into a series of results on various classes of modules, we need to make some easy but relevant observations about topological aspects. To avoid duplication, we steered away from certain topics covered by Matlis [5]; we touch only briefly upon R-completeness, but emphasize its homological aspects.

A delicate instrument for the investigation of certain modules appears in what we call filtrations and ultracompleteness. Their most rewarding applications lie in the study of RD-injectivity and torsion-ultracompleteness (Chapters XI-XII).

§1. THE R-TOPOLOGY

By a <u>topological module</u> is meant a module M which is at the same time a topological space such that the addition is a continuous function $M \times M \to M$, while multiplication by a ring element r is a continuous function $\bar{r} : M \to M$. (The underlying ring will carry the discrete topology in most of our considerations.)

1. THE R-TOPOLOGY

A topology is <u>linear</u> if it has a base of neighborhoods about 0 consisting of submodules U_i ($i \in I$) of M; in this case, for every $a \in M$, $\{a + U_i\}_{i \in I}$ will constitute a base of neighborhoods about a. In this way, M will carry a uniform structure.

As for every $r \in R$, $x \mapsto rx$ is a continuous map of M, it follows that M[r] is a closed submodule in every linear Hausdorff topology of M. Likewise, M[L] is closed being the intersection of closed submodules.

The most important linear topology on M is obtained by letting the submodules rM ($0 \neq r \in M$) of M form a subbase for the open neighborhoods of 0 in M; this is the so-called R-<u>topology</u> of M. The R-topology makes every R-module into a topological R-module and R itself into a topological ring.

It is clear that M is Hausdorff in its R-topology if and only if $M^1 = \bigcap_{0 \neq r \in R} rM$ vanishes. A torsion-free R-module is Hausdorff in its R-topology exactly if it is reduced.

A submodule N of M carries its own R-topology. If N is RD in M, then this is the same as the topology induced by the R-topology of M. For a submodule N of M, the closure of N in the R-topology of M is $\bigcap_{0 \neq r \in R} (N + rM)$. The density of N in M is equivalent to the divisibility of M/N.

An R-module M is R-<u>complete</u> if it is complete and Hausdorff in the uniformity of its R-topology, i.e. if every Cauchy net in M has a unique limit in M. It is convenient to consider a <u>Cauchy</u> <u>net</u> in M as a subset $\{a_r\}$, indexed by the non-zero elements r of R, subject to the condition that

$$a_r - a_{rs} \in rM \quad \text{for all nonzero } r,s \in R.$$

$a \in M$ is then a <u>limit</u> of this net provided that $a - a_r \in rM$ for all $0 \neq r \in R$.

The R-<u>completion</u> $\tilde{M}$ of M can be defined as the inverse limit

$$\tilde{M} = \varprojlim M/rM \quad (0 \neq r \in R)$$

with the natural maps $M/rM \leftarrow M/rsM$ given by $a + rM \mapsfrom a + rsM$ ($a \in M$). This $\tilde{M}$ can be regarded as a submodule of $\prod_{r \in R} M/rM$. The canonical homomorphism

$$\eta_M : a \longmapsto (\ldots, a+rM, \ldots) \in \widetilde{M} \qquad (a \in M)$$

maps M into $\widetilde{M}$ such that $\text{Ker } \eta_M = M^1$. Observe that all of M/rM are discrete in their respective R-topologies, thus $\Pi(M/rM)$ is complete in the product topology (as well as its R-topology which turns out to be a finer topology).

As is well known, the completion process works nicely for metrizable linear topologies, but can become unmanageable in the general case. Bearing this in mind, it is reasonable to confine our present study to two tractable cases: where Q is a countably generated R-module and where the modules considered are torsion-free, respectively. Both cases deserve particular attention in view of the rather surprising connection between R-completeness and homological properties. We delegate the discussion of the first case to the next section, and for the rest of this section, we shall be concerned with torsion-free modules. For a more complete discussion, we refer to Matlis [3] and [5].

LEMMA 1.1. [D] (Matlis [3]) Let M be a reduced torsion-free R-module, viewed as a submodule of $\widetilde{M}$. Then:

(i) M is an RD-submodule of $\widetilde{M}$;

(ii) $\widetilde{M}/M$ is torsion-free divisible;

(iii) the induced topology is the R-topology of $\widetilde{M}$, and $\widetilde{M}$ is R-complete.

   Proof. Let $0 \neq a \in M$ and suppose $r \in R$ does not divide a. Then $a + rM$ is not the trivial coset in M/rM, so it is not divisible by r either. Hence M is RD in $\Pi M/rM$, and a fortiori in $\widetilde{M}$.

   To check $\widetilde{M}/M$ torsion-free, let $\tilde{a} \in \widetilde{M}$ satisfy $r\tilde{a} = b \in M$, for some $0 \neq r \in R$. Write $\tilde{a} = (\ldots, a_s + sM, \ldots)$; then $ra_r - b \in rM$ implies $b = rc$ for some $c \in M$. From $r(a_{sr} - c) = ra_{rs} - b \in rsM$ and the torsion-freeness of M, $a_{sr} - c \in sM$ follows. Hence $\tilde{a} = c$. The divisibility of $\widetilde{M}/M$ will follow from part (iii).

1. THE R-TOPOLOGY

Now let U_s denote the product of all M/rM ($0 \neq r \in R$) with the exception of M/sM. Then $s\tilde{M} \leq U_s$ holds trivially. If $\tilde{a} = (\ldots, a_r + rM, \ldots) \in \tilde{M} \cap U_s$, then not only $a_s \in sM$, but even $a_{rs} \in sM$ for every $0 \neq r \in R$. Pick $b_r \in M$ such that $sb_r = a_{rs}$; then $\tilde{b} = (\ldots, b_s + sM, \ldots) \in \tilde{M}$ satisfies $s\tilde{b} = \tilde{a}$, establishing $\tilde{M} \cap U_s \leq s\tilde{M}$. □

If (1.1) is applied to $M = R$, we obtain a domain $\tilde{R}$ which is a torsion-free R-module. It is easy to check that the R- and $\tilde{R}$-topologies of $\tilde{R}$ coincide.

We can now prove a homological characterization of R-completeness.

PROPOSITION 1.2. [D] (Matlis [3]) Let M be a reduced torsion-free R-module. M is R-complete if and only if

$$\text{Ext}_R^1(Q, M) = 0.$$

Proof. Suppose $\text{Ext}_R^1(Q, M) \neq 0$, i.e. there is a non-split exact sequence $0 \to M \to N \to Q \to 0$. Here M is RD and dense in N, so M cannot be R-complete. On the other hand, if M is not complete, then $0 \to M \to \tilde{M} \to \tilde{M}/M \to 0$ is a non-split exact sequence where, because of (1.1)(ii), $\tilde{M}/M$ is torsion-free divisible $\neq 0$. As $\tilde{M}/M$ is a direct sum of copies of Q, $\text{Ext}_R^1(Q, M) \neq 0$ follows. □

By making use of the exact sequence obtained from $0 \to IM \to M \to M/IM \to 0$ via application of $\text{Ext}_R^1(Q, *)$, we conclude:

COROLLARY 1.3. [D] For every ideal $I \neq 0$ of R, and for every reduced torsion-free R-module M, M and IM are simultaneously R-complete or not. □

Before stating the main result on R-complete torsion-free modules, we shall prove two preparatory lemmas and a theorem. We set $K = Q/R$.

LEMMA 1.4. [D] A divisible torsion R-module is of the form $K \otimes_R M$ with a suitable reduced (R-complete) torsion-free R-module M if and only if it is h-divisible.

Proof. The "only if" part is obvious. To prove the converse, let H be an h-divisible torsion, D a divisible torsion-free module and $\alpha : D \to H$ an epimorphism. Without loss of generality, $M = \operatorname{Ker} \alpha$ may be assumed to be essential in D. Hence $Q \otimes M \cong D$, and from the exact sequence $0 \to M \to Q \otimes M \to K \otimes M \to 0$ the isomorphism $H \cong K \otimes M$ becomes evident. As $\widetilde{M}/M$ is divisible and torsion-free, we obtain $0 = \operatorname{Tor}_1(K, \widetilde{M}/M) \to K \otimes_R M \to K \otimes_R \widetilde{M} \to K \otimes_R \widetilde{M}/M = 0$, completing the proof. □

LEMMA 1.5. [D] For every reduced torsion-free R-module M, there is a natural monomorphism

$$\phi_M : M \to \operatorname{Hom}_R(K, K \otimes_R M).$$

ϕ_M is an isomorphism exactly if M is R-complete.

Proof. The exact sequence $0 \to R \to Q \to K \to 0$ implies that both $0 = \operatorname{Hom}_R(Q, M) \to \operatorname{Hom}_R(R, M) \cong M \to \operatorname{Ext}^1_R(K, M) \to \operatorname{Ext}^1_R(Q, M) \to 0$ and $0 \to M \to Q \otimes M \to K \otimes M \to 0$ are exact. From the last sequence, a fourth exact sequence is derived: $0 \to \operatorname{Hom}_R(K, K \otimes M) \to \operatorname{Ext}^1_R(K, M) \to \operatorname{Ext}^1_R(K, Q \otimes M) = 0$. Combining the second and fourth sequences, we can argue that M can be embedded by a natural monomorphism ϕ_M in $\operatorname{Hom}_R(K, K \otimes M)$. The cokernel of ϕ_M is $\operatorname{Ext}^1_R(Q, M)$ which vanishes exactly if M is R-complete; cf. (1.2).□

From the proof we can derive that for torsion-free modules the R-completion can be obtained homologically:

THEOREM 1.6. [D] (Matlis [3]) If M is a reduced torsion-free R-module, then there is a natural isomorphism

$$\widetilde{M} \cong \operatorname{Ext}^1_R(K, M).$$

Proof. ϕ_M embeds M in $\operatorname{Hom}_R(K, K \otimes M)$ with divisible cokernel. As $K \otimes M = K \otimes \widetilde{M}$, the second part of (1.5) shows that this Hom is naturally isomorphic to $\widetilde{M}$. The rest follows from the fourth exact sequence in the proof of (1.5). □

1. THE R-TOPOLOGY

We have arrived at the highlight of the theory; this is the so-called <u>Matlis Duality</u> (first discovered by D. K. Harrison for abelian groups):

THEOREM 1.7. [D] (Matlis [3]) There is an isomorphism between the category of h-divisible torsion R-modules T on one hand, and the category of reduced R-complete torsion-free R-modules M on the other hand. The correspondences

$$T \longmapsto \mathrm{Hom}_R(K,T) \quad \text{and} \quad M \longmapsto K \otimes_R M$$

are inverse to each other.

Proof. If T is as stated, then by (1.4) it can be written in the form $K \otimes M$ for a suitable R-complete torsion-free M. We obtain $\mathrm{Hom}(K,T) = \mathrm{Hom}(K, K \otimes M)$ which is, in view of (1.5), naturally isomorphic to M. Thus $T \cong K \otimes \mathrm{Hom}(K,T)$.

Next suppose M is of the stated kind. From (1.4) we infer that $K \otimes M$ is h-divisible torsion, and by (1.5) we obtain the desired $M \cong \mathrm{Hom}(K, K \otimes M)$.

Finally, we establish a natural isomorphism

$$\psi : \mathrm{Hom}_R(M,N) \to \mathrm{Hom}_R(K \otimes_R M, K \otimes_R N)$$

for R-complete torsion-free modules M, N. For every $\eta : M \to N$ there is a commutative diagram with exact rows:

$$\begin{array}{ccccccccc} 0 & \to & M & \to & Q \otimes M & \to & K \otimes M & \to & 0 \\ & & \downarrow \eta & & \downarrow 1 \otimes \eta & & \downarrow \eta^* & & \\ 0 & \to & N & \to & Q \otimes N & \to & K \otimes N & \to & 0 \end{array}$$

where η^* is uniquely determined by η. Now setting $\psi : \eta \longmapsto \eta^*$, it is clear that $\eta^* = 0$ means that $1 \otimes \eta$ carries $Q \otimes M$ into N. As N is reduced, $1 \otimes \eta = 0$ and $\eta = 0$ follows. Thus ψ is monic. ψ is surjective if $\mathrm{Hom}(Q \otimes M, Q \otimes N) \to \mathrm{Hom}(Q \otimes M, K \otimes N)$ is a surjective map. This is indeed the case as its cokernel is contained in $\mathrm{Ext}^1_R(Q \otimes M, N) = 0$; cf. (1.2). □

Anticipating results on cotorsion modules, it is easy to see which modules correspond, in this duality, to the injective torsion R-modules.

THEOREM 1.8. [D] (Warfield [4]) In the duality of (1.7), the duals of the injective torsion R-modules are exactly the RD-injective reduced torsion-free R-modules.

Proof. We begin with the observation that by (XII.3.6) RD-injectivity and cotorsionness are equivalent for torsion-free R-modules M. Furthermore, (XII.3.1) shows that M is cotorsion exactly if $Ext(Q,M) = 0$ and $i.d.M \leq 1$. The first condition is, by (1.2) equivalent to the R-completeness of M, while the second to the injectivity of $K \otimes M$, as is clear from the exact sequence $0 \to M \to Q \otimes M \to K \otimes M \to 0$. □

The next result points out a remarkable relationship between two endomorphism rings.

COROLLARY 1.9. [D] Let M be a reduced R-complete torsion-free R-module and E its injective hull. Then, canonically,
$$End_R M \cong End_R E/M .$$

Proof. The isomorphism ψ in the proof of (1.7) yields a group-isomorphism $End\, M \cong End(K \otimes M)$. Obviously, $K \otimes M \cong E/M$. That ψ preserves multiplication is readily checked by duplicating the diagram in the proof of (1.7) and verifying that $(\eta_1 \eta_2)^* = \eta_1^* \eta_2^*$ holds. □

Finally, we prove that completion commutes with localization. More precisely,

THEOREM 1.10. [D] For a domain R and a prime ideal J of R, we have a natural isomorphism
$$(\tilde{R})_J \cong \tilde{R}_J .$$

Proof. Before entering into the proof, we observe that the R- and R_J-topologies on R_J coincide.

1. THE R-TOPOLOGY

The canonical embedding $R \to R_J$ extends to a well-defined map $\tilde{R} \to \tilde{R}_J$ between the R-completions. This is monic. Tensoring by R_J (this is a flat R-module) yields a monic map between the localizations: $(\tilde{R})_J \to \tilde{R}_J$. This map has to be surjective, since the image of R_J in the second ring is dense. □

EXERCISES

1. [D] The ideal topology of an R-module M is defined by taking $\{IM \mid 0 \neq I \text{ an ideal of } R\}$ as a subbase of neighborhoods about 0. Show that the ideal topology is the same as the R-topology.

2. [D] If M is a torsion-free R-module, then $\tilde{M}$ is an RD-submodule of $\Pi(M/rM)$.

3. [D] Verify the isomorphism $\tilde{M}/M \cong \mathrm{Ext}^1_R(Q,M)$ for torsion-free R-modules M.

4. (a) If R is a domain, $\tilde{R}$ is always a ring but it need not be a domain.

   (b) The R- and $\tilde{R}$-topologies of $\tilde{R}$ coincide.

   (c) Prove that $\tilde{R}$ is an immediate extension of R whenever R is a valuation domain.

5. [D] (Matlis [3]) Prove that $K = Q/R$ satisfies

$$\mathrm{Hom}_R(K,K) \cong \tilde{R}.$$

6. [D] A torsion R-module can also be viewed, in a natural way, as an $\tilde{R}$-module.

7. [D] If M is a torsion-free R-module, $\tilde{M}$ has a unique $\tilde{R}$-module structure.

8. [D] (Matlis [3]) For an ideal $I \neq 0$ of R, we have:

   (a) $\tilde{I} = \tilde{R} \otimes I$;

   (b) $\tilde{R}I/I \cong \tilde{R}/R$;

   (c) $\tilde{R}/\tilde{R}I \cong R/I$.

From (a) conclude that I is R-complete exactly if R is R-complete.

9. [D] (Matlis [3]) Let $0 \to A \to B \to C \to 0$ be an exact sequence of h-reduced R-modules with C torsion-free. Then the induced sequence

$$0 \to \tilde{A} \to \tilde{B} \to \tilde{C}$$

is exact.

10. [D] Suppose Q is uncountably generated. If M_i ($i \in I$) are torsion-free R-complete R-modules, then $\oplus M_i$ is likewise R-complete.

11. [D] Show that, in the Matlis duality, for an ideal I of R, Q/I and $\tilde{I}$ correspond to each other.

12. [D] (Matlis [3]) There is a bijection between the direct decompositions of an h-divisible torsion module D and its dual $\text{Hom}_R(K,D)$.

§2. R-COMPLETE MODULES

As we have mentioned in §1, R-completeness can be dealt with successfully if either the modules are torsion-free or the field Q of quotients of R is a countably generated R-module. We now enter into the study of the second alternative, and concentrate on the homological aspects of R-completeness. Throughout this section, R will denote an arbitrary domain.

If Q is a countably generated R-module, then there is a sequence $t_1,\ldots,t_n,\ldots$ of non-zero elements of R such that Q is the union of a countable ascending chain of cyclic R-modules:

$$Rt_1^{-1} < Rt_1^{-1}t_2^{-1} < \ldots < Rt_1^{-1}\ldots t_n^{-1} < \ldots .$$

For every $0 \neq r \in R$, r^{-1} belongs to a member of this chain, i.e. $r | t_1 \ldots t_n$ for some n. Therefore, for every R-module M, the submodules

$$\{t_1 \ldots t_n M\}_{n<\omega}$$

2. R-COMPLETE MODULES

form a base of neighborhoods of 0 in the R-topology of M. Consequently,

LEMMA 2.1. [D] If Q is a countably generated R-module, then in every R-module M, the R-topology is pseudo-metrizable. □

By a standard result in topology, a pseudo-metrizable M has a completion $\tilde{M}$ which is a complete metric Hausdorff space:

$$\tilde{M} = \varprojlim (M/t_1 \ldots t_n M).$$

This $\tilde{M}$ can be viewed as a submodule of $\Pi(M/t_1 \ldots t_n M)$ consisting of all countable vectors of the form

$$\tilde{a} = (a_0, a_1 + t_1 M, \ldots, a_n + t_1 \ldots t_n M, \ldots) \qquad (1)$$

$(a_n \in R)$ such that $a_n - a_{n+1} \in t_1 \ldots t_n M$.

LEMMA 2.2. [D] If Q is countably generated, then the induced topology of $\tilde{M}$ coincides with its R-topology and $\tilde{M}$ contains ηM as an RD-submodule with $\tilde{M}/\eta M$ divisible.

Proof. The RD-character of ηM in $\Pi(M/t_1 \ldots t_n M)$ follows in the same way as in (1.1). Since a subbase of neighborhoods of 0 in the product topology consists of the submodules U_m = product of all $M/t_1 \ldots t_n M$ with the mth factor omitted, the inclusion $t_1 \ldots t_m \tilde{M} \leq U_m$ is obvious.

To verify the converse, suppose that $\tilde{a} \in \tilde{M} \cap U_m$. Writing $\tilde{a}$ in the form (1), necessarily $a_n \in t_1 \ldots t_m M$ holds for all $n \geq m$. Pick a $c_m \in M$ and a $c_{m+i} \in M$ ($i \geq 1$) so as to satisfy

$$t_1 \ldots t_m c_m = a_m \quad \text{and} \quad t_1 \ldots t_{m+i-1} c_{m+i} = a_{m+i} - a_{m+i-1}.$$

Setting $b_{m+i} = c_m + t_{m+1} c_{m+1} + \ldots + t_{m+1} \ldots t_{m+i} c_{m+i}$, it is clear that $\tilde{b} = (\ldots, b_n + t_1 \ldots t_n M, \ldots) \in \tilde{M}$ satisfies $t_1 \ldots t_m \tilde{b} = \tilde{a}$. Hence $\tilde{M} \cap U_m \leq t_1 \ldots t_m \tilde{M}$, establishing the converse inclusion.

The density of ηM in $\tilde{M}$ guarantees the divisibility of $\tilde{M}/\eta M$. □

After these easy preliminaries, we can now move towards a homological characterization of R-completeness.

LEMMA 2.3. [D] Supposing Q is countably generated, an R-module M is R-complete if and only if

(i) $M^1 = 0$;

(ii) $\text{RDext}^1_R(D,M) = 0$ for all divisible R-modules D.

Proof. Assume M is R-complete. By definition, (i) holds. Let $0 \to M \to A \to D \to 0$ be an RD-exact sequence with D divisible. First, suppose $A^1 = 0$. Then A has a Hausdorff topology, and M is a dense submodule in A. Complete submodules are closed, thus only $A = M$ is possible. If $A^1 \neq 0$, then $M \cap A^1 = 0$. The RD-property is evidently preserved under passage mod A^1, consequently, $0 \to (M+A^1)/A^1 \to A/A^1 \to A/(M+A^1) \to 0$ is again an RD-exact sequence. By the first part of the proof, we must have $M + A^1 = A$. Hence $A = M \oplus A^1$, i.e. the given sequence splits.

Conversely, assume M satisfies (i) and (ii). Because of (2.2), in the RD-exact sequence $0 \to M \to \widecheck{M} \to \widecheck{M}/M \to 0$ the last term is divisible, so the sequence splits by hypothesis. Therefore $\widecheck{M}/M = 0$, indeed. □

In order to simplify the condition formulated in the preceding result, we require a lemma.

LEMMA 2.4. [D] For every h-divisible R-module D, there is an (RD-)exact sequence

$$0 \to N \to H \to D \to 0 \qquad (2)$$

where H is a direct sum of copies of Q and K, while N is an h-divisible R-module.

Proof. It is readily seen that the proof can be confined to the torsion case, since by (VI.1.1), tD is a summand of D. The torsion case follows from (VI.2.3). □

Before stating our next result, the reader is reminded that if Q is countably generated, then all divisible R-modules are h-divisible; cf. (II.2.3).

2. R-COMPLETE MODULES

LEMMA 2.5. [D] Let Q be countably generated. An R-module M is R-complete if and only if
(i) $M^1 = 0$,
(ii) $\text{Ext}^1_R(Q,M) = 0$, and
(iii) $\text{RDext}^1_R(K,M) = 0$.

Proof. The necessity is a trivial corollary to (2.3). Conversely, suppose that M satisfies the stated conditions. Given a divisible D, by (VI.1.1), tD is a summand of D, thus the verification of (2.3)(ii) can be confined to the torsion case. We embed D in an RD-exact sequence like (2), and derive the exact sequence

$$0 \to \text{Hom}_R(N,M) \to \text{RDext}^1_R(D,M) \to \text{RDext}^1_R(H,M) \,.$$

The last term vanishes, being a product of R-modules $\text{RDext}^1_R(K,M) = 0$. A simple appeal to (2.3) concludes the proof. □

A further simplification in our condition is possible by making use of (III.2.8). If it is applied to our present situation, then it follows that — under the hypothesis that $p.d.K = 1$ — an exact sequence $0 \to M \to A \to K \to 0$ represents an element of $r\text{Ext}^1_R(K,M)$ exactly if the induced sequence $0 \to M/rM \to A/rM \to K \to 0$ is splitting exact. Bearing this in mind, we deduce:

LEMMA 2.6. [D] If Q is countably generated, then $\text{RDext}^1_R(K,M)$ is precisely the first Ulm submodule of $\text{Ext}^1_R(K,M)$ for any module M.

Proof. By Auslander's Lemma in IV. §2, Q countably generated implies $p.d.Q = 1$, thus $p.d._R K = 1$. Furthermore, M/rM is R-complete for every $0 \neq r \in R$, thus (2.5) indicates that the sequence $0 \to M/rM \to A/rM \to K \to 0$ splits if and only if M/rM is RD in A/rM. But $0 \to A \to B \to C \to 0$ is RD-exact precisely if $0 \to A/rA \to B/rA \to C \to 0$ is RD-exact for all $0 \neq r \in R$. We conclude that the sequence $0 \to M \to A \to K \to 0$ is RD-exact if and only if it represents an element of $\bigcap_{0 \neq r \in R} r\text{Ext}^1_R(K,M)$, as asserted. □

We are now ready to state and prove the central result for R-completeness.

THEOREM 2.7. (Matlis [3]) Suppose Q is countably generated, R an arbitrary domain. For an R-module M to be R-complete, it is necessary and sufficient that (i) and (ii) of (2.5) hold for M.

Proof. Only sufficiency requires a proof. So assume $M^1 = 0$ and $\text{Ext}_R^1(Q,M) = 0$. The exact sequence

$$0 = \text{Hom}(Q,M) \to \text{Hom}(R,M) \cong M \to \text{Ext}_R^1(K,M) \to \text{Ext}_R^1(Q,M) = 0$$

shows that $\text{Ext}_R^1(K,M) \cong M$. By the preceding lemma, $\text{RDext}_R^1(K,M) \cong M^1$ which vanishes by hypothesis. A reference to (2.5) completes the proof. □

It is worth pointing out that, under the hypotheses of (2.7), M is R-complete if and only if

$$M \cong \text{Ext}_R^1(K,M).$$

For an application of this duality, we refer to VI.§2.

Let us mention a few cases where R-complete modules appear "in nature".

PROPOSITION 2.8. [D] Let Q be countably generated. For all torsion R-modules A, $\text{Hom}_R(A,M)$ is R-complete.

Proof. Suppose η belongs to the first Ulm submodule of $\text{Hom}_R(A,M)$, and let $a \in A$. If $r \in R$ and $\chi \in \text{Hom}_R(A,M)$ are such that $ra = 0$ and $r\chi = \eta$, then $\eta a = r\chi a = \chi r a = 0$. Hence $\eta = 0$.

Let $0 \to M \to E \to E/M \to 0$ be exact with E injective. In the exact sequence $0 \to \text{Hom}_R(A,M) \to \text{Hom}_R(A,E) \to \text{Hom}_R(A,E/M)$, the second Hom is pure-injective (see (XI.2.4)), while the third Hom is reduced because of the first part of the proof. Hence $\text{Ext}_R^1(Q,\text{Hom}(A,M)) = 0$ follows, and (2.7) completes the proof. □

PROPOSITION 2.9. [D] If Q is countably generated and M is R-complete, then so is $\text{Hom}_R(A,M)$ for any A.

Proof. Hypothesis implies $\tilde{M} = \varprojlim M/rM$, thus $\text{Hom}_R(A,M) \cong \varprojlim \text{Hom}(A,M/rM)$. Each of these Hom's is bounded, thus the inverse

3. FILTRATION AND ULTRACOMPLETENESS

limit is R-complete. □

EXERCISES

1. [D] (Matlis [3]) Let Q be countably generated. If $0 \to A \to B \to C \to 0$ is an RD-exact sequence of R-modules, then so is the induced sequence $0 \to \tilde{A} \to \tilde{B} \to \tilde{C} \to 0$.

2. [D] (Matlis [3]) Let Q be countably generated. Then for an R-module M, $\tilde{M} \cong \text{Ext}_R^1(K,M)/E$ where E denotes the first Ulm submodule of $\text{Ext}_R^1(K,M)$.

3. [D] Verify (2.4) by first choosing an RD-exact sequence $0 \to A \to B \to D \to 0$ with B a direct sum of cyclically presented R-modules, and then embedding each of these in Q or K.

4. [D] Let Q be countably generated. The h-reduced part of the R-module

$$A = \Pi M_n / \oplus M_n$$

is R-complete for any countable set $\{M_n\}$ of R-modules.

§3. FILTRATION AND ULTRACOMPLETENESS

We come now to an idea which deserves to be considered as a substitute for topology-like considerations, though it is not a bona fide generalization. Its importance in our study begins to emerge with the next section.

By a <u>filtration</u> F_M of a module M will be meant a non-empty family of submodules of M, $\{N_i \mid i \in I\}$, which is directed downwards, i.e. for every pair N_i, N_j in F_M there is an $N_k \in F_M$ such that

$$N_k \leq N_i \cap N_j.$$

The most important types of filtration are those which are not only meaningful on a single R-module M, but are defined simultaneously on each object of the category R-Mod of R-modules. Such a collection of filtrations will be called a <u>functorial filtration</u> F if it originates from a downward directed partially ordered set Λ as follows: to every $\lambda \in \Lambda$ it assigns a submodule N_λ of M such that

(i) $\lambda \leq \mu$ in Λ implies $N_\lambda \leq N_\mu$;

(ii) for each R-homomorphism $\alpha : M \to M'$ in R-Mod, $\alpha N_\lambda \leq N'_\lambda$ holds for every $\lambda \in \Lambda$.

The simplest examples for filtrations are as follows.

1. F_M consists of all submodules of M; this may be called the <u>discrete</u> filtration. It is not functorial.

2. $F_M = \{0\}$, the <u>trivial</u> filtration. It is functorial.

3. The R-<u>filtration</u>: $R_M = \{rM \mid 0 \neq r \in R\}$; this is functorial where Λ is the set of principal ideals of R ordered by inclusion and assigns rM to Rr.

4. The <u>annihilator filtration</u> A_M is likewise functorial: $A_M = \{M[r]\}_{r \in R}$; here Λ is as before but with the dual ordering.

Manifestly, a filtration F_M defines a linear topology on M if the members of F_M are declared to be a subbase of open sets about 0 in M. In this way, both examples 1 and 2 define the discrete topology, while example 3 gives rise to the R-topology.

A filtration F_M is <u>topological</u> if $N \leq N'$ and $N \in F_M$ imply $N' \in F_M$, i.e. F_M is a filter in the usual sense. Evidently, every filtration on M is contained in a unique minimal topological filtration.

Consider cosets $a + N_i$ of submodules $N_i \in F_M$. It is a rather trivial, but an extraordinarily relevant fact that if $(a + N_i) \cap (b + N_j)$ is not empty, then it must be a coset of $N_i \cap N_j$ (which submodule need not belong to F). As customary, we say that a family of cosets

$$C = \{a_i + N_i\}_{i \in J} \quad (N_i \in F_M, \; J \text{ a subset of } I) \quad (1)$$

has the <u>finite intersection property</u> if, for every finite subset $\{i_1, \ldots, i_k\}$ of J, the intersection $(a_{i_1} + N_{i_1}) \cap \ldots \cap (a_{i_k} + N_{i_k})$ is not empty. Each $a \in M$ that belongs to all $a_i + N_i$ in C is called a <u>pseudo-limit</u> of C. If such an a does exist, C is

3. FILTRATION AND ULTRACOMPLETENESS

called <u>pseudo-convergent</u>. We shall talk simply of limits and convergence if C, in addition, satisfies $J = I$, or, more generally,

$$\bigcap_{i \in J} N_i = \bigcap_{i \in I} N_i. \qquad (2)$$

A module M is said to be <u>ultracomplete</u> in the filtration F_M (or, simply, F_M-<u>ultracomplete</u>) if every family of cosets C with the finite intersection property has a pseudo-limit in M (cf. Fleischer [2]). In case F_M is a topological filtration, ultracompleteness in F_M amounts to the linear compactness of M in the F_M-topology. In other words, if M is furnished with the topology induced by a filtration F_M, then it is F_M-ultracomplete whenever it is linearly compact.

A module M is F_M-<u>complete</u> if every family C of cosets in (1) with the finite intersection property satisfying (2) has a limit in M. Evidently, ultracompleteness implies completeness. On the other hand, we have:

LEMMA 3.1. A module M is F_M-ultracomplete if and only if for each non-empty subset J of I, the module $M/\bigcap_{i \in J} N_i$ is complete in its topology induced by F_M.

The proof is straightforward and may be left to the reader. □

Let M have a filtration F_M and let N be a submodule of M. We say N is f-<u>closed</u> in F_M if the following holds for each pseudo-convergent family (1): if, $\{N \cap (a_i + N_i) \mid i \in J\}$ has the finite intersection property, then $N \cap \bigcap_{i \in J}(a_i + N_i)$ is not empty either.

EXERCISES

1. Suppose F_1 and F_2 are two filtrations of the module M. If $F_1 \subset F_2$ and M is F_2-ultracomplete, then M is F_1-ultracomplete as well.

2. Let M be ultracomplete in a filtration F. Let F' be a filtration of M such that $F \subset F'$ and every $N \in F'$ is the intersection of submodules in F. Show that M is F'-ultracomplete.

§4. THE ANNIHILATOR FILTRATION

The objective of this section is to provide further insight into the nature of the annihilator filtration. It will allow us to place some results in a more general context.

Let R be any domain and L an ideal of R. A homomorphism $\phi : L \to M$ from L into an R-module M will be said to be <u>locally extendable</u> if for each finite subset $\{r_1, \ldots, r_k\}$ of L there is a $\psi : R \to M$ such that $\psi r_i = \phi r_i$ ($i = 1, \ldots, k$). Equivalently, there is an $x \in M$ such that $r_i x = \phi r_i$.

The fundamental connection between this concept and families of cosets in the annihilator filtration is as follows.

LEMMA 4.1. [D] Let $\{a_r + M[r]\}_{r \in Y}$ be a family of cosets, for some subset Y of R, and L the ideal of R generated by Y. This family has

(i) the finite intersection property exactly if the correspondence

$$\phi : r \longmapsto ra_r \qquad (r \in Y)$$

induces a locally extendable homomorphism $L \to M$;

(ii) a pseudo-limit in M if and only if ϕ has an extension $\psi : R \to M$.

<u>Proof.</u> To verify (i), first suppose that the given family of cosets has the finite intersection property. To assure that the correspondence ϕ induces a homomorphism, it is necessary and sufficient to ascertain that a relation of the form $\sum_{r \in Y} s_r r = 0$ (with almost all $s_r \in R$ vanishing) implies $\sum_{r \in Y} s_r r a_r = 0$. Let $r_1, \ldots, r_k$ be the indices for which $s_r \neq 0$, and let $x \in a_r + M[r]$ for $r = r_1, \ldots, r_k$. Then $ra_r = rx$ for these r's, thus $\sum s_r r a_r = \sum s_r (rx) = (\sum s_r r)x = 0$. This argument also shows that ϕ from $\sum_{i=1}^{k} Ra_{r_i}$ into M extends to a homomorphism $R \to M$ via $1 \longmapsto x$.

Conversely, if ϕ induces a locally extendable homomorphism, then for each finite set $\{r_1, \ldots, r_k\} \subset Y$ there is an $x \in M$ such

4. THE ANNIHILATOR FILTRATION

that $r_i x = r_i a_{r_i}$ ($i = 1,\ldots,k$). This simply means $x \in a_r + M[r]$ for $r = r_1,\ldots,r_k$.

The proof of (ii) is obvious from the first part of the proof. □

This lemma is relevant in the following characterization of injectivity.

THEOREM 4.2. [D] An R-module M is injective if and only if it satisfies the following two conditions:

(i) for each finitely generated ideal L of R, every homomorphism $\phi : L \to M$ extends to a $\psi : R \to M$.

(ii) M is ultracomplete in the annihilator filtration.

Proof. Suppose M is injective. Then (i) is obvious. To verify (ii), let $\{a_r + M[r]\}_{r \in Y}$ be a family of cosets with the finite intersection property. By the preceding lemma, it induces a map $\phi : L \to M$ where $L = \langle Y \rangle$. In view of the injectivity of M, it extends to a $\psi : R \to M$ which amounts to the existence of a pseudo-limit.

Conversely, let M satisfy (i) and (ii). Every homomorphism $\phi : L \to M$ from an ideal L of R is by (i) locally extendable. Therefore, for every $r \in L$ there is an $a_r \in M$ with $\phi r = ra_r$, and by (4.1), the cosets $\{a_r + M[r]\}_{r \in L}$ enjoy the finite intersection property. (ii) and (4.1) imply the existence of an extension $\psi : R \to M$ of ϕ. Hence M is injective, indeed. □

Observe that in the proof, ultracompleteness provided the step from finitely generated ideals to arbitrary ideals in the proof of extensibility. We borrow this idea to prove for Prüfer domains:

PROPOSITION 4.3. [PD] (Simmons [1]) Let N be an RD-submodule of M. Then N is cyclically pure in M if and only if N is f-closed in M in the annihilator filtration.

Proof. Suppose N cyclically pure in M, and let $\{a_r + M[r]\}_{r \in Y}$ with $a_r \in N$ be a family of cosets such that some $x \in M$ belongs to all of them. By hypothesis, there is a $b \in N$ such that $\mathrm{Ann}(x-b) = \mathrm{Ann}(x+N)$. For all $r \in Y$, $rx = ra_r \in N$, thus

$rb = rx$ and $b \in a_r + M[r]$. This shows N is f-closed in M.

Conversely, suppose N f-closed in M. For $x \in M$, set $L = \text{Ann}(x+N)$. Then for every $r \in L$, $rx \in N$, thus by the RD-property, $ra_r = rx$ for some $a_r \in N$. Consider the cosets $\{a_r + M[r]\}_{r \in L}$ each of which contains $x \in M$. Since R was assumed to be a Prüfer domain, N is pure in M, and thus $\{N \cap (a_r + M[r])\}$ has the finite intersection property. By hypothesis, some $b \in N$ satisfies $b \in a_r + M[r]$. We conclude $rb = ra_r = rx$ and $r(x-b) = 0$ for all $r \in L$. Hence $L \leq \text{Ann}(a-b)$ and N is cyclically pure in M. □

EXERCISE

1. [D] Show that condition (i) is satisfied whenever M is an absolutely pure module (for definition see II.§4, Ex. 4).

§5. R-ULTRACOMPLETE MODULES

Undoubtedly, the most important filtration is the R-filtration. This is the central theme of this section.

A module that is ultracomplete in its R-filtration is called R-<u>ultracomplete</u> or <u>maximal</u>. If M is R-ultracomplete, then M/rM is likewise R-ultracomplete for each $r \in R$.

If, for all $0 \neq r \in R$, the factor modules M/rM are R-ultracomplete, then M is said to be <u>almost maximal</u>. Equivalently, if a family of cosets mod $r_i M$ (for a subset $\{r_i\}$ of R) has the finite intersection property and $r_i M \geq rM$ for all r_i and some $r \neq 0$ in R, then the family is pseudo-convergent. The reader can easily convince himself that these definitions coincide with earlier definitions for $M = R$.

In order to state our condition on R-ultracompleteness, it seems desirable to introduce the following ad hoc definition. Let M be a submodule of A. If, for each coset $a + M$ ($a \in A$) there is a $b \in M$ such that, for each $r \in R$, $r | a + M$ implies $r | a + b$,

5. R-ULTRACOMPLETE MODULES

we then say: the cosets mod M <u>satisfy the divisibility condition</u>.

THEOREM 5.1. [D] A module M is R-ultracomplete if and only if the cosets mod M satisfy the divisibility condition in every R-module that contains M as an RD-submodule.

Proof. Let the R-ultracomplete module M be an RD-submodule in A. Given $a \in A \setminus M$, define $Y = \{r \in R \mid r \text{ divides } a + M\}$; in other words, $r \in Y$ exactly if $rb_r = a + c_r$ for some $b_r \in A$, $c_r \in M$. Consider the system of equations

$$ry_r = x + c_r \qquad (r \in Y)$$

whose unknowns are x, y_r ($r \in Y$). This system is solvable in A, and the finite subsystems are of the form which allows us to conclude, by (II.4.4), that every finite subsystem is solvable in M. This amounts to the finite intersection property of the family $\{c_r + rM\}_{r \in Y}$. By hypothesis, there is a $b \in M$ such that $b \in c_r + rM$ for all $r \in Y$. Hence $a + b \in rM$ for all $r \in Y$, as desired.

Conversely, assume M satisfies the stated condition and $\{c_r + rM\}_{r \in Y}$ is a family with the finite intersection property ($c_r \in M$). We embed M in its RD-injective hull $\hat{M}$ (cf. (XI.1.6)) which is by XI.§3, Ex.1, R-ultracomplete. As the cosets $c_r + r\hat{M}$ enjoy the finite intersection property, some $a \in \hat{M}$ satisfies $a \in c_r + r\hat{M}$ for every $r \in Y$. Hence $r \mid a + M$ for all $r \in Y$, thus there is a $b \in M$ such that $a - b \in r\hat{M}$ ($r \in Y$). Consequently, $b - c_r \in r\hat{M} \cap M = rM$ for every $r \in Y$. □

We prove the next theorem for valuation domains.

THEOREM 5.2. Let R be a valuation domain, M a torsion-free R-module and E the injective hull of M. Then the following conditions are equivalent:

(i) M is almost maximal;

(ii) E/M is ultracomplete in its annihilator filtration;

(iii) E/M is injective;

(iv) $\text{i.d.}_R M \leq 1$.

Proof. E is a direct sum of copies of Q and $T = E/M$ is an h-divisible torsion module. From (4.2) it is easy to deduce the equivalence of (ii) and (iii). The equivalence of (iii) and (iv) is obvious from the definition of injective dimension (cf. VI.§5).

To prove (i) $\Longleftrightarrow$ (ii), the basic observation is that for each $0 \neq s \in R$, we have $T[s] = s^{-1}M/M$, and so

$$\bar{a} + T[s] = \bar{a} + (s^{-1}M/M)$$

holds for each $a \in E$; here $\bar{a} = a + M$. Therefore, for a family of cosets $\{\bar{a}_s + T[s] \mid s \in Y\}$ to have a non-empty intersection in T, it is necessary and sufficient that the family $\{a_s + s^{-1}M \mid s \in Y\}$ have a non-empty intersection in E. As the set $\{T[s] \mid s \in Y\}$ is totally ordered by inclusion, no generality is lost if we confine ourselves to families $\{\bar{a}_s + T[s]\}$ such that for some fixed $0 \neq r \in R$ (depending on the family), both $s|r$ and $ra_s \in M$ hold for all $s \in Y$. Multiplication by r pushes the corresponding family $\{a_s + s^{-1}M\}$ in E down into M, and it is pretty clear that this family has a non-empty intersection in E exactly if $\{ra_s + rs^{-1}M\}$ has a non-empty intersection in M. As every family of cosets used to test the almost maximality of M comes from a family of cosets in T of the indicated type, the proof is completed. □

A consequence of (5.2) is an important condition for R to be almost maximal:

COROLLARY 5.3. (Matlis [1]) A valuation domain R is almost maximal exactly if Q/R is injective. □

EXERCISE

1. [D] Let J be an ideal of R with $\text{End}_R J \cong R$. Then J is R-ultracomplete if and only if R is R-ultracomplete.

NOTES

R-completeness is an extremely useful tool in the cases discussed in §§1-2. The difficulty in the general case can be better understood if we bring the RD-injective hulls into the picture.

For any R-module T, T is RD in its RD-injective hull $\hat{T}$ which is R-complete. Consequently, the topological closure of T in $\hat{T}$ coincides with the R-completion $\check{T}$ of T. In other words,

$$\check{T}/T = (\hat{T}/T)^1.$$

As $(\hat{T}/T)^1$ need not be RD in $\hat{T}/T$, $\check{T}$ will not in general be RD in $\hat{T}$. Therefore, the topology of $\check{T}$ need not be its own R-topology. (Observe that if T is torsion-free, then so is $\check{T}/T$, and $(\hat{T}/T)^1$ is the divisible part of $\hat{T}/T$.)

For additional results on R-completeness, the reader is advised to consult Matlis [3] and [5].

The following problem is worth while mentioning.

PROBLEM 4. [D] Is every summand of $\check{F}$, F a free R-module, of the form $\check{H}$ with H projective?

A systematic use of filtration and ultracompleteness in the study of modules seems to be new. The idea itself goes back to A. Ostrowski (valuation theory). Ultracompleteness in a more general setting was discussed by Fleischer [2]. Hopefully, soon much more will be known about ultracompleteness, so that this concept can be used more effectively in module theory.

VI. Divisibility and Injectivity

If our quest of a module theory is going to meet with any success, we must have a clear picture of modules for which we can get a pretty firm grip on the situation. With this principle in mind, we turn our attention to injective and divisible modules. We will uncover some of their striking features.

§1. DIVISIBLE MODULES

We resume the study of divisible and h-divisible modules which we started in II, §2. Our first objective is to clarify the relationship between divisibility and h-divisibility.

The starting point is the question as to when the torsion parts of divisible modules are summands. The h-divisible case is easy to settle:

LEMMA 1.1. [D] (Matlis [2]) The torsion submodule of an h-divisible module is always a summand.

1. DIVISIBLE MODULES

Proof. Let H be an h-divisible module. Suppose D is a direct sum of copies of Q and $\eta : D \to H$ an epimorphism. Select in H a maximal independent system $\{a_i\}$ mod tH, and for each i, pick an $x_i \in D$ with $\eta x_i = a_i$. Then $\{x_i\}$ is evidently an independent system in D; let X denote the Q-subspace they span. It is readily checked that $H = tH \oplus \eta X$. □

For arbitrary divisible modules, we verify:

LEMMA 1.2. (Matlis [2]) Let the domain R be such that in every divisible R-module the torsion submodule is a summand. Then

$$p.d._R Q = 1.$$

Proof. Given an arbitrary R-module M, let E be its injective hull. The R-module E/M is an h-divisible torsion module. Every extension of E/M by Q is divisible with the torsion part E/M, thus our hypothesis on R implies that $\text{Ext}^1_R(Q,E/M) = 0$. We deduce that in the exact sequence

$$\text{Ext}^1_R(Q,E/M) \to \text{Ext}^2_R(Q,M) \to \text{Ext}^2_R(Q,E) \tag{1}$$

the end terms vanish. Consequently, $\text{Ext}^2_R(Q,M) = 0$ for all R-modules M, i.e. $p.d._R Q \leq 1$. □

This leads us to the following theorem.

THEOREM 1.3. (Matlis [2], Hamsher [1]) For any domain R, the following conditions are equivalent:
    (i)   all divisible R-modules are h-divisible;
    (ii)   the torsion submodule of each divisible R-module is a summand;
    (iii) $p.d._R Q = 1$;
    (iv) $p.d._R Q/R = 1$.

Proof. The implications (i) ⟹ (ii) ⟹ (iii) follow from (1.1) and (1.2), while (iii) and (iv) are trivially equivalent in view of the exact sequence $0 \to R \to Q \to Q/R \to 0$ and Kaplansky's Lemma.

It is a rather long proof to verify (iii) $\implies$ (i); we refer to the article by Hamsher [1]. But let us observe that (i) follows from (iii) for valuation domains easily by making use of (IV.2.4) and (II.2.3). □

EXAMPLE 1.4. [VD] Suppose Q is uncountably generated as an R-module. Then p.d.$Q \geq 2$ and thus there is an R-module M with $\text{Ext}_R^2(Q,M) \neq 0$. With the injective hull E of M, we form the sequence (1) where the right end vanishes. Thus $\text{Ext}_R^1(Q, E/M) \neq 0$, and there is a non-splitting and hence not h-divisible extension of the h-divisible module E/M by Q. Moreover, there exists a torsion module which is the extension of an h-divisible module by another h-divisible module, but is itself not h-divisible. In fact, this follows at once from the exact sequence

$$\text{Ext}_R^1(Q/R, E/M) \to \text{Ext}_R^1(Q, E/M) \to \text{Ext}_R^1(R, E/M) = 0.$$

For a more general and useful method see §3.

EXERCISES

1. [VD] Let κ be an infinite cardinal. An R-module M will be called κ-<u>divisible</u>, if every homomorphism $\phi : R \to M$ can be extended to a homomorphism $\psi : J \to M$ where $R \leq J \leq Q$, J can be generated by $\leq \kappa$ elements. Show that $\aleph_0$-divisibility is equivalent to divisibility and $\aleph_\alpha$-divisibility to h-divisibility where $\aleph_\alpha$ is the cardinality of a generating set of the R-module Q of minimal cardinality.

2. [VD] Prove that, for every κ, a κ-divisible torsion-free R-module is a direct sum of copies of Q.

3. [VD] Generalize (1.3) by replacing divisibility by $\aleph_n$-divisibility for some $n > 0$.

4. For valuation domains R, give a full proof of the following: divisibility and h-divisibility are equivalent if and only if Q is a countably generated R-module.

§2. h-DIVISIBLE MODULES

5. There are (Dedekind) domains where $p.d.Q = 1$, but Q is not countably generated.

§2. h-DIVISIBLE MODULES

Our knowledge of h-divisible modules is very limited. Recently, de la Rosa and Fuchs [1] tried to classify them in terms of a dimension function which turned out to be closely related to the projective dimension. Here we discuss some of their results. As expected, the Matlis Duality (V.1.7) is a fundamental tool in the study.

An exact sequence of R-modules will be called <u>hd-exact</u> if the R-module $K = Q/R$ has the projective property relative to it. In other words, the exact sequence $0 \to U \to V \to W \to 0$ is hd-exact if the induced map $\text{Hom}_R(K,V) \to \text{Hom}_R(K,W)$ is surjective.

It is straightforward to check that hd-exactness gives rise to a proper class of exact sequences.

The definition of hd-exactness is motivated by the following simple lemma.

LEMMA 2.1. [D] (i) If $0 \to A \to B \to C \to 0$ is an exact sequence of torsion-free R-modules, then $0 \to K \otimes_R A \to K \otimes_R B \to K \otimes_R C \to 0$ is an hd-exact sequence of h-divisible torsion R-modules.

(ii) If $0 \to U \to V \to W \to 0$ is an hd-exact sequence of h-divisible torsion R-modules, then $0 \to \text{Hom}_R(K,U) \to \text{Hom}_R(K,V) \to \text{Hom}_R(K,W) \to 0$ is an exact sequence of R-complete torsion-free R-modules.

<u>Proof.</u> (i) The sequence of tensor products is evidently exact. To show that the induced map $\text{Hom}_R(K, K \otimes B) \to \text{Hom}_R(K, K \otimes C)$ is onto, it suffices to note that by (V.1.5) these Hom's are naturally isomorphic to B and C, respectively.

(ii) In view of (V.1.7), the Hom's are R-complete and torsion-free R-modules. The exactness is clear by the definition of hd-exactness. □

Next we verify:

LEMMA 2.2. [D] (de la Rosa-Fuchs [1]) Let $0 \to A \to D \xrightarrow{\beta} H \to 0$ be an hd-exact sequence of R-modules. D is h-divisible torsion if and only if both A and H are h-divisible torsion.

Proof. If D is h-divisible torsion, then so is H. By definition, for every $a \in A$ there is a $\phi : Q \to D$ such that $\phi 1 = a$. As $\beta \phi 1 = 0$, the map $\beta \phi : Q \to H$ factors through the canonical map $\gamma : Q \to Q/R = K$, i.e. $\beta \phi = \mu \gamma$ for some $\mu : K \to H$. By hd-exactness, there exists a $\nu : K \to D$ satisfying $\beta \nu = \mu$. We claim that $\phi - \nu \gamma$ maps Q into A and satisfies $(\phi-\nu\gamma)1 = a$. It suffices to show that $\beta(\phi-\nu\gamma) = 0$ which is evident from the definitions of μ and ν.

Conversely, let both A and H be h-divisible torsion, and $d \in D$. If $d \in A$, then there exists obviously a $\phi : Q \to D$ with $\phi 1 = d$. If $d \notin A$, then there exists a $\psi : Q \to H$ such that $\psi 1 = \beta d$. Since H is torsion, there is a $0 \neq q \in \text{Ker } \psi$; thus ψ factors as $Q \xrightarrow{\nu} Q/Rq \xrightarrow{\gamma} H$ with ν the canonical map. In view of $Q/Rq \cong K$ and hd-exactness, some $\chi : Q/Rq \to D$ satisfies $\beta \chi = \gamma$. Note that $d - \chi \nu 1 \in A$, because $\beta(d-\chi\nu 1) = \beta d - \gamma \nu 1 = 0$, so by hypothesis there exists a homomorphism $\psi' : Q \to A$ such that $\psi' 1 = d-\chi\nu 1$. Evidently, $\phi = \psi' + \chi\nu$ maps Q into D and carries 1 into d. □

The following result establishes the existence of 'enough' hd-projectives.

LEMMA 2.3. [D] (de la Rosa-Fuchs [1]) For every h-divisible torsion R-module D, there exists an hd-exact sequence

$$0 \longrightarrow N \longrightarrow H \xrightarrow{\phi} D \longrightarrow 0 \qquad (1)$$

where H is a direct sum of copies of K and N is h-divisible torsion.

Proof. It is readily seen that for every $d \in D$, there exists a homomorphism $\phi_d : K \to D$ such that $d \in \text{Im } \phi_d$. Define

$$H = \oplus K_\psi \quad (K_\psi \cong K \text{ for } \psi \in \text{Hom}_R(K,D)),$$

2. h-DIVISIBLE MODULES

and $\phi : H \to D$ acting on K_ψ as ψ. The arising exact sequence
(1) with $N = \text{Ker } \phi$ has to be hd-exact, since by design K has
the projective property relative to it. (2.2) shows that N is
h-divisible. □

Given an h-divisible torsion R-module D, we can proceed to
form a long hd-exact sequence

$$\cdots \longrightarrow H_1 \xrightarrow{\phi_1} H_0 \xrightarrow{\phi_0} D \longrightarrow 0$$

where each H_i is a summand of a direct sum of copies of K. If
there is a first index $n \geq 0$ such that $\text{Ker } \phi_n$ is a summand of a
direct sum of copies of K, then we define n to be the <u>hd-dimension</u>
of D,

$$\text{hd.d.}D = n.$$

Otherwise we set $\text{hd.d.}D = \infty$. A version of Schanuel's lemma shows
that the hd-dimension is well-defined. The global hd-dimension of
of R may be defined as the supremum of all hd.d.D with D
ranging over all h-divisible torsion R-modules.

The next result is a useful information about the hd-
dimension.

PROPOSITION 2.4. [D] (Matlis [3]) If T is an h-divisible torsion
R-module of projective dimension 1, then $\text{hd.d.}T = 0$.

<u>Proof.</u> By the Matlis Duality (V.1.7), there is an R-complete
torsion-free R-module M such that $T \cong K \otimes_R M$. The exact sequence
$0 \to R \to Q \to K \to 0$ tensored by M yields the exact sequence

$$0 \to M \to Q \otimes M \to T \to 0. \tag{2}$$

Let C be any R-complete torsion-free R-module. We have then the
induced exact sequence

$$0 = \text{Ext}^1_R(Q \otimes M, C) \to \text{Ext}^1_R(M,C) \to \text{Ext}^2_R(T,C).$$

Here the last term vanishes whenever $\text{p.d.}T = 1$. Therefore,
$\text{Ext}^1_R(M,C) = 0$.

Let $0 \to H \to F \to M \to 0$ be an exact sequence with F free. It is straightforward to check (e.g. Ex. 9 in V.§1) that the induced sequence $0 \to \widetilde{H} \to \widetilde{F} \to \widetilde{M}$ of R-completions is exact. From $\widetilde{M} = M$ we infer that $0 \to \widetilde{H} \to \widetilde{F} \to M \to 0$ is an exact sequence. By the preceding paragraph, $\operatorname{Ext}_R^1(M,\widetilde{H}) = 0$ whence $\widetilde{F} \cong \widetilde{H} \oplus M$. It follows that $K \otimes \widetilde{F} \cong (K \otimes \widetilde{H}) \oplus T$ where $K \otimes \widetilde{F} \cong K \otimes F$ is a direct sum of copies of K. □

In case p.d.$K = 1$, (2.4) shows that for any h-divisible torsion R-module T, hd.d.$T = 0$ exactly if p.d.$T = 1$. This generalizes in the following manner.

THEOREM 2.5. [D] (de la Rosa-Fuchs [1]) Suppose that p.d.$K = 1$. Then for an h-divisible torsion R-module T,

$$\text{hd.d.}T = \text{p.d.}T - 1.$$

The global hd-dimension of R is exactly gl.d.$R - 1$.

Proof. In view of the preceding observation, we may assume p.d.$T = n > 1$. As p.d.$K = 1$ is equivalent to p.d.$Q = 1$, the exact sequence (2) shows that p.d.$M = n-1$.

Let again $0 \to H \to F \to M \to 0$ be an exact sequence where F is a free R-module. By (2.1), $0 \to K \otimes H \to K \otimes F \to K \otimes M \to 0$ is hd-exact. $K \otimes F$ is a direct sum of copies of K, thus it is clear that hd.d.$K \otimes M = $ hd.d. $K \otimes H + 1$, the only possible exception being hd.d. $K \otimes M = 0$. This is, however, ruled out by the hypothesis p.d.$T > 1$. By induction, p.d.$H = n-2$ implies hd.d. $K \otimes H = n-2$, hence hd.d.$T = n-1$ follows.

To prove the second part, observe that for an ideal $J \neq 0$ of R we now have p.d.$Q/J = $ p.d.$J + 1$ (cf. IV.§2). In fact, the only exception p.d.$J = 1 = $ p.d.Q/J can be ignored after examining the projective dimensions in the exact sequence $0 \to R/J \to Q/J \to K \to 0$. Thus, by what has already been proved, hd.d.$Q/J = $ p.d.J. It is a well-known consequence of Auslander's lemma in IV.§2 that gl.d.R is $1 + \sup\{\text{p.d.}J \mid J \text{ ideals of } R\}$. Hence the second claim is immediate. □

EXERCISES

1. [D] Show that for every h-divisible torsion R-module T, the following inequality holds:

$$p.d.T \leq hd.d.T + p.d.K.$$

2. [D] (Matlis [3]) There always exists an h-divisible torsion R-module T with $p.d.T = gl.d.R$.

§3. DIVISIBLE MODULES OF PROJECTIVE DIMENSION ONE

The aim of this section is to exhibit a divisible R-module (over any domain R) with a number of remarkable features. Most importantly, it is a generator for the category of all divisible R-modules. Our discussion is based on Fuchs [6].

In this section R denotes an arbitrary domain, unless otherwise specified.

Let the R-module ∂ be generated by all k-tuples $(r_1,\ldots,r_k)$ of non-zero and non-unit elements r_i of R, for $k \geq 0$, subject to the defining relations

$$r_k(r_1,\ldots,r_k) = (r_1,\ldots,r_{k-1}) \qquad (k \geq 1).$$

The <u>length</u> of $(r_1,\ldots,r_k)$ is defined to be k. The element $w = (\emptyset)$ of length 0 generates a submodule $\cong R$, and ∂/Rw is a torsion module.

Let ∂_k be the submodule generated by the generators of lengths $\leq k$. It is clear that ∂ is the union of the ascending chain $\partial_0 = Rw < \partial_1 < \ldots < \partial_k < \ldots$ such that $\partial_{k+1}/\partial_k$ is a direct sum of cyclically presented R-modules, viz. generated by $(r_1,\ldots,r_{k+1}) + \partial_k$ with annihilators Rr_{k+1}. Hence for the Ulm submodule we have: $\partial_{k+1}^1 = \partial_k$ $(k \geq 0)$, and clearly, $p.d.\partial_{k+1}/\partial_k = 1$. We conclude:

LEMMA 3.1. [D] ∂ is a divisible R-module of projective dimension 1. □

A noteworthy property of ∂ is as follows:

LEMMA 3.2. [D] Let M be an R-module and $a \in M$. There exists a homomorphism $\eta : \partial \to M$ with $\eta w = a$ if and only if $a \in dM$.

Proof. As homomorphic images of divisible modules are divisible, this condition is necessary. To prove the converse, we construct successive homomorphisms $\eta_k : \partial_k \to M$. First, $\eta_0 : \partial_0 \to M$ is induced by $w \mapsto a$. If η_{k-1} has been defined to map ∂_{k-1} into dM, then let η_k carry $(r_1,\ldots,r_k)$ into an element $x \in dM$ such that $r_k x = \eta_{k-1}(r_1,\ldots,r_{k-1}) \in dM$. This defines $\eta_k : \partial_k \to dM$ as desired. □

The module ∂ has numerous summands. For instance, ∂/Rrw (for any $r \in R$) is isomorphic to a summand of ∂. As a matter of fact, it is readily checked that the correspondence

$$(r_1,\ldots,r_k) \mapsto (r,r_1,\ldots,r_k) - s(rs,r_1,\ldots,r_k)$$

gives rise to a homomorphism of ∂ whose image is a summand of ∂ isomorphic to ∂/Rrw; here s is a non-zero, non-unit element of R.

From what has been shown in (III.3.2) it follows that $\mathrm{Tor}_n(R/Rr,C) = 0$ holds for $n \geq 1$ and for every torsion-free R-module C. Using this, one can easily derive the canonical isomorphism $\mathrm{Tor}_n(\partial_k,C) \cong \mathrm{Tor}_n(\partial_{k-1},C)$ for every $k \geq 1$ and torsion-free C. As $\mathrm{Tor}_n(\partial_0,C) = 0$ trivially, we conclude (note that Tor_n commutes with direct limits):

PROPOSITION 3.3. [D] $\mathrm{Tor}_n^R(\partial,C) = 0$ for torsion-free modules C and for all $n \geq 1$. □

We proceed to prove:

PROPOSITION 3.4. [D] $\mathrm{Ext}_R^n(\partial,D) = 0$ holds for all divisible R-modules D and for all $n \geq 1$.

Proof. The modules $\partial_{k+1}/\partial_k$ are direct sums of cyclically presented R-modules, thus $\mathrm{Ext}^1(\partial_{k+1}/\partial_k,D) = 0$ by virtue of (II.2.2). The claim now follows from (IV.2.1) for $n = 1$, while for $n > 1$ it is a trivial consequence of (3.1). □

3. DIVISIBLE MODULES OF DIMENSION ONE

The following two results tell us about the extreme cases where ∂ is h-divisible, resp. h-reduced.

LEMMA 3.5. [D] The R-module ∂ is h-divisible if and only if $p.d._R Q = 1$.

Proof. From (3.2) it is clear that ∂ is h-divisible exactly if all divisible R-modules are h-divisible. As is shown by (1.3), this happens if and only if $p.d.Q = 1$. □

LEMMA 3.6. Let R be a valuation domain and $p.d._R Q \geq 2$. Then ∂ is h-reduced.

Proof. Q is now uncountably generated. We may assume $Q = \cup Ra_\alpha$ with α running over the ordinals less than the initial ordinal Ω corresponding to the minimal cardinality of generating systems of Q. Suppose $\phi : Q \to \partial$ is a homomorphism $\neq 0$. As $\partial = \cup \partial_k$, it is clear that there is an index m such that ∂_m contains ϕa_α for an index set of cardinality Ω. By our choice, ∂_m contains ϕQ entirely. As ∂_m is h-reduced, $\phi Q = 0$, as claimed. □

Call an exact sequence ∂-exact if ∂ has the projective property relative to it. By our remark above, then all factor modules of the form ∂/Rrw (for all $r \in R$) have the projective property relative to this sequence. The same holds for $\partial^0 = \partial/Rw$.

LEMMA 3.7. [D] (Fuchs [6]) For every divisible R-module M, there is a ∂-exact sequence

$$0 \longrightarrow N \longrightarrow D \xrightarrow{\eta} M \longrightarrow 0 \qquad (1)$$

of divisible R-modules such that D is a direct sum of modules each of which is isomorphic to some ∂/Rrw ($r \in R$). If M is torsion, $D = \oplus \partial^0$ can be chosen.

Proof. Given a divisible R-module M, for each pair (a,r) with $a \neq 0$ in M, $r \in R$ and $ra = 0$, select a copy D_a of ∂/Rrw along with a homomorphism $\eta_a : D_a \to M$ mapping the coset of w onto a. That this is possible should be clear from (3.2). The η_a's induce a map $\eta : \oplus D_a = D \to M$ which is evidently surjective. It is readily seen that the arising exact sequence (1) is ∂-exact. For the divisibility of N, observe that ∂-exactness implies that all cyclically presented R-modules have the projective property relative to (1). Therefore (1) is an RD-exact sequence, and N as an RD-submodule in a divisible module is itself divisible. □

An obvious consequence of (3.7) is that ∂ is a generator for the category of divisible R-modules. (Here the maps can be restricted to those corresponding to ∂-exact sequences.)

Before specializing the ring R, let us point out the following interesting fact:

PROPOSITION 3.8. [D] (Fuchs [6]) Every R-module of projective dimension k can be embedded in a divisible R-module whose projective dimension is $\leq \max\{k,1\}$.

Proof. Let F be a free R-module and $\phi : F \to M$ an epimorphism. The exact sequence $0 \to R \to \partial \to \partial/\partial_0 \to 0$ induces the exact sequence in the top row:

$$\begin{array}{ccccccccc} 0 & \to & F & \to & F \otimes \partial & \to & F \otimes \partial/\partial_0 & \to & 0 \\ & & \downarrow \phi & & \downarrow \psi & & \| & & \\ 0 & \to & M & \to & \overline{M} & \to & F \otimes \partial/\partial_0 & \to & 0 \end{array}.$$

By push-out construction we obtain this commutative diagram with exact bottom row. As ψ is surjective, $\overline{M}$ is divisible. $F \otimes \partial/\partial_0$ has projective dimension 1, so it is clear that p.d. $\overline{M} \leq \max\{p.d.M, 1\}$, as claimed. □

We could prove the next result only for Prüfer domains. It shows that modules behave towards divisible modules as if their projective dimensions were one less.

PROPOSITION 3.9. Let R be a Prüfer domain and M an R-module with $p.d._R M = m$. Then

3. DIVISIBLE MODULES OF DIMENSION ONE

$$\text{Ext}_R^m(M,D) = 0$$

for all divisible R-modules D.

Proof. We induct on m. If m = 1, then by (IV.4.7) M is the union of a well-ordered continuous ascending chain of submodules, $0 = M_0 < M_1 < \ldots < M_\alpha < \ldots$ ($\alpha < \tau$) such that $M_{\alpha+1}/M_\alpha$ is finitely presented cyclic for each $\alpha < \tau$. Hence (II.2.2) implies $\text{Ext}_R^1(M_{\alpha+1}/M_\alpha, D) = 0$ for $\alpha < \tau$. An appeal to (IV.2.1) shows that our claim holds for m = 1.

If m > 1, let $0 \to N \to F \to M \to 0$ be an exact sequence with F projective. Evidently, p.d.N = m−1. The exact sequence

$$\text{Ext}^{m-1}(N,D) \to \text{Ext}^m(M,D) \to \text{Ext}^m(F,D) = 0$$

together with the induction hypothesis completes the proof. □

It is not difficult to identify the divisible modules of projective dimension one in the Prüfer case.

THEOREM 3.10. [PD] (Fuchs [6]) A divisible module has projective dimension 1 if and only if it is a summand of a direct sum of copies of δ.

Proof. It suffices to show that if p.d.D = 1 holds for a divisible R-module D, then D is a summand of some $\oplus \delta$. Using (3.7) and the fact that δ/Rrw is isomorphic to a summand of δ, we conclude that there is an exact sequence $0 \to N \to \oplus \delta \to D \to 0$ where N is divisible. In view of p.d.D = 1, (3.9) implies that this sequence splits; thus the claim follows. □

We conclude this section with a structure theorem that characterizes divisible modules of projective dimension 1 over valuation domains by two cardinal invariants. First, we prove a relevant lemma.

LEMMA 3.11. [VD] Let D and D' be divisible R-modules of projective dimension 1, T and T' tight submodules containing D[r] and D'[r], respectively, for some $0 \neq r \in P$, and $\psi: T \to T'$ an isomorphism. For $a \in D \setminus T$ with $\text{Ann}(a+T) \neq 0$, there is an isomorphism

$$\psi' : T + Ra \to T' + Ra'$$

extending ψ for some $a' \in D'$. Here $T + Ra$ and $T' + Ra'$ are again tight submodules.

Proof. By (IV.4.3), $\mathrm{Ann}(a+T) = Rs$ ($s \in R$). Pick an $a' \in D'$ such that $\psi(sa) = sa' \in T'$. If there existed a $t \in R$ with $Rt > Rs$ and $ta' \in T'$, then $b = \phi^{-1}(ta') \in T$ would imply $sa = st^{-1}b$ whence $st^{-1}(ta - b) = 0$. If $Rr \leq Rst^{-1}$, then $ta - b \in D[r] \leq T$ would lead to $ta \in T$ contradicting $\mathrm{Ann}(a + T) = Rs$. If $Rr > Rst^{-1}$, then $r^{-1}st^{-1}(ta - b) \in D[r]$ would lead to $r^{-1}sa \in T$, a similar contradiction. Consequently, $\mathrm{Ann}(a' + T') = Rs$, and $a \longmapsto a'$ induces an extension ψ' of ψ between the submodules $T + Ra$ and $T' + Ra'$. Evidently, $\mathrm{Ker}\,\psi' = 0$. The last assertion follows from (IV.4.3). □

Observe that in (3.11) the tightness of T' was irrelevant.

Now we can prove our structure theorem; for 2) see (XIII.5.4).

THEOREM 3.12. [VD] For a divisible R-module D with p.d.D = 1, the following two cardinals constitute a complete set of invariants:

1) the rank $r(D)$ of D;

2) the cardinality of a basis of $D[r]$ (which is a direct sum of modules $\cong R/Rr$) for some $0 \neq r \in P$.

Proof. We show that two divisible R-modules D, D' of projective dimension one are isomorphic if the indicated two invariants are the same for D and D'. Let $\{a_i \mid i \in I\}$ and $\{b_j \mid j \in J\}$ be a maximal independent set in D and a basis for $D[r]$, respectively. Let $\{a'_i \mid i \in I\}$ and $\{b'_j \mid j \in J\}$ have the same meaning for D'. Thus $\mathrm{Ann}\,a_i = 0 = \mathrm{Ann}\,a'_i$ and $\mathrm{Ann}\,b_i = Rr = \mathrm{Ann}\,b'_j$. By hypothesis, we can use the same index sets I, J for D and D'.

Since $D/(\Sigma Ra_i + \Sigma Rb_j)$ is the quotient of $D/\Sigma Rb_j \cong D$ modulo the free R-module ΣRa_i, its projective dimension is ≤ 1. By making use of (IV.5.1), we can find in D a system $\mathcal{T}$ of tight submodules, all containing $T_0 = \Sigma Ra_i + \Sigma Rb_j$ such that $\mathcal{T}$ mod T_0 is

3. DIVISIBLE MODULES OF DIMENSION ONE 129

a tight system in D/T_0. Let T' be a similar system in D'.

The correspondence $a_i \mapsto a_i'$, $b_j \mapsto b_j'$ gives rise to an isomorphism $\psi_0 : T_0 \to T_0' = \Sigma R a_i' + \Sigma R b_j'$. Consider pairs (T_α, ψ_α) such that

(i) $T_\alpha \in T$, and ψ_α is an isomorphism of T_α with some T_α' in T';

(ii) $\psi_\alpha | T_0 = \psi_0$.

The set of all these pairs is partially ordered in the natural way. Zorn's lemma yields a maximal pair (T^*, ψ^*) in this set. If T^* is not all of D, then pick an $a_1 \in D \setminus T^*$ and, with the aid of (3.11), extend ψ^* to an isomorphism $\phi_1 : A_1 = T^* + R a_1 \to A_1' = \psi^* T^* + R a_1'$ for some $a_1' \in D'$. There exist submodules $B_1 \in T$, $B_1' \in T'$ which are countably generated over A_1 and A_1', respectively; say, $B_1 = \langle T^*, a_{11}, \ldots, a_{1n}, \ldots \rangle$ and $B_1' = \langle \psi^* T^*, a_{11}', \ldots, a_{1n}', \ldots \rangle$. Using (3.11) again, we can find an extension ϕ_2 of ϕ_1 which is an isomorphism between $A_2 \leq D$ and $A_2' \leq D'$ such that $\langle A_1, a_{11} \rangle \leq A_2$ and $\langle A_1', a_{11}' \rangle \leq A_2'$. There exist submodules $B_2 \in T$, $B_2' \in T'$ such that $B_1 + A_2 \leq B_2$, $B_1' + A_2' \leq B_2'$ and they are of the form $B_2 = \langle T^*, a_{21}, \ldots, a_{2n}, \ldots \rangle$, $B_2' = \langle \psi^* T^*, a_{21}', \ldots, a_{2n}', \ldots \rangle$. In the next step, we extend ϕ_2 to an isomorphism $\phi_3 : A_3 \to A_3'$ where A_3/A_2 is finitely presented and $a_{ij} \in A_3$, $a_{ij}' \in A_3'$ for $i,j \leq 2$. Then we select $B_3 \in T$, $B_3' \in T'$ to contain $B_2 + A_3$ and $B_2' + A_3'$, respectively, such that $B_3 = \langle T^*, a_{31}, \ldots, a_{3n}, \ldots \rangle$, $B_3' = \langle \psi^* T^*, a_{31}', \ldots, a_{3n}', \ldots \rangle$. In this way, the union A of the chain $A_1 \leq \ldots \leq A_n \leq \ldots$ will be equal to the union of the chain $B_1 \leq \ldots \leq B_n \leq \ldots$, and $A' = \cup A_n' = \cup B_n'$. Thus $A \in T$, $A' \in T'$ and the ϕ_n define an isomorphism $\phi : A \to A'$, contradicting the choice of (T^*, ψ^*). Hence $T^* = D$ and ψ^* is an isomorphism of D into D'.

It remains to show that ψ^* is surjective. If not, then $\psi^* D \in T'$ implies that there is a $d' \in D'$ such that $\mathrm{Ann}(d' + \psi^* D) = Rr$. If $d \in D$ satisfies $r\psi^* d = rd'$, then $d' - \psi^* d \in D'[r] \leq \psi^* D$ leads to the contradiction $d' \in \psi^* D$. □

EXERCISES

1. [D] Keeping $r \in R$, $0 \neq s \in P$ fixed, show that the elements $(r, r_1, \ldots, r_k) - s(rs, r_1, \ldots, r_k)$ generate a summand of ∂; a complement is generated by all $(r_1, \ldots, r_k)$ with $r_1 \neq r$.

2. [D] For every m with $1 \leq m \leq \mathrm{gl.d.}R$, there is a divisible R-module whose projective dimension is m.

3. [D] Following the pattern of §2, define $\underline{\partial\text{-dimensions}}$ of divisible R-modules. For Prüfer domains, verify the analogues of (2.5).

§4. INJECTIVE MODULES

This section is devoted to the study of injective modules. $E_R(M)$, or simply $E(M)$ will denote the injective hull of M.

Injective modules are always divisible. The converse holds in the torsion-free case.

THEOREM 4.1. [D] (Cartan-Eilenberg [1]) A torsion-free R-module is injective if and only if it is divisible. It is then a direct sum of copies of Q.

Proof. Assume M is a torsion-free divisible R-module and $\phi : L \to M$ is a homomorphism of an ideal $L \neq 0$ of R into M. For each $0 \neq r \in L$ there exists a unique $a_r \in M$ such that $\phi r = r a_r$. For another element $0 \neq s \in L$,

$$r(sa_s) = r(\phi s) = \phi(rs) = s(\phi r) = s(ra_r).$$

Hence $a_s = a_r$ is independent of $r \in L$; say $a_r = a$ for all $r \in L$. Consequently, the correspondence $1 \longmapsto a$ induces a homomorphism $R \to M$ extending ϕ.

We see that a torsion-free divisible R-module is a vector space over Q, thus it has the indicated structure. □

Injectives are moreover h-divisible, so by (1.1) their torsion submodules are summands. Hence, we can reduce the study of injectives to that of torsion injectives:

4. INJECTIVE MODULES   131

THEOREM 4.2. [D] An injective R-module M is the direct sum of an injective torsion module (viz. tM) and an injective torsion-free module ($\cong$ M/tM). □

Turning our attention to torsion injectives, we prove what is a far-reaching generalization of (4.1). This is implicitly in Fleischer [1] and Matlis [1].

THEOREM 4.3. An R-module over a Prüfer domain is injective if and only if
   (i) it is divisible, and
   (ii) it is ultracomplete in the annihilator filtration.

Proof. This is an immediate consequence of (V.4.2) and (II.2.2). □

In the rest of this section, we confine ourselves to injectives over valuation rings.

Over a valuation ring R, a cyclic R-module Ra is uniserial, therefore its injective hull E(Ra) is indecomposable. We infer that every element in an injective R-module M can be embedded in an indecomposable summand of M. Evidently, a maximal direct sum of indecomposable injectives in M has to be an essential submodule of M. This leads to

THEOREM 4.4. (Warfield [2]) Every injective module over a valuation ring is the injective hull of a direct sum of indecomposable injectives. □

As one expects, this direct sum is unique up to isomorphism. For valuation domains, this is a consequence of our later theorems (cf. (XI.4.9)).

Needless to say, the indecomposable injectives are precisely the injective hulls of cyclic R-modules R/L, each ideal L being irreducible. The isomorphy question is easy to settle:

LEMMA 4.5. [VR] (Nishi [1]) For two ideals, J and L, of R, $E(R/J) \cong E(R/L)$ holds if and only if $J \sim L$.

Proof. Suppose J = Lr. It is clear that for every $0 \neq r \in R$, multiplication followed by the natural map induces a surjective homomorphism $R \to Rr/Lr$. Its kernel is evidently $Lr:r$ which is equal to L, because of (I.4.1). Consequently, $R/L \cong Rr/Lr$ whence $E(R/L) \cong E(R/J)$ is evident.

Conversely, let $E(R/J) = M = E(R/L)$ and $a, b \in M$ such that Ann $a = J$, Ann $b = L$. Pick an element $0 \neq c \in Ra \cap Rb$, say $ra = c = sb$ with $r, s \in R$. Then $J:r =$ Ann $c = L:s$, thus $J \sim$ Ann $c \sim L$ follows. □

EXAMPLE 4.6. Let R be a valuation domain with value group $\cong \mathbb{Q}$. There exist continuously many pairwise non-isomorphic indecomposable injective R-modules.

EXAMPLE 4.7. If the value group of R is $\cong \mathbb{R}$, then the indecomposable injectives are isomorphic either to Q, $E(Q/R)$ or $E(Q/P)$.

Manifestly, $E(R/J)$ contains a copy of the uniserial module Q/J. The case where they coincide is of utmost interest. In the next result, linear compactness is understood in the discrete topology.

THEOREM 4.8. (Matlis [1], Gill [1], Vámos [1]) For a valuation domain R, the following are equivalent:

    (i) R is almost maximal;

    (ii)   every ideal J of R is almost maximal;

    (iii) Q/R is injective;

    (iv) Q/J is injective for every ideal J of R;

    (v) Q/R is linearly compact;

    (vi) Q/J is linearly compact for every ideal $J \neq 0$ of R;

    (vii) $E(Q/R)$ is uniserial;

    (viii) all indecomposable injectives are uniserial.

Proof. (i) ⟺ (ii) Note that submodules of linearly compact modules are likewise linearly compact. Since $J < R$ and J contains a copy of R, R/L is linearly compact for all ideals $L \neq 0$ of R if and only if J/L is linearly compact for all ideals $0 \neq L < J$.

(v) ⟺ (vi) Linear compactness being preserved under epimorphic images, the equivalence of (v) and (vi) is obvious.

4. INJECTIVE MODULES

(i) $\Longleftrightarrow$ (vi) Since (i) implies that all cyclic submodules of Q/J are linearly compact, it is clear that Q/J itself has to be linearly compact. The converse is trivial.

(vi) $\Longrightarrow$ (iv) As Q/J is divisible, by (4.3) it suffices to ascertain that it is ultracomplete in the annihilator filtration. This is guaranteed by (vi).

(iv) $\Longrightarrow$ (iii) and (viii) trivially.

(i) $\Longleftrightarrow$ (iii) has been shown in (V.3.3).

(viii) $\Longrightarrow$ (vii) is again trivial.

(vii) $\Longrightarrow$ (iii) For $x \in E(Q/R)$, there is an $r \in R$ such that $0 \neq rx \in Q/R$. By the divisibility of Q/R, some $a \in Q/R$ satisfies $ra = rx$. Hence $r(x-a) = 0$, and (vii) implies $x-a = pa$ for a suitable $p \in P$. Consequently, $x = (1+p)a \in Q/R$, proving $Q/R = E(Q/R)$. □

The proof can somewhat be shortened by making use of (V.5.2).

EXERCISES

1. [D] Show that a domain R is Dedekind if and only if all divisible R-modules are injective if and only if all h-divisible R-modules are injective.

2. [VD] (a) The endomorphism ring of $E(R/L)$, $L \neq 0$ an ideal of R, is a local ring.

   (b) (Nishi [1]) If R is not almost maximal, then this ring is not commutative.

3. Construct a valuation domain R such that there exist exactly κ pairwise non-isomorphic indecomposable injective R-modules where κ is an arbitrary infinite cardinal.

4. [VD] An indecomposable injective R-module is either uniserial or it contains elements a, b such that $Ra \cong Rb$, but neither of Ra, Rb contains the other.

5. [VD] Call an R-module H <u>homogeneous</u> if the annihilators of all non-zero elements of H belong to the same equivalence class of ideals. Show that

(a) in an injective R-module E a maximal homogeneous submodule is a summand;

(b) E is the injective hull of $\oplus H_i$ for suitable homogeneous H_i such that for $i \neq j$, $H_i \oplus H_j$ is not homogeneous;

(c) every homogeneous injective R-module is the injective hull of an R-module of the form $\oplus(R/J)$ with a fixed ideal J of R.

6. For a valuation domain R, $E(R/P)$ is an injective cogenerator of the category of R-modules.

7. Let S be a maximal immediate extension of R. Show that all S-injective modules are injective as R-modules.

§5. THE INJECTIVE DIMENSION

The dual of projection dimension, the injective dimension of modules also deserves attention. We pause for a short while to prove a couple of results on it.

Recall that M has injective dimension $\leq n$ means that the functor

$$\operatorname{Ext}_R^{n+1}(*, M) = 0.$$

An easy criterion for this to happen is given by Matlis:

LEMMA 5.1. [D] For an R-module M, $\mathrm{i.d.}_R M \leq n$ if and only if, for every ideal J of M,

$$\operatorname{Ext}_R^{n+1}(R/J, M) = 0.$$

<u>Proof</u>. To verify sufficiency, let $0 \to M \to E \to H \to 0$ be an exact sequence with E injective. Then we have $\operatorname{Ext}_R^{n+1}(X,M) \cong \operatorname{Ext}_R^n(X,H)$ for every R-module X. By Baer's criterion for injectivity, H is injective whenever $\operatorname{Ext}_R^1(R/J,H) = 0$ for every ideal J of R. Our assertion is now obvious by induction. □

5. THE INJECTIVE DIMENSION

In view of the exactness of

$$0 = \text{Ext}_R^n(R,M) \to \text{Ext}_R^n(J,M) \to \text{Ext}_R^{n+1}(R/J,M) \to \text{Ext}_R^{n+1}(R,M) = 0,$$

the preceding lemma can be rephrased:

LEMMA 5.2. [D] $\text{i.d.}_R M \leq n$ exactly if $\text{Ext}_R^n(J,M) = 0$ for all ideals J of R. □

The following is a noteworthy connection between a torsion-free module M and its factors M/rM.

LEMMA 5.3. [D] A torsion-free R-module M has injective dimension $\leq n$ if and only if

$$\text{i.d.}_R M/rM \leq n \quad \text{for all} \quad 0 \neq r \in R.$$

Proof. Start off with the exact sequence

$$0 \longrightarrow M \xrightarrow{r} M \longrightarrow M/rM \longrightarrow 0$$

induced by multiplication by r. We derive the exact sequence $\text{Ext}^{n+1}(X,M) \to \text{Ext}^{n+1}(X,M/rM) \to \text{Ext}^{n+2}(X,M)$. Now if $\text{i.d.}_R M \leq n$, then the two ends vanish, thus $\text{i.d.}_R M/rM \leq n$ follows.

Conversely, if $\text{i.d.}_R M/rM \leq n$ for all $0 \neq r \in R$, then $\text{Ext}_R^{n+1}(X,M) \xrightarrow{r} \text{Ext}_R^{n+1}(X,M) \longrightarrow \text{Ext}_R^{n+1}(X,M/rM) = 0$ implies $\text{Ext}_R^{n+1}(X,M)$ divisible for every X. This Ext is annihilated by the ideal J of R if $X = R/J$, hence it vanishes. (5.1) completes the proof. □

The case in which the injective dimension is equal to 1 is of particular interest. Recall that (V.5.3) asserts that for a valuation domain R, $\text{i.d.}_R R = 1$ exactly if R is almost maximal. More generally, we can prove:

PROPOSITION 5.4. Let M be a torsion-free module of finite rank over an almost maximal valuation domain R. Then

(a) $\text{i.d.}_R M \leq 1$, and

(b) $\text{Ext}_R^1(J,M) = 0$ for every ideal J of R.

Proof. If M is of rank 1, then M is isomorphic either to Q or to an ideal I of R. From almost maximality we conclude that Q/I is injective, i.e. (a) holds for M. We now induct on the rank n of M. Let $n \geq 2$ and N a pure submodule of rank $n-1$ in M. In the exact sequence $\text{Ext}_R^2(*,N) \to \text{Ext}_R^2(*,M) \to \text{Ext}_R^2(*,M/N)$ the end terms vanish by induction hypothesis, so (a) holds true.

Assertion (b) follows from (a) and from (5.2). □

EXERCISES

1. A valuation domain R is almost maximal if and only if $\text{Ext}^1(J,R) = 0$ for the ideals J of R. It is maximal if and only if $\text{Ext}^1(J,R) = 0$ for all submodules J of Q.

2. [VD] All factor modules of indecomposable injective R-modules are again injective exactly if R is almost maximal.

3. [VD] For an ideal I of R, $\text{i.d.}_R R/I \leq 1$ if and only if $I_{\tilde{F}} \leq I$ (see I.§5).

4. Prove the analogues of Kaplansky's and Auslander's Lemmas in IV.§2 for injective dimensions.

5. [VD] Find an R-module whose injective dimension is $\text{gl.d.}R$.

§6. QUASI-INJECTIVITY

A module M (over any ring) is called quasi-injective if every homomorphism of a submodule of M into M is induced by an endomorphism of M. It is well-known that a module is quasi-injective exactly if it is a fully invariant submodule of its injective hull. Our objective here is to find a more explicit characterization of quasi-injectives over valuation rings. Our discussion presents us with no surprises: everything which is expected to be true (in view of the close relationship with injectives) holds true.

First we give examples. Owing to our previous remark, they have to be fully invariant submodules of injectives. Ignoring the

6. QUASI-INJECTIVITY

injectives themselves as trivial examples, we have:

EXAMPLE 6.1. Let E be any injective R-module, and $L \neq 0$ an ideal of R. Then both $E[L]$ and $E[L^+]$ are quasi-injectives. In fact, homomorphisms can never decrease annihilators.

If R is a valuation ring, then there are no more examples as it is clear from

THEOREM 6.2. Let M be an R-module, R a valuation ring, and E its injective hull. Then M is quasi-injective if and only if

$$M = E[\text{Ann } M].$$

Proof. In the proof of necessity, we distinguish two cases according as $\text{Ann } M = L$ or L^+.

In the first case, let $a \in M$ have annihilator L. If b is any element in E whose annihilator is $\geq L$, then the correspondence $a \mapsto b$ extends to a homomorphism $Ra \to Rb$ which is obviously induced by an endomorphism of E. By the full invariance of M in E, $b \in L$ follows. Thus $E[L] \leq M$. The reverse inclusion is a consequence of the definition of L.

In the second alternative, we argue that to every $b \in E$ with $\text{Ann } b > L$ there is an $a \in M$ such that $\text{Ann } b \geq \text{Ann } a > L$. The rest of the argument is unchanged, and we are led to $M = E[L^+]$. □

In other words, the only fully invariant submodules of an injective R-module E are the submodules $E[L]$ and $E[L^+]$, for the various ideals L of R. (The choice $L = 0$ yields E and tE, while the choice $L = R$ yields 0 and the socle of E.)

Observe that the non-torsion quasi-injectives are necessarily injective.

It is immediate to raise the question as to which uniserial torsion modules are quasi-injective. In view of Ex. 1, it is natural to restrict ourselves to almost maximal valuation domains.

PROPOSITION 6.3. A uniserial module J/I ($0 < I < J \leq Q$) over an almost maximal valuation domain R is quasi-injective if and only if

$$\text{End } I \leq \text{End } J.$$

Proof. From (III.2.1) we conclude that the generator 1+I of R/I can be mapped by a suitable homomorphism R/I → Q/I upon an element q+I of Q/I exactly if $q \in I:I$. It follows at once that the quasi-injective hull of J/I is the submodule (I:I)J/I of the injective R-module Q/I. Hence, for J/I to be quasi-injective, it is necessary and sufficient that $(I:I)J \leq J$. This is equivalent to $I:I \leq J:J$, and so to the stated condition. □

Notice that an easy corollary to (6.3) is that, over almost maximal domains, J/I is quasi-injective if I is an archimedean ideal, and R/I is quasi-injective precisely if I is archimedean.

We close this section with a result which will be an easy consequence of (XI.2.1); however, the following proof is simple enough for inclusion here.

PROPOSITION 6.4. [VD] Let N be a quasi-injective submodule of an R-module M such that Ann N = Ann M. Then N is a summand of M.

Proof. The identity of N extends to a homomorphism $\eta : M \to E$ where E is the injective hull of N. But by (6.2), N = E[Ann N], thus the hypothesis implies Im η = N. It follows that η is a map of M onto N which is the identity on N. □

EXERCISES

1. [VD] A quasi-injective R-module is ultracomplete in the annihilator filtration.

2. [VD] The direct sum of two quasi-injective R-modules is again quasi-injective if they have the same annihilator.

3. [VD] Prove that the quasi-injective hull of M is precisely E[Ann M] where E = E(M).

4. [VD] All α-basic submodules of a quasi-injective module are basic (cf. Chapter X).

5. [VD] If E is an injective R-module, E[L] is an injective R/L-module, for any ideal L of R.

6. [VD] Give a counterexample to show that (6.4) fails if Ann $N = L^+$ and Ann $M = L$.

7. [VD] Every element a of a quasi-injective module M is contained in an indecomposable summand of M which is unique up to isomorphism.

8. [VD] Let $E = E(R/J)$ where J is an ideal of R. For the ideals K and L of R, $E[K] = E[L]$ holds if and only if $J:K = J:L$.

NOTES

In this chapter, we surveyed a number of theorems on divisible, h-divisible and injective modules over domains in general, and over Prüfer and valuation domains in particular. However, we still lack a satisfactory way of describing them via invariants. At this stage, the search for such a description looks hopeless, unless severe restrictions are imposed, like in (3.12).

There are indications which make the authors believe that divisible modules of projective dimension 1 over Prüfer domains will admit a characterization by numerical invariants, generalizing (3.12). In order to prove such a result, one must cope with a specific difficulty that occurs in extending isomorphisms between submodules to the divisible modules themselves, just as in the valuation domain case.

PROBLEM 5. Characterize by numerical invariants the divisible modules of projective dimension 1 over arbitrary Prüfer domains.

We also state formally as a problem:

PROBLEM 6. Are all divisible modules of projective dimension 1 over any domain necessarily direct summands of direct sums of copies of ∂?

Recently, B. de la Rosa and G. Viljoen have studied quasi-injective modules and extended results in §6 to more general rings.

VII. Uniserial Modules

In our discussions, we have frequently encountered uniserial modules. It is time to study them more thoroughly.

An important point that emerges from the study of uniserial modules is that their behavior depends heavily on their annihilators which can be of the form I or I^+.

Direct sums of uniserials will also be discussed briefly. No satisfactory condition is known under which a module over a valuation domain is a direct sum of uniserials.

§1. UNISERIAL MODULES

Recall that we called a module U over a ring R <u>uniserial</u> (in the literature, it is also called serial or chain module) if its submodules are totally ordered by inclusion. Equivalently, given $x, y \in U$, either $y \in Rx$ or $x \in Ry$ holds.

Submodules and quotients of uniserial modules are likewise uniserial. The simplest examples of uniserial modules are the

1. UNISERIAL MODULES

valuation rings R themselves, their rings Q of quotients, their cyclic modules, and more generally, the R-modules of the form J/I with R-submodules $I < J$ of Q. Uniserial modules of the latter type will be called **standard**, for easy reference.

From now on we assume R is a valuation ring.

We start off with associating with a uniserial module U its annihilator set

$$A(U) = \{\text{Ann } a \mid 0 \neq a \in U\}.$$

By (II.1.1), this is a set of equivalent ideals of R. Setting

$$A = \cap \{L \mid L \in A(U)\},$$

it is pretty obvious that

$$\text{Ann } U = \begin{cases} A & \text{if } A \in A(U), \\ A^+ & \text{otherwise.} \end{cases}$$

The two alternatives are discussed separately.

LEMMA 1.1. [VR] (Shores-Lewis [1]) If $\text{Ann } U = A$ for a uniserial module U over a valuation ring R, then U is isomorphic to an R-submodule of $Q(R/A)$.

Proof. No generality is lost in assuming $A = 0$.

Let $u \in U$ satisfy $\text{Ann } u = 0$. The inclusion $Ru \to U$ gives rise to a monomorphism $\phi : Q \otimes Ru \to Q \otimes U$ where Q is the ring of quotients of R. To see that ϕ is surjective, note that given $v \in U \setminus Ru$, $u = tv$ for some $t \in R$ which ought to be regular in view of $\text{Ann } u = 0$. Hence if $q \in Q$, then $q \otimes v = t(t^{-1}q) \otimes v = t^{-1}q \otimes u \in \text{Im } \phi$ and ϕ is an isomorphism. The canonical map $\psi : U \to Q \otimes U$ is monic, for if $v \in \text{Ker } \psi$, then $tv = 0$ for a regular $t \in R$. Setting $v = su$ ($s \in R$), $tsu = 0$ implies $ts = 0$ and $s = 0$. We conclude that $\phi^{-1}\psi$ maps U isomorphically into $Q \otimes Ru$ which is clearly isomorphic to $Q \otimes R \cong Q$. □

LEMMA 1.2. [VR] (Shores-Lewis [1]) Let $\text{Ann } U = A^+$ for a uniserial module U. For each pair $L_j \leq L_i$ with $L_i, L_j \in A(U)$, there is an element $r_{ij} \in R$ such that

(i) $L_i = L_j : r_{ij}$,
(ii) for $L_k \leq L_j \leq L_i$, $r_{ij}r_{jk} \equiv r_{ik} \mod L_k$,
(iii) $U \cong \varinjlim R/L_i$ where $L_i \in \mathsf{A}(U)$ and $\pi_{ij} : R/L_i \to R/L_j$ acts as multiplication by r_{ij}.

Conversely, if A is a totally ordered set of all equivalent ideals $L_i > A$ with intersection A, and if there are elements $r_{ij} \in R$ satisfying (i)-(ii), then U as defined in (iii) will be a uniserial module.

Proof. If U is uniserial, then for each $L_i \in \mathsf{A}(U)$ fix an element $u_i \in U$ with $\mathrm{Ann}\, u_i = L_i$. Note that $L_j \leq L_i$ implies $u_i \in Ru_j$, so we can choose $r_{ij} \in R$ with $u_i = r_{ij}u_j$. Then (i)-(iii) follow readily. The converse is easy to check. □

In view of the simple form of standard uniserial modules, it is natural to look for criteria under which a uniserial module is of standard form.

LEMMA 1.3. [VD] A uniserial R-module U is standard if either
(a) U is countably generated; or
(b) R is almost maximal.

Proof. (a) Let U be the union of the ascending chain $0 < Ru_0 \leq Ru_1 \leq \ldots \leq Ru_n \leq \ldots$ ($n < \omega$). Setting $r_n u_n = u_{n-1}$ ($r_n \in R$, $n \geq 1$), and $I = \mathrm{Ann}\, u_0$, we define a map

$$\psi : U \to Q/I$$

via $\psi u_n = r_n^{-1} \ldots r_1^{-1} + I$. As $U \cong \psi U = J/I$ for an R-submodule J of Q, U is standard.

(b) Choose $0 \neq u \in U$ and set $I = \mathrm{Ann}\, u$. The isomorphism $Ru \to R/I$ (with $u \mapsto 1+I$) extends to a monomorphism of U into Q/I, as Q/I is injective (see (VI.4.8)). □

The following result fully classifies standard uniserial modules over valuation domains.

THEOREM 1.4. [VD] Two standard uniserial R-modules, J/I and J'/I', are isomorphic if and only if there exists a $0 \neq q \in Q$ such that

1. UNISERIAL MODULES

$$I = qI' \quad \text{and} \quad J = qJ'.$$

Proof. The "if" part being obvious, assume $\phi : J/I \to J'/I'$ is an isomorphism. If $I = 0$, then also $I' = 0$, and the claim follows from the equivalence of $J \cong J'$ to $J \sim J'$. Suppose $I \neq 0$. Clearly, $E(J/I) \cong E(J'/I')$ implies by (VI.4.5) that $I \sim I'$, thus $I = qI'$ for some $0 \neq q \in Q$. Replacing J' by qJ', the proof is reduced to the case $I = I' \neq 0$.

If $J = Q$, then clearly $J' = Q$ too; so we can assume that $J', J < R$.

First, let R be maximal. Then Q/I is injective, therefore ϕ extends to an endomorphism ψ of Q/I. By (2.5) in the next section, $\text{End } Q/I$ is isomorphic to a subring of Q, therefore ϕ is induced by multiplication by some $0 \neq q \in Q$. The case when R is maximal is thus settled.

In the general case, when R is an arbitrary valuation domain, let S be a maximal immediate extension of R. Then $J/I \cong J'/I$ implies $JS/IS \cong J'S/IS$ as S-modules. Because of the settled case, there exists a $0 \neq \tilde{q} \in Q(S)$ such that $\tilde{q}IS = IS$ and $\tilde{q}J'S = JS$. By (I.1.10), $\tilde{q} = q\varepsilon$ where $q \in Q$ and ε is a unit of S. Thus we have $qIS = IS$ and $qJ'S = JS$. By I.§1, these equalities imply $qI = I$ and $qJ' = J$. □

EXERCISES

1. Determine all uniserial modules (a) over a P.I.D.; (b) over a Dedekind domain.

2. [VD] A divisible uniserial R-module U is maximal uniserial in $E(U)$.

3. [VR] A uniserial R-module U is standard if $\text{Ann } U$ contains the ideal I_F defined in I.§5.

4. [VR] Let U, V be uniserial R-modules, and

$$\{\phi_r : U \to V \mid r \in R\}$$

a family of homomorphisms satisfying

$$\phi_{rs} = \phi_s r + \phi_r s \qquad (r, s \in R).$$

Convert the additive group $U \oplus V$ into an R-module M by setting

$$r(u,v) = (ru, \phi_r u + rv) \quad (u \in U, v \in V).$$

Show that M is uniserial and there is an exact sequence $0 \to V \to M \to U \to 0$.

    5. [VR] Let V be a proper submodule of a uniserial module U. Then $V = Lu$ for some $u \in U$ and ideal L of R.

    6. [VR] (Shores-Lewis [1]) There exists a faithful uniserial R-module U with $\text{Ann } U = I^+$ for some ideal I of R if and only if $\text{Ann } P = 0$.

    7. [VD] If $R < J \leq Q$, then $\text{Ann } J/R$ is J^{-1} or $(J^{-1})^+$ according as J is principal or not.

§2. ENDOMORPHISM RINGS OF UNISERIAL MODULES

We shall now embark on the study of the structure of $\text{End } U$ for uniserial modules U over valuation rings R. In this way, we shall also learn about $\text{End } L$ for arbitrary ideals L of R (cf. I.§4).

    Preliminary definitions are required, generalizing $L^\#$ and $L_\#$, introduced in I.§4 for ideals L.

    Let R be a valuation ring. For a uniserial R-module $U \neq 0$, we set

$$U^\# = \{r \in R \mid rU < U\}$$

and

$$U_\# = \{r \in R \mid ra = 0 \text{ for some } 0 \neq a \in U\}.$$

It is easy to verify that these are indeed ideals of R. Moreover, $U^\#$ is a prime ideal containing $\text{Ann } U$, while $U_\#$ is simply $\cup A(U)$, again a prime ideal.

    For $r \in R$, $\bar{r}$ will denote the multiplication by r viewed as an endomorphism of U. It is clear that $\bar{r}$ is surjective if and only if $r \notin U^\#$ and injective if and only if $r \notin U_\#$. Hence,

LEMMA 2.1. [VD] For $r \in R$, $\bar{r}$ is an automorphism of U if and only if $r \notin U^\# \cup U_\#$. □

2. ENDOMORPHISM RINGS OF UNISERIAL MODULES

Evidently, one of $U^\#$, $U_\#$ contains the other. It is somewhat surprising that the form of Ann U determines which can be larger.

PROPOSITION 2.2. [VD] For a uniserial module $U \neq 0$,

(i) $U_\# \leq U^\#$ if Ann $U = I$;

(ii) $U^\# \leq U_\#$ if Ann $U = I^+$.

Proof. (i) Let $a \in U$ satisfy Ann $a = I$. By way of contradiction, assume $r \in R$ satisfies $rU = U$ and $ru = 0$ for some $0 \neq u \in U$. Obviously, $u = sa$ for a suitable $s \in R \backslash I$. Then $rsa = ru = 0$ shows $rs \in I$ whence $sU = rsU = 0$. We obtain the contradiction $s \in I$.

(ii) Working again toward contradiction, let $r \in R$ be such that $rU < U$ and $ru = 0$ implies $u = 0$. Pick an $a \in U \backslash rU$. As Ann $U = I^+$, there must exist an $s \in R$ such that $sa = 0$ but $sU \neq 0$. Now we have $srU \leq saR = 0$ whence $sU = 0$ follows. □

The discussion of End U is divided into two cases according as Ann $U = I$ or Ann $U = I^+$. We start with the first alternative.

THEOREM 2.3. [VR] (Shores-Lewis [1]) Let U be a uniserial R-module and Ann $U = I$. Then

(i) U carries the natural structure of an S-module where S is the ring R/I localized at $U^\#/I$;

(ii) $\text{End}_R U \cong S$.

Proof. Without loss of generality, $I = 0$ can be assumed.

(i) Suppose $r \in R \backslash U^\#$; then by (2.1) and (2.2), $\bar{r}$ is an automorphism of U. Therefore, multiplication by r^{-1} makes sense and, in the natural way, U becomes an S-module. (Now (1.1) guarantees that r is a regular element in R.)

(ii) Because of (1.1) and part (i), U may be regarded as an R-submodule of Q containing 1 and S as a subring of $\text{End}_R U$. We claim that $\alpha \in \text{End}_R U$ is nothing else than multiplication by $\alpha(1)$. This action is obvious if $u \in U$ belongs to R. If it does not, then $u = t^{-1}$ for a regular element $t \in R$, and $\alpha(1) = t\alpha(t^{-1})$ shows $\alpha(t^{-1}) = \alpha(1)t^{-1}$. □

We proceed to the second alternative.

THEOREM 2.4. [VR] (Shores-Lewis [1]) Let U be a uniserial R-module and $\text{Ann } U = I^+$. Then

(i) U has a natural structure of a T-module where T denotes the ring R/I localized at $U_\#/I$;

(ii) $\text{End}_R U$ is isomorphic to the completion of T in the T-topology.

Proof. Again $I = 0$ can be assumed.

(i) The proof is just the repetition of the proof of (i) in (2.3).

(ii) By virtue of (i), T can be viewed as the subring of $\text{End}_R U$ via $t \longmapsto \bar{t}$ ($t \in T$). It is then enough to show that the finite topology on $\text{End}_R U$ induces the T-topology on T and that T is dense in $\text{End}_R U$. The proofs of these run parallel to those given in (I.4.10), and are left to the reader. □

A special case is of particular interest.

COROLLARY 2.5. [VR] If R is R-complete, then $\text{End } Q/I$ is the localization of R at $I^\#$.

Proof. Observe that $\text{Ann } Q/I = 0^+$ and $(Q/I)_\#$ is just $I^\#$, and apply (2.4). By (V.1.10), localizations are complete. □

As localizations of valuation rings are again valuation rings, while their completions are always local rings, we are led to:

COROLLARY 2.6. [VR] (Shores-Lewis [1]) Endomorphism rings of uniserial modules are (commutative) local rings.

It is worthwhile noting that $\text{End } U$ is moreover a valuation ring if either $\text{Ann } U = I$ or if $\text{Ann } U = I^+$ with a prime ideal I.

From (2.6) we deduce, in view of (II.7.3):

COROLLARY 2.7. [VR] Uniserial modules have the exchange property. □

EXERCISES

1. [VR] If $U = J/I$ is a standard uniserial module, then
$$U^\# = J^\# \text{ and } U_\# = I^\#.$$

3. NON-STANDARD UNISERIALS   147

2. [VD] Let U be uniserial and $\operatorname{Ann} U = I$ or I^+. Then $I\# \leq U_\#$ and, if $\operatorname{Ann} U = I$, then equality holds.

3. [VR] (Shores-Lewis [1]) Let U be a uniserial R-module such that $\operatorname{Ann} U = 0^+$. Show that every regular ideal of $\operatorname{End} U$ is comparable to every other ideal. Moreover, $\operatorname{End} U$ is a valuation domain if and only if R is a domain.

4. [VR] Let U be a uniserial module with $\operatorname{Ann} U = I$.

   (i) Every surjective endomorphism of U is an automorphism.

   (ii) Prove that $\operatorname{End} U$ is local by showing that the endomorphisms with image $\neq U$ form the maximal ideal in $\operatorname{End} U$.

5. [VR] Let U be uniserial such that $\operatorname{Ann} U = I^+$.

   (i) Show that every injective endomorphism of U is an automorphism.

   (ii) The endomorphisms of U whose kernels are $\neq 0$ form the maximal ideal in the local ring $\operatorname{End} U$.

§3. NON-STANDARD UNISERIAL MODULES

One of the most challenging problems on modules over valuation domains is concerned with the existence of non-standard uniserial modules. In fact, for quite a while it was not known if such modules could exist at all. Very recently, S. Shelah proved that there are models of ZFC in which such modules do exist. Here we present a more algebraic approach to settle this question.

Throughout, R will denote a valuation domain. We start with a simple lemma which will make our considerations somewhat easier.

LEMMA 3.1. [VD] Let J be a submodule of Q containing R. Then every submodule of J/R is fully invariant.

Proof. A cyclic submodule of J/R is of the form Rr^{-1}/R. Its generators are the only elements whose annihilators are Rr. As endomorphisms can never decrease annihilators, the claim follows. □

A crucial lemma is as follows.

LEMMA 3.2. [VD] Let $J < J'$ be submodules of Q containing R and η an epimorphism $J \to J/R$ with kernel R. Suppose J is countably generated and J/R has an automorphism which is not induced by multiplication by a unit of R. Then there exists an R-module V such that

(i) V contains J/R as a submodule;
(ii) $V \cong J'/R$;
(iii) η does not extend to a homomorphism of J' into V.

Proof. We represent J as the union of an ascending chain $Rx_0 < \ldots < Rx_n < \ldots$ in Q and choose $r^{-1} \in J' \setminus J$ ($r \in R$). Set $r_n = rx_n \in R$ ($n < \omega$). As $\operatorname{Ann} J/R = (J^{-1})^+$, $\operatorname{End} J/R$ is by (2.4) the completion of R/J^{-1} in its R-topology. Let $u_0, \ldots, u_n, \ldots$ be units of R converging to an automorphism χ of J/R which is not induced by a multiplication by a unit of R. We can choose them such that

$$\chi(\eta x_n) = u_n(\eta x_n) \quad \text{for } n < \omega.$$

Define a module M by adjoining to J/R a symbol y subject to the relations

$$r_n y = u_n^{-1} \eta x_n \quad \text{for } n < \omega.$$

As $\{\eta x_n \mid n < \omega\}$ generates J/R, $M = Ry$ follows. Moreover, J/R is in fact a submodule of M. Clearly, $M \cong Rr^{-1}/R$ and $y \longmapsto r^{-1} + R$ induces an embedding of M in an R-module V isomorphic to J'/R.

By way of contradiction, assume η extends to a homomorphism η' of J' into V. (3.1) implies that $\eta' r^{-1} = vy$ for some unit v of R. Thus $vu_n^{-1} \eta x_n = vr_n y = \eta' r_n r^{-1} = \eta' x_n = \eta x_n$ for $n < \omega$. Consequently, $\chi \eta x_n = u_n \eta x_n = v \eta x_n$ for $n < \omega$, contrary to the choice of χ. □

Before entering into the discussion of the actual construction, we formulate a well-known combinatorial principle due to R. B. Jensen which we shall need:

3. NON-STANDARD UNISERIALS

$\diamondsuit_{\omega_1}$: there is a sequence of functions $h_\nu : \nu \to \nu$ ($\nu < \omega_1$) such that, for every function $f : \omega_1 \to \omega_1$, the set

$$\{\nu < \omega_1 \mid f_\nu = h_\nu\}$$

(where $f_\nu = f|\nu$) is stationary in ω_1.

It is known that this principle holds in the constructible universe and that it implies the Continuum Hypothesis.

THEOREM 3.3. [VD] Let R have cardinality $\aleph_1$ and assume

$$R < J_0 < J_1 < \ldots < J_\nu < \ldots < J_{\omega_1} = J \qquad (1)$$

is a continuous, well-ordered ascending chain of R-submodules of Q such that
    (i) J_ν is countably generated for $\nu < \omega_1$;
    (ii) R/J_ν^{-1} is not complete for any limit ordinal $\nu < \omega_1$.
Then $\diamondsuit_{\omega_1}$ implies the existence of a continuous well-ordered ascending chain of uniserial R-modules

$$U_0 < U_1 < \ldots < U_\nu < \ldots < U_{\omega_1} = U \qquad (2)$$

such that
    (iii) $U_\nu \cong J_\nu/R$ for each $\nu < \omega_1$;
    (iv) U is not isomorphic to J/R.

<u>Proof.</u> Let $\{\varepsilon_\alpha \mid \alpha < \omega_1\}$ be a list of units of R and $\{g_\alpha \mid \alpha < \omega_1\}$ a generating set of J. It is straightforward to see that there is a club C of limit ordinals in ω_1 such that for each ν in this club, $\{g_\alpha \mid \alpha < \nu\}$ generates J_ν.

Let $\{h_\nu \mid \nu < \omega_1\}$ be a set of functions as stated above.

We proceed to define a chain (2) by transfinite induction. Let $\phi_0 : J_0/R \to U_0$ be an arbitrary isomorphism. Suppose that, for a given $\nu < \omega_1$, uniserial modules U_μ and isomorphisms $\phi_\mu : J_\mu/R \to U_\mu$ have been defined for each $\mu < \nu$ such that the U_μ form a continuous chain. If ν is a limit ordinal $< \omega_1$, we simply set $U_\nu = \cup\{U_\mu \mid \mu < \nu\}$. As U_ν is countably generated, (1.3) implies the existence of an isomorphism $\phi_\nu : J_\nu/R \to U_\nu$. If $\nu = \lambda + 1$ is a successor ordinal, we distinguish two cases, depending on the function h_λ.

Case I: h_λ is such that $g_\alpha \longmapsto \varepsilon_{h_\lambda(\alpha)} g_\alpha + R$ $(\alpha < \lambda \in C)$ induces an epimorphism $\eta_\lambda : J_\lambda \to J_\lambda/R$ with kernel R. In this case, we embed J_λ/R - with the aid of (3.2) - in a module $V \cong J_{\lambda+1}/R$ so that η_λ does not extend to a homomorphism $J_{\lambda+1} \to V$. For the sake of convenience, identify V with $J_{\lambda+1}/R$ (keeping in mind that we are not using the canonical embedding $J_\lambda/R \to J_{\lambda+1}/R$). Choose an isomorphism $\phi_{\lambda+1} : J_{\lambda+1}/R \to U_{\lambda+1}$ and define the embedding ψ_λ so as to make the diagram

$$\begin{array}{ccc} J_\lambda/R & \longrightarrow & J_{\lambda+1}/R \\ \downarrow \phi_\lambda & & \downarrow \phi_{\lambda+1} \\ U_\lambda & \xrightarrow{\psi_\lambda} & U_{\lambda+1} \end{array}$$

commute. We may identify U_λ with its image under ψ_λ.

Case II: If h_λ is not as in Case I, then define ϕ_λ and $U_{\lambda+1}$ by using the same diagram with the canonical (or any other) embedding in the first row.

The union U of the chain $\{U_\nu \mid \nu < \omega_1\}$ is evidently a uniserial R-module. Assume that there is an epimorphism $\xi : J \to U$ with kernel R. From (3.1) we conclude that, for each $\nu < \omega_1$, ξ maps g_ν onto $\phi_\nu(\varepsilon_{\alpha(\nu)} g_\nu + R)$ for some unit $\varepsilon_{\alpha(\nu)}$ of R. This yields a function $f : \nu \longmapsto \alpha(\nu)$ $(\nu < \omega_1)$, and $\diamondsuit$ shows that there is a $\lambda \in C$ (as a matter of fact, stationarily many of them) such that $f|\lambda = h_\lambda$. Thus $g_\nu \longmapsto \varepsilon_{\alpha(\nu)} g_\nu + R$ $(\nu < \lambda)$ gives rise to an epimorphism $\eta_\lambda : J_\lambda \to J_\lambda/R$ with kernel R. But then $f|\lambda+1$ would induce an extension of η_λ to $J_{\lambda+1} \to J_{\lambda+1}/R$ which is, by construction, impossible. □

We proceed to construct valuation domains R satisfying the hypotheses of (3.3).

Let Γ be a totally ordered group of cardinality $\aleph_1$ with the following property: it has a continuous well-ordered ascending chain of convex subgroups

$$0 < \Gamma_1 < \ldots < \Gamma_\nu < \ldots < \Gamma = \Gamma_{\omega_1} \qquad (\nu < \omega_1)$$

such that Γ_ν is countable for each $\nu < \omega_1$. E.g. $\Gamma = \oplus \mathbb{Z}$,

3. NON-STANDARD UNISERIALS

lexicographically ordered over an index set which carries the inverse order of ω_1.

Let K be any countable field and $K[[\Gamma]]$ the formal power series ring over Γ with coefficients in K. The monomials x^γ ($\gamma \in \Gamma$) generate a subfield in the quotient field of $K[[\Gamma]]$ which we shall call Q. The valuation v on $K[[\Gamma]]$ makes Q into a field with valuation. Its valuation ring is our choice for R.

For $\nu < \omega_1$, define J_ν as the R-submodule of Q generated by elements $x \in Q$ with $v(x) \in \Gamma_\nu$. In this way, we obtain a continuous well-ordered ascending chain

$$R = J_0 < J_1 < \ldots < J_\nu < \ldots < J_{\omega_1} = Q$$

Evidently, each J_ν ($\nu < \omega_1$) is a countably generated R-module.

It is straightforward to check that

$$J_\nu^{-1} = \{x \in R \mid v(x) \notin \Gamma_\nu\} \qquad (\nu < \omega_1)$$

and that this is a prime ideal of R. Furthermore, the factor rings R/J_ν^{-1} are countable valuation domains with value groups Γ_ν ($\nu<\omega_1$). They cannot be complete in their R-topologies (otherwise they would have the power of the continuum). Consequently, from (3.3) we obtain:

THEOREM 3.4. There exist valuation domains R of cardinality $\aleph_1$ such that $\diamondsuit_{\omega_1}$ implies the existence of non-standard (divisible) uniserial R-modules. □

The last theorem makes it possible to answer a question raised by I. Kaplansky some time ago: Do there exist valuation rings that are not quotients of valuation domains?

THEOREM 3.5. Let R be a valuation domain and U a non-standard divisible uniserial R-module. Then the ring R_U of (I.1.9) is a valuation ring which is not a quotient of any valuation domain.

Proof. Without loss of generality, we can assume that the annihilators of elements in U are principal ideals. Suppose, by way of contradiction, that there is an epimorphism $\phi: S \to R_U$ for some valuation domain S. In view of the natural map $\psi: R_U \to R$,

U and Q/R become both R_U- and S-modules. We now have $R_U \cong S/St$ for some $t \in S$ and $U \cong I/St$ for some ideal I of S. Hence $U \cong It^{-1}/S$. The last module is contained in Q'/S where Q' is the field of quotients of S, and divisibility relations show that it must be $\cong Q/R$. Thus $U \cong Q/R$ as S-modules, and so as R-modules - a contradiction. □

EXERCISES

1. [VD] If there is a non-standard uniserial R-module where the annihilators of elements are principal ideals, then for every isomorphy class of ideals $\neq 0$ of R there are non-standard uniserial R-modules with annihilators in this class.

2. [VD] Check that in the proof of (3.3), U can be chosen in non-isomorphic ways.

3. [VD] If R is defined as before (3.4), then $\text{End } U \simeq R$ for the U of (3.3).

4. [VD] Relax condition (ii) in (3.3) by stipulating it only for ν's in a stationary subset of ω_1.

5. [VD] Verify (3.2) for uncountably generated J.

§4. DIRECT SUMS OF UNISERIAL MODULES

Before entering into the discussion of direct sums of uniserials, we show that uniserial modules (over any ring) are countably small. This is evident for countably generated modules. Otherwise, even a stronger result can be established.

PROPOSITION 4.1. (Fuchs-Salce [1]) *Uniserial modules are countably small. Uncountably generated uniserial modules are small.*

Proof. If a module U is not small, then there exist a homomorphism $\phi : U \to \oplus_{i \in I} A_i$ and a suitable countably generated submodule V of U such that ϕV has nonzero projections in countably many A_i's. If U happens to be an uncountably generated uniserial module, then there is a $u \in U \setminus V$. As $V \leq Ru$, ϕu has nonzero coordinate in each A_i in which the projection of ϕV fails

4. DIRECT SUMS OF UNISERIALS

to vanish. This is absurd, so U is small. □

By (2.6), uniserial modules have local endomorphism rings. In view of this and (4.1), (II.7.3) yields:

THEOREM 4.2. (Fuchs-Salce [1]) Let R be a valuation ring and $M = \bigoplus_{i \in I} U_i$ with U_i uniserial R-modules. Every decomposition of M into the direct sum of uniserial modules is isomorphic to the given decomposition, and any summand of M is isomorphic to $\bigoplus_{i \in J} U_i$ for a suitable subset J of I. □

This theorem gives a satisfactory description of direct sums of uniserial modules. In (X.1.7) we shall see that, for standard U_i, they admit complete sets of numerical invariants.

We have not said as yet anything about criteria for a module to be the direct sum of uniserials. Unfortunately, no general criteria are known. We shall partially remedy the situation by proving:

PROPOSITION 4.3. (Fuchs-Salce [1]) Let R be an almost maximal valuation domain and M a torsion R-module. Then M is a direct sum of countably many uniserial modules if and only if it is the union of a countable ascending chain of pure submodules of finite Goldie dimensions.

The same holds if R is a maximal valuation ring and M is any R-module.

Proof. The necessity being trivial, let $M = \bigcup_{n<\omega} M_n$ where $M_1 \leq \ldots \leq M_n \leq \ldots$ is a chain of pure submodules of finite Goldie dimensions. Because of the hypothesis on R, M_n is by (IX.5.5) polyserial. By (XI.4.2), we have $M_n = M_{n-1} \oplus A_n$ for all $n \geq 1$, $M_0 = 0$ and all the A_n's are direct sums of uniserial modules. Hence $M = \bigoplus_n A_n$ is a direct sum of countably many uniserial modules. □

We interject here two examples which may or may not be relevant in understanding the enormous difficulties in generalizing (4.3), but which certainly shed light on how badly submodules in direct sums of uniserials can behave.

EXAMPLE 4.4. We show that a pure submodule of a direct sum of uniserial R-modules need not be a direct sum of uniserials. Let R be a valuation domain with value group Γ isomorphic to a dense subgroup of $\mathbb{R}$. Let $0 < \gamma \in \Gamma$ and, for every $\beta \in \Gamma$, with $0 < \beta \leq \gamma$, choose an $r_\beta \in R$ with $v(r_\beta) = \beta$. Define H_γ as the R-module generated by $\{x, x_\beta$ for all $\beta \in \Gamma$ with $0 < \beta < \gamma\}$ subject to the relations

$$\text{Ann } x = P, \qquad r_\beta x_\beta = x \qquad (0 < \beta < \gamma).$$

If we adjoin a new generator y subject to the relation

$$r_\gamma y = x,$$

then the R-module $M = H_\gamma + Ry$ will be a direct sum of cyclics:

$$M = Ry \oplus \bigoplus_\beta R(x_\beta - r_\gamma r_\beta^{-1} y).$$

The submodule $N = Py \oplus \bigoplus_\beta R(x_\beta - r_\gamma r_\beta^{-1} y)$ is a direct sum of uniserials, and it is readily seen that H_γ is a pure submodule of N. If we had $H_\gamma = \bigoplus_{i \in I} U_i$ with U_i uniserial, then because of $N[P] = H_\gamma[P] = Rx$, we would have $Rx \leq U_j$ for some $j \in I$. As H_γ/Rx is a direct sum of cyclic modules, each U_i would be cyclic. Since there is no $r \in R$ with maximal value such that $r|x$, this is impossible.

EXAMPLE 4.5. We can improve on (4.4) and show: A pure submodule of a direct sum of cyclics need not be a direct sum of uniserials. Let R be an almost maximal valuation domain with value group $\mathbb{Q}$. Let

$$M = Rx \oplus Ry \oplus \bigoplus_{n \geq 1} Rx_n$$

where Ann $x = P$, Ann $y = Rr^2$, Ann $x_n = Rrr_n$; here $r, r_n \in R$ satisfy: $v(r) = 1$ and $v(r_n)$ form a strictly decreasing sequence in $[0,1]$ with limit 0. Define N as the submodule generated by

$$a = x - ry \quad \text{and} \quad a_n = x_n + y \qquad (n \geq 1).$$

Evidently, Ann $a = Rr$, Ann $a_n = Rr^2$ and $r_n a = rr_n a_n$ for all $n \geq 1$. It is easy to check that N is pure in M and $N[P] = 0$. However, N fails to be a direct sum of uniserial modules. In

NOTES

fact, N is not smooth (see VIII.§5), since $a \notin N[P] + rN = rN$ and $pa \in prN$ for all $p \in P$, so $i(a)$ is not smooth at P.

EXERCISES

1. [VD] Give details of proof for the purity of N in M and for $N[P] = 0$ in (4.5).

2. Verify (4.2) for arbitrary domains R.

NOTES

The study of uniserial modules has been initiated by Shores-Lewis [1]; in this paper, the reader will find further information on uniserial modules over commutative rings in general. Additional results were obtained by Fuchs-Salce [1].

Just before the completion of the manuscript of this volume, the existence of non-standard uniserial modules has been established. S. Shelah used forcing to verify the existence of models of ZFC such that, over suitable valuation domains, non-standard uniserial modules exist. The approach presented in §3 is entirely different and is due to one of the present authors. Currently, Fuchs and Shelah are investigating valuation domains of cardinality $\aleph_1$ over which non-standard uniserial modules exist.

There is an interesting connection between certain non-standard uniserial modules and Aronszajn trees. This is, however, not relevant at this time.

The class of direct sums of uniserials has not received as yet much attention. Their significance will become apparent in the study of α-basic submodules in Chapter X.

PROBLEM 7. Give criteria for a module to be a direct sum of uniserial modules.

VIII. Heights and Indicators

If one wishes to pinpoint the concepts which seem to be the most relevant for the theory of modules over valuation domains, then among those to be mentioned prominent place must be awarded to the notions of heights and indicators. In fact, as we proceed with our discussion, it will become increasingly clear that they are indispensable tools.

The first thought is to borrow their definitions from their counterparts in abelian group theory. However, the situation in the general valuation domain case is considerably more complicated. As a consequence, more delicate definitions are warranted (which are basically equivalent to the usual definitions for discrete rank one valuation domains).

§1. HEIGHTS

Throughout, R will denote a valuation ring and M an R-module.

For $a \in M$, let $D(a)$ denote the set of all R-submodules J of Q containing R such that there is a homomorphism $\alpha : J \to M$

1. HEIGHTS

with $a1 = a$ ($1 \in R$). Define the __height ideal__ of a in M as

$$H_M(a) = \cup\{J \mid J \in D(a)\}.$$

It is evident that, for r regular in R, $r^{-1} \in H_M(a)$ exactly if $a \in rM$. (Roughly speaking, $H_M(a)$ measures the divisibility of a in M by the regular elements of R.)

Let $I = H_M(a)$ be the height ideal of a in M, and set $U = I/R$. Define the __height__ $h_M(a)$ of a in M as:

$$h_M(a) = \begin{cases} U & \text{if } I \in D(a), \\ U^- & \text{otherwise.} \end{cases} \qquad (1)$$

(We write $h(a)$ instead of $h_M(a)$ if there is no danger of confusion.) Accordingly, we distinguish between __non-limit heights__ U and __limit heights__ U^-. If I is cyclic, then U^- can not occur as height. But this is the only exception as is shown by the two examples below.

The set of all heights will be denoted by Σ. Thus Σ consists of all uniserial modules of the form $U = I/R$ with $R \leq I \leq Q$ and of symbols U^- for non-cyclic U. We adjoin a symbol ∞ to Σ and consider it as the height of 0. The set Σ can be totally ordered by inclusion for non-limit heights U, V, while if $U < V$ and V^- exists, we set $U < V^- < V$. Evidently, ∞ is declared to be the largest element of Σ. Every non-void subset of Σ has an infimum and a supremum. Manifestly, $h_M(a) = Q/R$ exactly if $a \in hM$, the maximal h-divisible submodule of M, and $h_M(a) = (Q/R)^-$ if and only if a belongs to the first Ulm submodule M^1 of M. The elements of M^1 are said to be of __infinite height__. Furthermore, if $\{M_i \mid i \in I\}$ is a family of R-modules and $x = (x_i)_{i \in I} \in \prod M_i$, then

$$h(x) = \inf_i \{h_M(x_i) \mid i \in I\}.$$

Let $0 \neq r \in R$ and $U = I/R$. The multiplication of the heights U and U^- by r is defined by the rules

$$rU = (rI + R)/R \quad \text{and} \quad rU^- = ((rI + R)/R)^-.$$

Division by a regular element $t \in R$ can be defined via

$$t^{-1}U = (t^{-1}I)/R \quad \text{and} \quad t^{-1}U^- = ((t^{-1}I)/R)^-.$$

EXAMPLE 1.1. Let R be a valuation domain. Let $U = J/R$, where J is an R-submodule of Q containing R, and let L be an ideal of R. Consider the standard uniserial module $V(U,L) = J/L$. The element $a = 1 + L \in V(U,L)$ satisfies $h(ra) = r^{-1}U$ for $r \in R \setminus L$, and $h(ra) = \infty$ for $r \in L$.

EXAMPLE 1.2. Let U and L be as in (1.1), and assume that J is not cyclic; say, it is generated by an infinite set $\{q_i\}_{i \in I}$. Here $q_i \notin R$ can be assumed, thus $q_i^{-1} \in R$ for all $i \in I$. Define the R-module

$$V(U^-,L) = \langle Ra, Ra_i \mid i \in I \rangle$$

such that the defining relations between the generators are:

$$q_i^{-1}a_i = a \ (i \in I); \quad ra = 0 \text{ if } r \in L.$$

It is straightforward to check that $h(ra) = r^{-1}U^-$ for $r \notin L$, and $h(ra) = \infty$ for $r \in L$.

The following inequalities satisfied by the heights are easy to prove.

LEMMA 1.3. [VD] Let M be an R-module, $x,y \in M$ and r a regular element of R. Then

$$h(x+y) \geq \min(h(x), h(y)) \qquad (2)$$

$$h(rx) \geq r^{-1}h(x). \qquad \square \qquad (3)$$

If $h(x) \neq h(y)$, then (2) becomes an equality. Equality holds in (3) whenever r is a unit in R. This can happen even if r is not a unit. In fact, if $r \notin (h(a))^\#$, then $rh(a) = h(a)$ and $h(ra) = h(a) = r^{-1}h(a)$. More generally, we deduce:

LEMMA 1.4. [VR] Let U be a uniserial module. If $a \in U$ and r is a regular element in R such that $ra \neq 0$, then $h(ra) = r^{-1}h(a)$.

Proof. If there exists a homomorphism $\phi : J \to U$ ($R \leq J \leq Q$) such that $\phi 1 = ra$, then, by (1.1), there exists a homomorphism $\psi : rJ \to U$ such that $\psi 1 = a$ (clearly $rJ \geq R$). This shows that

1. HEIGHTS

$rh(ra) \leq h(a)$, therefore $h(ra) \leq r^{-1}h(a)$. The converse inequality is always true. □

The subsequent result is concerned with the situation where $h(ra) > r^{-1}h(a)$; this can happen only if M is not uniserial.

LEMMA 1.5. [VR] Let M be an R-module, $a \in M$ and r regular in R. If $h(ra) > r^{-1}h(a)$, then there exists a $b \in M$ such that $rb = ra$ and $h(b) > h(a)$.

Proof. Let $h(a) = J/R$ or $(J/R)^-$ and $h(ra) = L/R$ or $(L/R)^-$. Hypothesis implies that there exist an R-module I in Q with $r^{-1}J \leq I \leq L$ and a homomorphism $\phi : I \to M$ with $\phi 1 = ra$, such that $a \notin \phi I$. There is a $b \in \phi I$ such that $rb = ra$, and clearly $h(b) \geq r(I/R) > h(a)$. □

The following result compares heights in a module over different rings.

LEMMA 1.6. [VD] Let J be a prime ideal of the valuation domain R, A an R_J-module and $a \in A$. Let L be an R_J-submodule of Q containing R_J. Then $h_A(a) = L/R_J$ if and only if the height of a in A, regarded as an R-module, is L/R.

Proof. It is enough to notice that, if K is an R-submodule of Q containing R, any R-homomorphism $\phi : K \to A$, such that $\phi 1 = a$ extends uniquely to an R_J-homomorphism $\phi' : K_J \to A$. □

If $\sigma \in \Sigma$ and M is an R-module, we can form the submodules

$$M^\sigma = \{a \in M \mid h(a) \geq \sigma\}$$

and

$$M^{\sigma+} = \{a \in M \mid h(a) > \sigma\}.$$

These are fully invariant submodules of M. The following facts are easily proved.

(a) $M^{\sigma+} = \cup \{M^\tau \mid \tau > \sigma\}$.
(b) $\sigma < \tau$ in Σ implies $M^{\sigma+} \geq M^\tau$.
(c) $h(a) = \sigma$ exactly if $a \in M^\sigma \setminus M^{\sigma+}$.
(d) If $\sigma = r^{-1}R/R$, r regular in R, then $M^\sigma = rM$ and $M^{\sigma+} = rPM$.

(e) If $\sigma = J/R$ or $(J/R)^-$, then $J^\#M^\sigma \leq M^{\sigma+}$.

(f) If $\phi : M \to N$ is a homomorphism, then $h_N(\phi a) \geq h_M(a)$ for all $a \in M$. Therefore, $\phi M^\sigma \leq N^\sigma$ and $\phi M^{\sigma+} \leq N^{\sigma+}$ for all $\sigma \in \Sigma$.

Evidently, (a) and (b) show that, for any module M, we have a chain of submodules

$$M \geq \ldots \geq M^\sigma \geq M^{\sigma+} \geq \ldots \geq M^{(Q/R)^-} = M^1 \geq M^{Q/R} = hM \ .$$

This will be called the <u>height filtration</u> of M.

In order to distinguish between elements of infinite heights with different divisibility behavior, transfinite heights can be introduced, by using the heights as defined above in the higher Ulm submodules. As we do not need transfinite heights, we do not give a formal definition at this time.

EXERCISES

1. Let A be a $\mathbb{Z}_p$-module. $a \in A$ has height $n \in \mathbb{Z}$ in the usual sense of abelian group theory if and only if $h_A(a) = p^{-n}\mathbb{Z}_p/\mathbb{Z}_p$.

2. Let R be a valuation domain and U a uniserial R-module. Show that the following are equivalent:
   (i) there exists a $0 \neq a \in U$ such that $h(a)$ is non-limit;
   (ii) every element $\neq 0$ in U has non-limit height;
   (iii) U is standard uniserial.

3. [VD] Let U be a uniserial module. Show that U is divisible whenever $U^1 \neq 0$.

4. [VD] Let M be an R-module, $a \in M$ and $h_M(a)$ non-limit. If $r^{-1}R/R \leq h_M(a)$, there exists a $b \in M$ such that $rb = a$ and $h_M(b) = rh_M(a)$. Show that this is not necessarily true if $h_M(a)$ is a limit height.

2. EQUIHEIGHT SUBMODULES

5. Let M be an R-module, σ a non-limit height and $N \leq M^\sigma$. If $a \in M$, $h_M(a) < \sigma$, then $h_M(a) = h_{M/N}(a+N)$.

6. Give an example contradicting the statement in Exercise 5, for σ a limit height. (A divisible module D such that $0 \neq hD = tD$ and $D/hD \cong Q$; see (IV.1.4).)

§2. EQUIHEIGHT SUBMODULES

Next we introduce a new kind of submodule which will be useful in our subsequent discussions. It strengthens the notion of relative divisibility.

Let R be a valuation ring and M an R-module. A submodule N of M is called __equiheight__ (cf. Fuchs-Salce [2]) if

$$N^\sigma = N \cap M^\sigma \quad \text{for all } \sigma \in \Sigma. \tag{1}$$

It is clear that (1) implies

$$N^{\sigma+} = N \cap M^{\sigma+} \quad \text{for all } \sigma \in \Sigma.$$

Hence, N is equiheight in M if and only if

$$h_N(a) = h_M(a) \quad \text{for all } a \in N.$$

This explains the choice of terminology.

It is obvious that equiheight submodules are necessarily pure. The converse fails as evidenced by the following example.

EXAMPLE 2.1. [VR] Let H_γ be the R-module considered in (VII.4.4). It is a pure submodule of $H_\gamma + Py = N$. The height of x in H_γ is $(r_\gamma^{-1}P/R)^-$, while its height in N is $r_\gamma^{-1}P/R$. Thus H_γ is not equiheight in N.

It should be pointed out that for a pure submodule N of M, (1) holds automatically for every height σ which is either cyclic or limit. Thus, in order that a pure submodule N be equiheight in M, it is necessary and sufficient that (1) holds for all non-cyclic, non-limit heights σ.

The following properties of equiheight submodules are straightforward to verify.

(A) Direct summands are equiheight.

(B) If N is a pure submodule of M and all the elements of N have non-limit heights in N, then N is equiheight in M.

(C) If R is a domain and M is an R-module, every submodule N of M with M/N torsion-free is equiheight.

(D) The equiheight property is transitive.

(E) The equiheight property is inductive.

We will find it convenient to call a module M <u>cohesive</u> if no element in M has a limit height. In this terminology, (B) asserts that a cohesive pure submodule is equiheight.

EXERCISES

1. [VD] If N is an equiheight submodule of a module M, then
   (a) $N^1 = M^1 \cap N$,
   (b) $hN = N \cap hM$.

2. [VD] Torsion-free modules are cohesive.

3. [VD] A uniserial module is standard if and only if it is cohesive.

4. [VD] If all the elements $\neq 0$ of M have cyclic heights, then the pure submodules of M are equiheight.

§3. INDICATORS

The notion of height introduced in the first section can be used to define indicators which are most useful tools in investigating certain classes of modules (pure-injective and separable modules, Chapters XI, XIII). The indicators of an element a in the module M gives an excellent picture of the manner Ra is embedded in M.

We are assuming that R is a valuation domain. Let Γ be the value group of R. For $\gamma \in \Gamma^+$, we let r_γ denote an element of R whose value is γ; we may think of r_γ as being fixed.

Let M be an R-module and $a \in M$. The <u>indicator</u> $i_M(a)$ of a in M is a function from Γ^+ to Σ defined by:

$$i_M(a) : \gamma \longmapsto r_\gamma h_M(r_\gamma a) \qquad (\gamma \in \Gamma^+).$$

3. INDICATORS

As usual, we drop the index M in $i_M(a)$, if there is no danger of confusion. From (1.3) we infer that $i(a)$ is a non-decreasing function,

$$i(a)(\gamma) \leq i(a)(\delta) \qquad \text{if } \gamma \leq \delta \text{ in } \Gamma^+.$$

This inequality expresses the most relevant property of indicator functions. Actually, it is the only restriction to which the indicators are subject, as will be seen from (3.3) below.

EXAMPLE 3.1. [VD] Let U be a uniserial R-module and $0 \neq a \in U$. By (1.4), for every $r \in R \setminus \text{Ann } a$, $h(ra) = r^{-1}h(a)$. So, for every $\gamma \in \Gamma^+$ such that $r_\gamma a \neq 0$, we have

$$i_M(a)(\gamma) = r_\gamma h_M(r_\gamma a) = r_\gamma(r_\gamma^{-1} h_M(a)) = h_M(a).$$

It follows that $i_M(a)$ is constant up to Ann a. In particular, if $V(U,L)$ is the module considered in (1.1) and $a = 1+L \in V(U,L)$, then $i(a)(\gamma) = U$ for all $r_\gamma \notin L$.

EXAMPLE 3.2. [VD] Let $V(U^-,L)$ be the module considered in (1.2). Then $i(a)(\gamma) = U^-$ for all $r_\gamma \notin L$.

We can now characterize indicators.

THEOREM 3.3. [VD] (Fuchs-Salce [2]) Let $f : \Gamma^+ \to \Sigma$ be any function such that $\gamma \leq \delta$ in Γ^+ implies $f(\gamma) \leq f(\delta)$. Then there exists an R-module M and an element $a \in M$ such that $i_M(a) = f$.

Proof. For every $\gamma \in \Gamma^+$ define the module

$$V_\gamma = V(f(\gamma), r_\gamma P)$$

using (1.1) or (1.2) according as $f(\gamma)$ is a non-limit or a limit height. In each V_γ select an element a_γ of height $f(\gamma)$ and with annihilator $r_\gamma P$. If $f(\gamma) = \infty$, we let $V_\gamma = 0$ and $a_\gamma = 0$. We clearly have

(i) if $\gamma \leq \delta$, then $h(a_\gamma) \leq h(a_\delta)$;

(ii) if $\gamma < \delta$, then Ann $a_\gamma >$ Ann a_δ, provided that $f(\gamma) \neq \infty$.

Define the module M as the cartesian product of all V_γ, for $\gamma \in \Gamma^+$, and $a \in M$ with coordinates $a_\gamma \in V_\gamma$. In view of (ii), $r_\gamma a_\delta = 0$ for $\delta < \gamma$, hence from (i) we conclude that

$$i_M(a)(\gamma) = r_\gamma h_M(r_\gamma a) = r_\gamma h_{v_\gamma}(r_\gamma a_\gamma) = f(\gamma) . \quad \square$$

Figure 1 on p.167 shows the indicator function of an element.

In view of (3.1) and §2(B), it is evident that, given a module M, if an element $a \in M$ belongs to a pure standard uniserial submodule U of M, then $i_M(a)$ is constantly equal to a non-limit height up to Ann a. The converse also holds.

LEMMA 3.4. [VD] An element a of an R-module M is contained in a pure standard uniserial submodule U of M if and only if $i_M(a)$ is constant non-limit up to Ann a.

Proof. Only sufficiency is needed. Let $h_M(a) = J/R$ and $\phi : J \to M$ a homomorphism such that $\phi 1 = a$. To show that ϕJ is a pure submodule of M, suppose $rx = \phi q$ for $r \in R$, $x \in M$ and $q \in J$. If $q \in R$, then $rx = q\phi 1 = qa$, and as $h_M(qa) = q^{-1}h_M(a)$, we have

$$r^{-1}R/R \leq h_M(rx) = h_M(qa) = q^{-1}h_M(a) = q^{-1}J/R.$$

Thus $r^{-1}R \leq q^{-1}J$, $qr^{-1} \in J$ and $\phi q = r\phi(r^{-1}q) \in r\phi J$. If $q \notin R$, then $h_M(rx) = q^{-1}J/R$ likewise implies $r^{-1}R \leq q^{-1}J$, and we again obtain $\phi q \in r\phi J$. $\square$

EXERCISES

1. [VD] Let $a \in M$ where M is either torsion-free or injective. Show that $i_M(a)$ is constant up to Ann a.

2. [VD] Compare the indicators $i_M(a)$ and $i_M(ra)$, for $a \in M$ and $r \in R\backslash$Ann a.

§4. IRREGULARITIES OF INDICATORS

By (3.3), indicator functions can have arbitrarily irregular behavior, as they are subject only to the condition of being non-decreasing functions. We shall analyze in detail what kinds of irregularities may actually occur. We keep assuming that R is a valuation domain.

4. IRREGULARITIES OF INDICATORS

Given a module M and an element $a \in M$, we say that the indicator $i_M(a)$ is <u>irregular</u> at the ideal $L < R$ if

$$rh_M(ra) < sh(sa)$$

holds for all $r \in R \backslash L$, $s \in L$.

Since $i_M(a)$ is always irregular at $L = \text{Ann } a$, it suffices to consider irregularities only at ideals properly containing $\text{Ann } a$.

By definition, the absence of irregularity at L means that there exist $r_0 \in R \backslash L$ and $s_0 \in L$ such that $r_0 h(r_0 a) = s_0 h(s_0 a)$, i.e. $i(a)$ is constant in the interval $[v(r_0), v(s_0)]$. An indicator that has no irregularities at all is constant.

LEMMA 4.1. [VD] Let $a \in M$. If $h(a) < th(ta)$ for some $t \notin \text{Ann } a$, then $i(a)$ is irregular at some ideal $L \geq Rt$.

<u>Proof</u>. Let L be the ideal generated by all $s \in R$ such that $h(a) < sh(sa)$. Then $L \leq P$ and $h(a) = rh(ra)$ for all $r \in R \backslash L$. It is clear that $i(a)$ is irregular at L and $L \geq Rt$. □

In order to distinguish between various kinds of irregularities, we set

$$i_M(a)_L = \sup \{rh_M(ra) \mid r \in R \backslash L\}$$

and

$$i_M(a)^L = \inf \{rh_M(ra) \mid r \in L\}.$$

Obviously, $i_M(a)_L \leq i_M(a)^L$ holds.

The irregularities of $i_M(a)$ that can occur at an ideal $L > \text{Ann } a$ are clearly of one of the following kinds:

(i) $i_M(a)$ <u>increases on the right</u> at L, i.e. there is no $s_0 \in L$ such that $i_M(a)^L = s_0 h(s_0 a)$.

(ii) $i_M(a)$ <u>increases on the left</u> at L, i.e. there is no $r_0 \in R \backslash L$ such that $i_M(a)_L = r_0 h(r_0 a)$.

(iii) $i_M(a)$ has a <u>gap</u> at L, i.e. $i_M(a)_L < i_M(a)^L$. A gap is <u>proper</u> if it occurs at an ideal $L \neq \text{Ann } a$.

(iv) Simultaneous occurrence of two or all of (i)-(iii).

Observe that $i_M(a)_L$ can be a non-limit height only if $i_M(a)$ does not increase on the left at L, while $i_M(a)^L$ can be a limit height only if $i_M(a)$ does not increase on the right at L.

Figures might be of some help in recognizing irregularities (Figure 2 on p.167).

Here are some explicit examples of the various kinds of irregularities.

EXAMPLE 4.2. [VD] Consider $0 < \gamma < \delta$ in Γ^+ and let

$$M = (R/Rr_\gamma) \oplus (R/Rr_\delta).$$

The elements $b = 1+Rr_\gamma$ and $c = 1+Rr_\delta$ have constant indicator up to Rr_γ and Rr_δ, respectively, by (3.1). The indicator $i_M(b+c)$ has constant value equal to 0 up to Rr_γ. It has a gap at Rr_γ, and then constant value $r_\gamma r_\delta^{-1} R/R$ from Rr_γ up to Rr_δ.

EXAMPLE 4.3. [VD] We shall construct in X. §4 a torsion module A satisfying the following property: for every $0 \neq a \in A$, there exists an $r \in R$ such that $ra \neq 0$ and, for all $s,t \in R\backslash\text{Ann}\, a$ such that $v(s) > v(t) \geq v(r)$, $h(sa) > ts^{-1}h(ta)$. It follows that $i(ra)$ increases on both sides at every ideal $L > \text{Ann}\, ra$.

At ideals of some particular forms certain irregularities cannot occur.

PROPOSITION 4.4. Let $a \in M$ and $R > L > \text{Ann}\, a$. If $L = Rr$ for some $r \in R$, then $i_M(a)$ cannot increase on the right at L. If $L = Pr$, then $i_M(a)$ cannot increase on the left at L.

Proof. It is enough to notice that $i_M(a)^L = rh_M(ra)$ if $L = Rr$, while $i_M(a)_L = rh_M(ra)$ if $L = Pr$. □

EXERCISES

1. Let R be a discrete rank one valuation domain and M an R-module. Show that, for all $a \in M$, the only irregularities of $i_M(a)$ that can occur at any ideal are gaps.

2. An indicator has a gap at L if it increases on both sides at L.

4. IRREGULARITIES OF INDICATORS

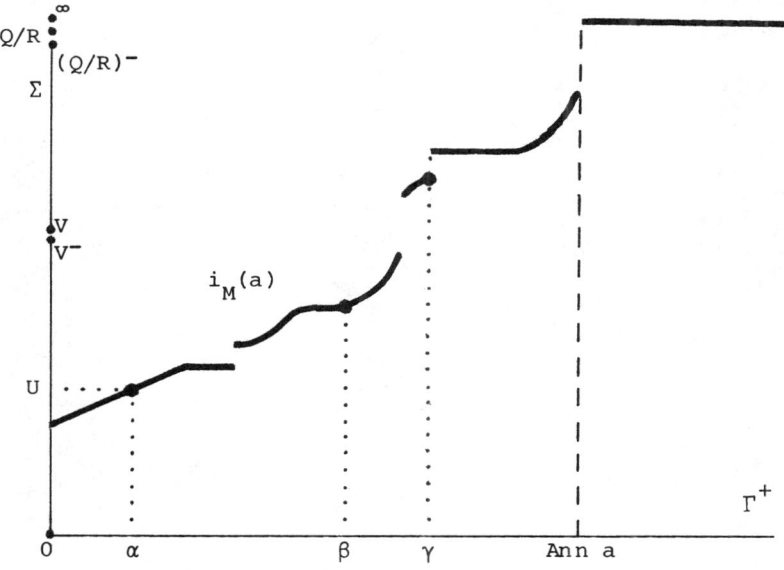

Figure 1. Indicator of an element a in a module M.

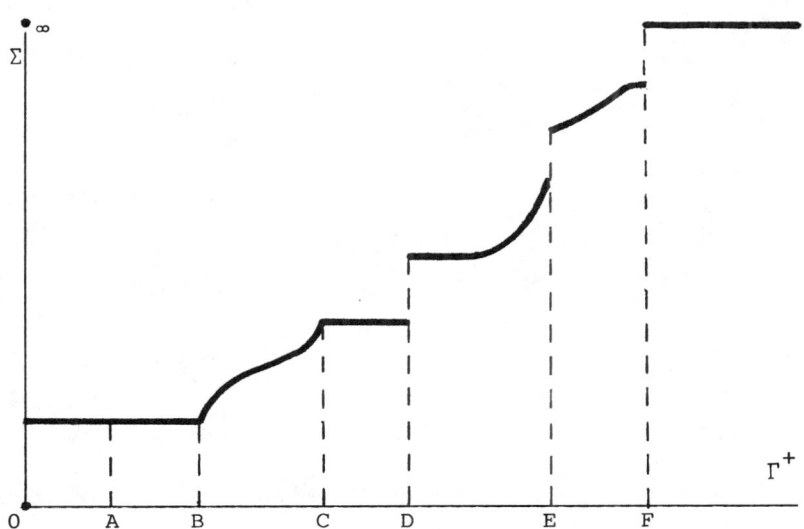

Figure 2. Irregularities. A: no irregularity; B, C: increase on one side; D: proper gap; E: all possible irregularities; F: improper gap.

§5. SMOOTHNESS

We now shift our focus to the consideration of a property of modules which allows us to split elements into pieces whose indicators have less irregularities. This property is motivated by modules in several important classes (separable, pure-injective, etc.). R will denote throughout a valuation domain.

Let M be an R-module, $a \in M$ and L an ideal of R. We say that $i_M(a)$ is <u>smooth at</u> L if

$$a \in M[L] + M^\sigma \quad \text{where} \quad \sigma = i_M(a)^L,$$

i.e. a is of the form $a = b + c$ with $Lb = 0$ and $h(c) \geq i_M(a)^L$. It is to be emphasized that σ depends both on L and a.

The indicator $i_M(a)$ is <u>smooth</u> if it is smooth at every ideal $L > \text{Ann } a$; M itself is <u>smooth</u> if $i_M(a)$ is smooth for each $a \in M$.

LEMMA 5.1. [VD] Let $a = b + c$ ($a,b,c \in M$) and $R > L > \text{Ann } a$. Assume that $b \in M[L]$ and $c \in M^\sigma$, with $\sigma = i_M(a)^L$. Then
 (i) $i(c)$ is constant up to L and has no gap at L;
 (ii) if $i(a)$ increases on the left or has a gap at L, then
   $i(a) = i(b)$ up to L.

<u>Proof.</u> (i) Assume, by way of contradiction, that there exists an $r \in R\setminus L$ such that $rh(rc) > h(c)$. For each $s \in L$, $h(c) < rh(rc) \leq sh(sc)$ holds which implies $h(c) < i(a)^L$, a contradiction. If there is a gap at L, then for each $r \in R\setminus L$, $h(c) = rh(rc) = i(a)_L < i(a)^L \leq h(c)$, again a contradiction.
 (ii) For every $r \in R\setminus L$, $rh(ra) < i(a)^L$, thus $h(ra) < h(rc)$, whence $h(rb) = h(ra)$ follows. □

Our discussion of smoothness begins with simple observations.
 a) If $i(a)$ is smooth, then $i(ra)$ is smooth for all $r \in R$.
 b) If $s \in L\setminus\text{Ann } a$, and if $i(a)$ has no irregularities up to Rs, then $i(a)$ is smooth at L.

For, $i(a)^L = h(a)$, so $a \in M^\sigma$ where $\sigma = i(a)^L$.

5. SMOOTHNESS

c) The following are classes of smooth modules: torsion-free modules, uniserial modules, h-divisible modules and h-reduced divisible modules.

In fact, in each of the listed modules the elements have constant indicators up to the annihilators.

d) The class of smooth modules is closed under formations of direct sums, direct products and direct summands.

The proof is straightforward and is left to Exercises.

Here is an example of a smooth divisible module which is neither h-divisible nor h-reduced.

EXAMPLE 5.2. Let R be a valuation domain such that $p.d.Q \geq 2$. By (VI.1.4) there exists a divisible module D such that $0 \neq hD = tD$ and $D/tD \cong Q$. If $a \in hD$, then $i(a)$ is obviously smooth. If $a \in D\setminus hD$, then $Ra \cap hD = 0$, therefore $i(a)$ is constantly equal to $(Q/R)^-$. Thus $i(a)$ is again smooth.

We analyze smoothness at principal ideals. The definition yields:

e) If $R > Rr > \text{Ann } a$, $i(a)$ is smooth at Rr if and only if the equation $rx = ra$ has a solution of height $rh(ra)$. This definitely holds if $h(ra)$ is non-limit.

f) Let $i(a)$ be non-increasing on the right at $L > \text{Ann } a$. If $\sigma = i(a)^L$ is non-limit and M^σ is ultracomplete in the annihilator filtration, then $i(a)$ is smooth at L.

Let $s_0 \in L\setminus\text{Ann } a$ such that $\sigma = s_0 h(s_0 a)$. For each $s \in L\setminus Rs_0$ the equation $sx = sa$ has a solution $b_s \in M^\sigma$. The family of cosets $\{b_s + M^\sigma[s] \mid s \in L\setminus Rs_0\}$ has the finite intersection property. By hypothesis, there is a $b \in M^\sigma$ in the intersection of the family. Clearly $a-b \in M[L]$, therefore $a = (a-b)+b \in M[L] + M^\sigma$.

If in f) we drop the hypothesis of σ being non-limit, the statement is no longer true, as is shown by the following example.

EXAMPLE 5.3. Let H_γ be the module constructed in (VII.4.4). Let $0 < \beta < \gamma$ and consider $x_\beta \in H_\gamma$. Then

$$i(x_\beta)^{Rr_\beta} = h(r_\beta x_\beta) = h(x) = (r_\gamma^{-1} P/R)^-.$$

It is easily seen, by using e), that $i(x_\beta)$ is not smooth at Rr_β. An example for a non-smooth module is as follows.

EXAMPLE 5.4. [VD] Let N be the R-module defined in (VII.4.5) as a submodule in a larger module, and $a \in N$ as defined there. $i_N(a)$ is not smooth at P, since $i_N(a)^P = r^{-1}R/R$, $N[P] = 0$ and $a \notin rN$.

We close this section with a result which we will need later on.

LEMMA 5.5. [VD] If M is a cohesive smooth module and its elements have principal annihilators, then equiheight submodules of M are smooth.

Proof. Let N be an equiheight submodule of M and $R > L > \operatorname{Ann} a$, where $a \in N$. If L is principal, then $i_N(a)$ is, because of e), smooth at L; so assume L is not principal. We can write $a = b + c$, where $b \in M[L]$ and $c \in M^\sigma$, $\sigma = i(a)^L$. As $\operatorname{Ann} b > L$, there exists an $r \in R\setminus L$ such that $rb = 0$. Then $ra = rc \in M^{r^{-1}\sigma} \cap N = N^{r^{-1}\sigma}$, therefore there exists a $d \in N^\sigma$ with $ra = rd$. Thus $a-d \in N[L]$, and N is smooth. □

EXERCISES

1. Show that an abelian p-group G is smooth if and only if $G^1 = dG$.

2. [VD] Let $a \in M$ and $L > \operatorname{Ann} a$; assume that $i(a)$ is smooth at L. Show that a has a proper decomposition $a = b + c$ ($0 \neq b \in M[L]$, $0 \neq c \in M^\sigma$), provided that one of the following conditions holds:
    a) $i(a)$ has a gap at L.
    b) $i(a)$ increases on the left at L.
    c) $h(a)$ has an immediate successor and $i(a)$ increases on the right at L.

3. Give an explicit example of an element a of an R-module M such that $i_M(a)$ is not smooth, but $i_M(ra)$ is smooth for some $r \in R\setminus\operatorname{Ann} a$.

4. Verify all statements in d) above.

5. Show that, in (5.4), rN is not ultracomplete in the annihilator filtration and conclude that ultracompleteness cannot be dropped in f).

NOTES

For valuation domains in general, heights were defined by Fuchs-Salce [1], and more fully developed in [2], where indicators and smoothness were introduced.

Heights and indicators give major information on the way elements are embedded in modules. In this respect, they play similar roles as their counterparts in abelian group theory, and we believe that their importance will increase as the theory develops.

As noted in §1, there is no difficulty in introducing transfinite heights. So far, there has not been any demand for them, as no systematic investigation has taken place for modules with elements of infinite heights. Once transfinite heights will be introduced, the notion of indicators has to be extended accordingly.

Over discrete rank one valuation domains, torsion modules share a number of useful, equivalent properties. These properties fail to hold over general valuation domains and are no longer equivalent either. Among them, smoothness and separability (to be discussed in Chapter XIII) are the most significant ones. Smoothness differs from traditional concepts, and its real meaning is not yet fully understood, except its implications for finitely generated modules (see IX, §2, Ex.3), but its scope is far wider than this particular case. §5 f suggests that it might have close connection with ultracompleteness.

Guided by Kaplansky's definition of transitivity and full transitivity of abelian p-groups, we introduce the analogous concepts for modules over valuation domains. M is <u>transitive</u> (<u>fully transitive</u>), if for every pair a, b of elements of M with $i(a) = i(b)$ (with $i(a) \leq i(b)$) there exists an automorphism (endomorphism) α of M such that $\alpha a = b$.

PROBLEM 8. Investigate the properties of transitive (fully transitive) modules over valuation domains.

Even restriction to divisible modules seems interesting.

IX. Finitely Generated and Polyserial Modules

The study of finitely generated modules over domains is still in a highly embryonic state, though the commutative rings over which these modules are direct sums of cyclics have been identified; see Brandal [2]. Here we concern ourselves exclusively with valuation rings, and the basic themes are: what can be stated of the structures of finitely generated modules? and what conditions force them to be direct sums of cyclics? Once the appropriate theorems are proved, we will have a better picture of what finitely generated modules are like.

Some of the results on finitely generated modules generalize to polyserial modules. This is an interesting class which includes the finite direct sums of uniserials.

Throughout the chapter, R will denote a valuation ring. A restriction to domains would essentially simplify our discussions, but the methods employed and the results obtained seem sufficiently rewarding to warrant the extra effort.

§1. FINITELY GENERATED MODULES

Needless to say, our point of departure is (II.3.6). This motivates the following definition.

A <u>pure-composition series</u> for an R-module M is a finite chain of pure submodules of M:

$$0 = M_0 < M_1 < \ldots < M_{n-1} < M_n = M \tag{1}$$

such that M_i/M_{i-1} is uniserial for each i. The sequence of ideals

$$A_i = \operatorname{Ann} M_i/M_{i-1} \qquad (i = 1,\ldots,n)$$

is said to be the <u>annihilator sequence</u> of (1). In this terminology, (II.3.6) states that a finitely generated module M over a valuation ring R has a pure-composition series (1) with cyclic factors M_i/M_{i-1} and with a non-decreasing annihilator sequence

$$\operatorname{Ann} M = A_1 \leq A_2 \leq \ldots \leq A_n. \tag{2}$$

Unquestionably, for almost maximal valuation rings we have a most satisfactory result:

THEOREM 1.1. (Kaplansky [1], Matlis [1], Lafon [1], Gill [1]) All finitely generated modules over a valuation ring R are direct sums of cyclic modules if and only if R is almost maximal.

<u>Proof.</u> Assume R is almost maximal. If R is a domain, then by (IV.1.2) and (IV.1.3) a finitely generated M decomposes as $M = tM \oplus F$ with F free of finite rank and tM finitely generated torsion. As cyclic torsion R-modules are pure-injective (cf. (XI.4.2)), (II.3.6) leads to the desired conclusion via induction. If R is not a domain, then it is maximal in view of (I.5.6), so all cyclic R-modules are pure-injective, and again (II.3.6) yields the desired result.

Conversely, assume R is not almost maximal. There exists a nonzero ideal I of R such that R/I is not complete in the R-topology. Hence we can find a family of elements (which can be assumed to be units in R), say $\{u_a \mid a \in R\setminus I\}$ such that

1. FINITELY GENERATED MODULES

(i) $u_a - u_b \in aR$ if $b \in aR$,

(ii) no $u \in R$ satisfies $u - u_a \in aR$ for all $a \in R\setminus I$.

Choose a $0 \neq c \in I$ and define $J = \{r \in R \mid c \notin rI\}$. Clearly, $Rc < J \leq P$. For each $r \in J\setminus Rc$, let $c = rp_r$ with $p_r \in R\setminus I$. Define M as the R-module generated by two elements x, y, subject to the relations

$$cx = 0, \quad rx = ru_{p_r} x \quad \text{for all } r \in J\setminus Pc.$$

Assume, by way of contradiction, that M is a direct sum of cyclic modules. As pure submodules of direct sums of cyclics are summands (see (5.6)) and as Rx is easily seen to be pure in M, we obtain $M = Rx \oplus R(y-sx)$ for a suitable $s \in R$. Now $\text{Ann}(y+Rx) = J$, for every $r \in J$ we have $ry = rsx$. Consequently,

$$s - u_{p_r} \in p_r R \quad \text{for all } 0 \neq r \in J.$$

This contradicts (ii), so M is not a direct sum of cyclics. □

We may add to (1.1) that the uniqueness of the decompositions into cyclic summands follows from (VII.4.2).

The minimal number of elements required to generate a finitely generated module M will be called the <u>length</u> of M and denoted by $\ell(M)$. From Nakayama's lemma it follows that a family of elements of M generates M if and only if their cosets mod PM generate M/PM as an R/P-vectorspace. Consequently,

$$\ell(M) = \dim_{R/P} M/PM.$$

Our next goal is to show that $\ell(M)$ coincides with the lengths of pure-composition series of M.

LEMMA 1.2. [VR] (Salce-Zanardo [2]) Let (1) be a pure-composition series of the finitely generated R-module M with $M_i/M_{i-1} = R(x_i + M_{i-1})$ ($1 \leq i \leq n$). Then $\ell(M) = n$.

<u>Proof</u>. The elements $x_i \in M$ ($i = 1, \ldots, n$) generate M, thus $n \leq \ell(M)$. To verify the converse inequality, it is enough to establish the independence of $x_1, \ldots, x_n$ mod PM. Assume, by way of contradiction, that

$$\sum_{i=1}^{n} r_i x_i = p \sum_{i=1}^{n} t_i x_i$$

where $r_i, t_i \in R$, $p \in P$ and not all of the r_i belong to P. No generality is lost in supposing $p = 0$ and $r_j = 1$ for the largest index j with $r_j \notin P$. Then

$$r_1 x_1 + \ldots + r_{j-1} x_{j-1} + x_j \in M_j \cap PM = PM_j.$$

Hence $r_1 x_1 + \ldots + r_{j-1} x_{j-1} + x_j = t(s_1 x_1 + \ldots + s_j x_j)$ with suitable $s_i \in R$, $t \in P$. We infer that $x_j \in M_{j-1}$, a contradiction. □

Next we establish the isomorphy of any two pure-composition series, i.e. the existence of a bijection between the two sets of cyclic factors such that corresponding factors are isomorphic. We need three preliminary lemmas.

LEMMA 1.3. [VR] Let N be a pure submodule of an R-module such that $M/N = R(x+N)$ is cyclic. If $\text{Ann } M/N \leq \text{Ann } N$, then $M = N \oplus Rx$.

Proof. Obviously, $M = N + Rx$. If $rx \in N$ for $r \in R$, then by the purity of N, $rx \in rN$. But $rx \in N$ implies $rN = 0$ in view of our hypothesis. We deduce $rx = 0$, $Rx \cap N = 0$, thus $M = N \oplus Rx$. □

An easy consequence of (1.3) is the following technical lemma.

LEMMA 1.4. [VR] (Salce-Zanardo [2]) Every pure-composition series of a finitely generated R-module is isomorphic to one with a non-decreasing annihilator sequence.

Proof. Assume that $A_i > A_{i+1}$ in the pure-composition series (1) of M. Setting $M_{i+1}/M_i = R(x_{i+1} + M_i)$, (1.3) implies

$$M_{i+1}/M_{i-1} = R(x_{i+1} + M_{i-1}) \oplus M_i/M_{i-1}.$$

Therefore, in (1) M_i can be replaced by $M'_i = M_{i-1} + Rx_{i+1}$ where $\text{Ann } M'_i/M_{i-1} = A_{i+1} < A_i = \text{Ann } M_{i+1}/M'_i$. In a finite number of similar steps, we obtain from (1) an isomorphic pure-composition series

1. FINITELY GENERATED MODULES

whose annihilator sequence is non-decreasing. □

Another noteworthy consequence of (1.3) is that if (1) is a pure-composition series for M with $M_i/M_{i-1} = R(x_i + M_{i-1})$ for all i, and if $A_1 = \ldots = A_k$ for some $k \leq n$, then $M_k = Rx_1 \oplus \ldots \oplus Rx_k$. In particular, if the annihilator sequence is constant, then M is a direct sum of cyclic modules.

We now examine the submodules rM ($r \in R$).

LEMMA 1.5. [VR] Let M be a finitely generated module with pure-composition series (1) with $M_i/M_{i-1} = R(x_i + M_{i-1})$ for all i and a non-decreasing annihilator sequence (2). If $r \in A_{k+1} \setminus A_k$ ($1 \leq k \leq n$) (where $A_{n+1} = R$), then $0 < rM_1 < \ldots < rM_k = rM$ is a pure-composition series of rM with annihilator sequence $A_i : r$ ($1 \leq i \leq k$) and

$$M_k/rM = \bigoplus_{i=1}^{k} R(x_i + rM) \cong \bigoplus_{i=1}^{k} R/rR.$$

Proof. The first claim is straightforward to verify. It is also easy to prove that, setting $\overline{M}_i = (M_i + rM)/rM$,

$$0 < \overline{M}_1 < \ldots < \overline{M}_k \qquad (3)$$

is a pure-composition series of $\overline{M}_k = M_k/rM$ such that $\overline{M}_i/\overline{M}_{i-1} = R(\overline{x}_i + \overline{M}_{i-1}) \cong R(x_i + rM)$ where $\overline{x}_i = x_i + rM$ for $1 \leq i \leq k$. The annihilator sequence of (3) is constantly equal to Rr, therefore, the second claim follows from the preceding remark. □

We can now prove what our main goal was.

THEOREM 1.6. [VR] (Salce-Zanardo [2]) Any two pure-composition series of a finitely generated module M are isomorphic. Thus the annihilators of factors in a pure-composition series of M are uniquely determined by M (up to order).

Proof. In view of (1.4), we can assume that the two pure-composition series of M have non-decreasing annihilator sequences: A_i ($1 \leq i \leq n$) and B_j ($1 \leq j \leq m$). Note that $A_1 = \text{Ann } M = B_1$ and, by (1.2), necessarily $n = m$. Assume $A_k = B_k$ for some k,

$1 \leq k < n$. If there is an $r \in A_{k+1} \setminus A_k$, then $\ell(rM) = k$ by (1.5), thus again by (1.5), $r \in B_{k+1}$. This shows $A_{k+1} \leq B_{k+1}$, and by symmetry, $A_{k+1} = B_{k+1}$. An obvious induction completes the proof. □

EXERCISES

1. [VD] Simplify the second part of the proof of (1.1) for the domain case as follows. If R is not almost maximal, there is an a in $E(Q/R)$ which is not in Q/R and there is a $b \in Q/R$ not in Ra. Show that $Ra + Rb$ is not cyclic and cannot be a proper direct sum.

2. [VD] The heights of elements in a finitely generated R-module are cyclic.

3. [VD] The indicator of an element in a finitely generated R-module assumes but a finite number of different values. [Use (XI.5.7).]

4. [VR] A pure submodule of a finitely generated module M is likewise finitely generated. The annihilators of factors in its pure-composition series form a subset of the annihilators of factors in a pure-composition series for M.

§2. THE GOLDIE DIMENSION

There is another numerical invariant of finitely generated modules whose comparison with the length sheds light on the complexity of the structure.

Recall that a module M (over any ring R) has <u>Goldie dimension</u> n if it has an independent set of uniform submodules $M_1, \ldots, M_n$ such that $M_1 \oplus \ldots \oplus M_n$ is essential in M. Equivalently, if the injective hull $E(M)$ is the direct sum of n indecomposable injective modules. $g(M)$ will stand for the Goldie dimension of M. Clearly, a cyclic module over a valuation ring has Goldie dimension 1, and g is an additive function.

2. THE GOLDIE DIMENSION

The crucial result in this section is the next lemma.

LEMMA 2.1. [VR] (Salce-Zanardo [2]) Let $0 = M_0 < M_1 < \ldots < M_n = M$ be a pure-composition series of the finitely generated module M, with non-decreasing annihilator sequence A_i ($i = 1, \ldots, n$). If M_{n-1} is not essential in M, then M has a cyclic summand $\neq 0$.

Proof. Set $M_i/M_{i-1} = R(x_i + M_{i-1})$ for $1 \leq i \leq n$, and suppose $Ry \cap M_{n-1} = 0$ for some $0 \neq y \in M$. Without loss of generality, $y \in M\backslash PM$ may be assumed (see Ex. 1); i.e. $y = \sum_{i=1}^{n} a_i x_i$ ($a_i \in R$) and there is a largest index $j \geq 1$ such that $a_j \notin P$. If $j = n$, then $M = Ry \oplus M_{n-1}$, and we are done. If $j < n$, then we claim $M = Ry \oplus N$ with $N = \sum_{i \neq j} Rx_i$. Since $Ry + N = M$ is obvious, we need only to show that $Ry \cap N = 0$. Suppose

$$0 \neq ry = \sum_{i \neq j} b_i x_i \qquad (r, b_i \in R). \qquad (1)$$

The choice of y implies $r \notin A_n = \text{Ann } M_n/M_{n-1}$. Passing mod rM in (1), we deduce from (1.5) that $b_i x_i \in rM$ for each i. Hence $b_i = rc_i$ for some $c_i \in R$, $1 \leq i \leq n$. We obtain $\sum_{i=1}^{n} (a_i - c_i) rx_i = 0$ with $c_j = 0$. As $a_j \notin P$, the elements $rx_1, \ldots, rx_n$ are dependent in rM mod PrM which contradicts the fact that $0 < rM_1 < \ldots < rM_n = rM$ is a pure-composition series of rM (cf. (1.5)). □

We can draw from this lemma a couple of corollaries.

COROLLARY 2.2. [VR] Every finitely generated R-module M contains a pure submodule B which is a direct sum of cyclic modules such that $g(B) = g(M)$.

Proof. We induct on $\ell(M) = n$. The case $n = 1$ is obvious; so let $n > 1$. By induction, M_{n-1} has a pure submodule B' that is a direct sum of $g(M_{n-1})$ non-zero cyclic submodules. If M_{n-1} is essential in M, then we are done by setting $B = B'$. If not, then by (2.1) $M = Ry \oplus N$ for some $0 \neq y \in M$ and submodule N of M. Since $\ell(N) = n-1$, N has a pure submodule B'', a direct sum of $g(N)$ non-zero cyclic submodules. Put $B = Ry \oplus B''$ to conclude the proof. □

The uniqueness of B up to isomorphism will follow from (X.3.2).

COROLLARY 2.3. [VR] For a finitely generated module M, we always have

$$g(M) \leq \ell(M).$$

Proof. Choose B as described in (2.2). Then $PB = B \cap PM$ implies that B/PB is isomorphic to a submodule of M/PM. We infer that $g(M) = g(B) = \dim B/PB \leq \dim M/PM = \ell(M)$. □

We now formulate a criterion for the decomposability into direct sums of cyclics.

THEOREM 2.4. [VR] (Salce-Zanardo [2]) A finitely generated module M is a direct sum of cyclics if and only if $g(M) = \ell(M)$.

Proof. Only sufficiency requires a proof. Assume $g(M) = \ell(M)$. By virtue of (2.2), M contains a pure submodule B which is the direct sum of $g(M) = \ell(M)$ non-zero cyclic submodules. If $B < M$, then $\ell(M) > \ell(B) = g(B) = g(M)$, which is absurd. Hence $B = M$, which completes the proof. □

For a valuation domain R, (2.2) asserts that a finitely generated R-module M has an essential α-basic submodule. From the isomorphy of α-basics we can conclude that every α-basic submodule of M is essential, and so basic in M. (See X, §3.)

EXERCISES

1. [VR] If M is a finitely generated R-module, then for every $0 \neq x \in M$ there exist $y \in M$ and $r \in R$ such that $x = ry$ and $y \notin PM$.

2. [VD] Let M be a finitely generated R-module with pure composition series and annihilator sequence like (1) and (2) in §1. If $A_k < A_{k+1} = \ldots = A_n$ and $r \in A_{k+1} \setminus A_k$, then $0 \neq rx_n = r(t_1 x_1 + \ldots + t_k x_k)$ $(t_i \in R)$. Show that if the indicator of $z = x_n - t_1 x_1 - \ldots - t_k x_k$ is smooth at A_n, then M_{n-1} is a

summand of M.

3. [VD] (Salce-Zanardo [2]) Use Ex. 2 to verify: a finitely generated R-module is a direct sum of cyclics if and only if it is smooth.

4. [VD] Let $M = Rx + Ry$ be the module in the proof of (1.1). Show that $i_M(y)$ is constant up to Ann y, but the indicator of $y - u_{p_r} x$ $(0 \neq r \in J)$ is not smooth at J.

5. [VD] The indicator of an element in a finitely generated R-module M can assume at most $\ell(M)$ different values $\neq \infty$.

§3. INDECOMPOSABLE FINITELY GENERATED MODULES

This section centers around issues related to the indecomposability of finitely generated modules over valuation domains R. The remarkable feature here is that they are dependent on the ideal structure of R and on the extent R fails to be almost maximal.

In pursuing this subject, interesting questions to keep in mind are concerned with the lengths and the Goldie dimensions of the modules: do finitely generated uniform R-modules of large lengths exist? and are there indecomposable finitely generated R-modules of large Goldie dimensions?

The answer to the first question is tied in with the <u>Fleischer rank</u> r of the injective hull $E(R/P) = E$; this r is defined to be the minimum rank of torsion-free R-modules having E as an epimorphic image (see Fleischer [1]).

PROPOSITION 3.1. [VD] There exist finitely generated uniform R-modules of length n if and only if n does not exceed the Fleischer rank of $E(R/P) = E$.

<u>Proof.</u> If r is the rank of E, then E is an epimorphic image of a direct sum of r copies of Q, say, $E = \bigoplus_{i=1}^{r} Q_i / N$ with $Q_i \cong Q$ for each i. Let n be an integer such that $1 < n \leq r$. We can choose $Q_1, \ldots, Q_n$ such that Q_i is not contained in $A_i = (Q_1 \oplus \ldots \oplus Q_{i-1} \oplus Q_{i+1} \oplus \ldots \oplus Q_n) + N$ for $i = 1, \ldots, n$. Pick

$x_i \in Q_i \setminus A_i$ and define

$$M = \langle x_i + N \mid i = 1,\ldots,n\rangle < E.$$

Obviously, M is uniform and $\ell(M) \leq n$. If we had $\ell(M) < n$, then - after permuting indices if necessary - we would get

$$x_1 = \sum_{i=2}^{n} s_i x_i + y \qquad (s_i \in R,\ y \in N),$$

in contradiction to the choice of x_1.

Conversely, if M is a uniform finitely generated R-module of length n, then M embeds in $E(Rx)$ for any $0 \neq x \in M$. The conclusion now follows from $r(M) = \ell(M)$ (see Ex. 2) and the inequalities $r(M) \leq r(E(Rx)) \leq r(E)$; cf. Ex. 1. □

We proceed to study indecomposable finitely generated R-modules M with Goldie dimension > 1, with R not almost maximal.

Let S denote a maximal immediate extension of R and $U(S)$ the multiplicative group of its units. For $u \in U(S)\setminus R$, the set

$$I(u) = \{a \in R \mid u \notin aS + R\}$$

is an ideal of R. For each $r \notin I(u)$, there is a unit u_r of R such that $u - u_r \in rS$. Clearly, $u \notin SI(u) + R$. The equality $S = PS + R$ implies $PI(u) = I(u)$, thus the non-zero submodules of $R/I(u)$ give rise to a Hausdorff topology. $R/I(u)$ is incomplete in this topology, because $u + I(u) \notin R/I(u)$ and $u + SI(u)$ is the limit of the Cauchy net $\{u_r + SI(u) \mid r \notin I(u)\}$ in $S/SI(u)$.

As u lies in the completion of $R/I(u)$, but not in $R/I(u)$ itself, the R-topology of $R/I(u)$ is not discrete. Hence $I(u)$ is not of the form Pa $(a \in R)$, and it follows that

$$R : (R:I(u)) = I(u).$$

We deduce that, for $0 \neq a \in I(u)$,

$$I(u) = \bigcap_{r \in J} Rr^{-1}a \qquad \text{with } J = Pa : I(u). \tag{1}$$

Suppose $u_1,\ldots,u_n \in U(S)\setminus R$ with $I(u_i) = I$ for all i. We say that $u_1,\ldots,u_n$ are <u>u-independent over I</u> if $r_0 + r_1 u_1 + \ldots + r_n u_n \in SI$ $(r_i \in R)$ implies that $r_0, r_1,\ldots,r_n \in P$.

3. INDECOMPOSABLE MODULES

THEOREM 3.2. [VD] (Zanardo [2]) Assume that there exist $u_1, u_2, \ldots, u_n \in U(S) \setminus R$ which are u-independent over the ideal $I = I(u_i)$ of R. Then there exists an indecomposable finitely generated module M such that $g(M) = n$ and $\ell(M) = n+1$.

Proof. Pick a non-zero $a \in I$, and select units $\{u_i^r \mid r \in J = Pa : I\}$ in R such that $u_i - u_i^r \in Sr^{-1}a$ for each i. Notice that such u_i^r obviously exist if $r \notin Ra$, while, for $r \in Ra$, their existence is guaranteed by the hypothesis $I(u_i) = I$.

An R-module M is now defined to be generated by symbols $x_0, \ldots, x_n$ subject to the relations

$$ax_1 = \ldots = ax_n = 0, \quad rx_0 = ru_1^r x_1 + \ldots + ru_n^r x_n$$

for all $r \in J$. Thus $B = Rx_1 \oplus \ldots \oplus Rx_n$ is a pure submodule of M such that $M/B \cong R/J$. Evidently, $\ell(M) = n+1$ and $g(M) \geq n$. Hence, it suffices to establish the indecomposability of M.

By way of contradiction, assume M is decomposable. From $g(M) \geq n$ and (2.4) it follows that M must have a non-zero cyclic summand, $M = Ry_0 \oplus \langle y_1, \ldots, y_n \rangle$ for suitable $y_i \in M$. Set

$$y_i = \sum_{j=0}^{n} a_{ij} x_j \quad (a_{ij} \in R; \ i = 0, \ldots, n),$$

and note that, in view of $\ell(M) = n+1$, the matrix $A = [a_{ij}]$ is invertible mod P. Thus A is invertible as a matrix over R. For each $r \in J$, there exist $b_i^r \in R$ ($i = 0, \ldots, n$) such that

$$\sum_{i=0}^{n} a_{ij} b_i^r = u_j^r \quad (j = 0, 1, \ldots, n) \tag{2}$$

where $u_0 = u_0^r = -1$ for each $r \in J$. Hence

$$r \sum_{i=0}^{n} b_i^r y_i = r \sum_{i=0}^{n} b_i^r \sum_{j=0}^{n} a_{ij} x_j = r \sum_{j=0}^{n} u_j^r x_j = 0$$

which implies $rb_0^r y_0 = 0$, $rb_0^r \in \text{Ann } y_0$. From $J > Ra$ and earlier results it is not difficult to derive that $\text{Ann } y_0 = Ra$ can be assumed (by the way, this is an immediate consequence of the isomorphy of basic submodules, see (X.3.2)). Hence $b_0^r \in Rr^{-1}a$.

Cramer's rule applied to (2) yields

$$b_0^r = \sum_{j=0}^{n} t_j u_j^r$$

for suitable $t_j \in R$. Therefore, $\sum_{j=0}^{n} t_j u_j^r \in Rr^{-1}a$ implies $\sum_{j=0}^{n} t_j u_j \in Sr^{-1}a$ for each $r \in J$. We conclude from (1) that

$$\sum_{j=0}^{n} t_j u_j \in SI.$$

u-independence implies $t_j \in P$ for each j. But the t_j are algebraic minors of the first row of A divided by $\det A$. Since $\det A$ is a unit of R, a contradiction is reached. Thus M is indecomposable. □

The next example exhibits arbitrarily large finite sets of u-independent units.

EXAMPLE 3.3. Let R be a valuation domain, I a non-zero prime ideal such that $R/I \cong \mathbb{Z}_p$. For a maximal immediate extension S of R, S/SI contains a copy of the ring J_p of p-adic integers. For any integer $n > 0$, we can find units $u_1, \ldots, u_n$ of S such that $1 + SI, u_1 + SI, \ldots, u_n + SI$ are linearly independent units of J_p over $\mathbb{Z}_p$ and $I(u_i) = I$ for all i. In this case, $a_0 + a_1 u_1 + \ldots + a_n u_n \in SI$ ($a_i \in R$) implies $a_i \in I$ for each i, so the u-independence of $u_1, \ldots, u_n$ is obvious.

EXERCISES

1. [VD] For the Fleischer ranks, $r(N) \leq r(M)$ and $r(M/N) \leq r(M)$ hold whenever N is a submodule of M.

2. [VD] For a finitely generated M, the Fleischer rank equals $\ell(M)$.

3. [VD] Let S be a maximal immediate extension of R and $u, v \in U(S) \setminus R$. Show that $I(v) > I(u)$ if and only if $u + I(v) \in R/I(v)$.

4. [VD] (a) An ideal I of R satisfies $I = \bigcap_{a \notin I} Ra$ exactly if I is not of the form Px ($x \in R$).

(b) If I is not of the form Px, then $R : (R:I) = I$.

5. [VD] (Zanardo [2]) Using the notations of (3.2), show that the module M generated by $x_0, x_1, \ldots, x_n$, subject to the defining relations

$$ax_0 = 0, \quad rx_i = ru_i^r x \quad \text{for all } r \in J \quad (i = 1, \ldots, n)$$

is uniform of length $n+1$.

§4. DECOMPOSITIONS OF FINITELY GENERATED MODULES

Since the length of finitely generated modules is an additive function, it follows at once that over a valuation ring R, a finitely generated module decomposes into a finite direct sum of indecomposable modules. However, the question as to the uniqueness of such decompositions has to be answered. We will give an answer to this question in the affirmative for valuation rings with a unique non-zero prime ideal (i.e. in the domain case, the value group Γ is archimedean). The general case remains an open problem.

We start with a result whose proof is an immediate consequence of the exchange property of cyclic modules over valuation rings.

PROPOSITION 4.1. [VR] Let

$$M = C_1 \oplus \ldots \oplus C_m \oplus A = D_1 \oplus \ldots \oplus D_n \oplus B$$

be two decompositions of a finitely generated module M where C_i and D_j are cyclic and neither A nor B has a cyclic summand. Then $m = n$, $C_i \cong D_i$ for each i (after a suitable rearrangement), and $A \cong B$. □

As a consequence of (4.1), we derive:

COROLLARY 4.2. [VR] Suppose M is a finitely generated module with $g(M) + 1 = \ell(M)$. Then any two direct decompositions of M into indecomposable summands are isomorphic.

Proof. Note that both g and ℓ are additive functions. Hence $M = C_1 \oplus \ldots \oplus C_n \oplus A$ with C_i cyclic and A indecomposable with $g(A) + 1 = \ell(A)$, and any indecomposable decomposition of M is of this kind. An appeal to (4.1) concludes the proof. □

Valuable information as to direct decompositions can frequently be obtained from the endomorphism ring.

THEOREM 4.3. [VR] (Zanardo [2]) If the annihilator of a finitely generated indecomposable R-module M is a P-primary ideal of R, then End M is a local ring.

Proof. Let ϕ be an endomorphism of $M = \sum_{i=1}^{n} Rx_i$. Then $\phi x_i = \sum_{j=1}^{n} r_{ij} x_j$ $(r_{ij} \in R)$. We infer that the determinant of the matrix $\|r_{ij} - \delta_{ij}\phi\|$ (where δ_{ij} is the Kronecker delta) represents the zero endomorphism of M. The expansion of the determinant leads to a relation in End M:

$$\phi^n + a_{n-1}\phi^{n-1} + \ldots + a_0 = 0$$

for suitable $a_i \in R$. If a_0 happens to be a unit in R, then $\phi(\phi^{n-1} + a_{n-1}\phi^{n-2} + \ldots) = -a_0$ shows that ϕ is an automorphism of M. Suppose a_0 is not a unit, and let $a_0, \ldots, a_{j-1} \in P$, $a_j \notin P$ for some $j \geq 1$. As Ann M was assumed to be P-primary, there is a power of $a_{j-1}\phi^{j-1} + \ldots + a_0$ which vanishes. This leads to a relation

$$\phi^m + \ldots + b_k \phi^k = 0 \qquad (m \geq k \geq 1) \qquad (1)$$

where b_k is a unit of R. Hence we conclude that Im $\phi^k \leq$ Im ϕ^h for each $h \geq k$, thus Im $\phi^k =$ Im $\phi^{k+1} = \ldots =$ Im ϕ^{2k}. Therefore, for every $x \in M$ there is a $y \in M$ such that $\phi^k x = \phi^{2k} y$. This implies

$$x = (x - \phi^k y) + \phi^k y \in \text{Ker } \phi^k + \text{Im } \phi^k,$$

i.e. $M = \text{Ker } \phi^k + \text{Im } \phi^k$. Going back to (1), we obtain Ker $\phi^k \geq$ Ker ϕ^h for $h \geq k$ whence Ker $\phi^k =$ Ker ϕ^{2k} follows. Hence if $x \in \text{Ker } \phi^k \cap \text{Im } \phi^k$, then $x = \phi^k z$ for some $z \in M$, whence $0 = \phi^k x = \phi^{2k} z$, $z \in \text{Ker } \phi^{2k} = \text{Ker } \phi^k$ and $\phi^k z = 0$. Consequently,

$$M = \text{Ker } \phi^k \oplus \text{Im } \phi^k .$$

M is indecomposable, thus either Ker $\phi^k = 0$ or Im $\phi^k = 0$. In the first case, ϕ^k (and so ϕ) is an automorphism of M. In the

4. DECOMPOSITIONS OF FINITELY GENERATED MODULES

second case, ϕ is nilpotent, so $1-\phi$ is an automorphism. We conclude that $\text{End } M$ is a local ring. □

From the preceding theorem we can derive:

COROLLARY 4.4. [VD] (Zanardo [2]) Let R be a valuation ring with P as unique non-zero prime ideal. Then the decompositions of finitely generated R-modules into indecomposable summands are unique up to isomorphism.

<u>Proof.</u> If R is a domain, then an indecomposable finitely generated R-module is either isomorphic to R or has a P-primary annihilator. If R is not a domain, then all its proper ideals are P-primary, so every finitely generated R-module has a P-primary annihilator. (4.3) shows that the indecomposable summands have local endomorphism rings whence the claim follows from (II.7.3). □

We exhibit an example of an indecomposable finitely generated module over a valuation domain whose endomorphism ring is not local.

We need the following preliminary result; as usual, $U(J_p)$ denotes the multiplicative group of the p-adic units.

LEMMA 4.5. For a prime number p, the polynomial $X^2 + X - p^2$ has a root in $U(J_p) \setminus \mathbb{Z}_p$.

<u>Proof.</u> $X^2 + X - p^2 \mod p$ has two different roots in the field with p element. By Hensel's Lemma, the polynomial has two different roots $a_0, a_1 \in J_p$. Obviously, $a_0 a_1 = -p^2$ and $a_0 + a_0 = -1$. It follows that one of these roots, say a_0, is a unit in J_p. Assume by way of contradiction, that $a_0 \in \mathbb{Z}_p$; then $a_0 \in \mathbb{Z}$ and we obtain $a_0(1+a_0) = p^2$. Note that $a_0(1+a_0)$ is even; so $p = 2$. But $a_0(1+a_0) = 4$ has no solution in $\mathbb{Z}$; therefore $a_0 \in J_p \setminus \mathbb{Z}_p$. □

EXAMPLE 4.6. (Zanardo [2]) Let R be a valuation domain with a non-zero prime ideal I such that $R/I \cong \mathbb{Z}_p$. Thus $P = Rp$ is a principal ideal. Let $0 \neq c \in I$ and consider the module $M = Rx + Ry$ constructed in the proof of (1.1), subject to an additional

condition. By (4.5), there exists a $u \in U(S) \setminus R$ such that $I(u) = I$ and $u^2 + u - p^2 \in IS$. For every $r \in R \setminus I$, there exists a $u_r \in R$ such that $u - u_r \in rS$. As in (1.1), for all $r \in (Pc:I) \setminus Pc$ let $s_r = u_{r-1_c}$. Then the relations on the generators of M are:

$$\text{Ann } x = Rc; \quad ry = r_{s_r} x \qquad (r \in (Pc:I) \setminus Pc).$$

We know from (1.1) that M is indecomposable. We show now that the map $\phi : M \to M$ defined by

$$\phi(y) = p^2 x \; ; \; \phi(x) = y + x$$

is an endomorphism of M. It is enough to show that

$$r\phi(y) = r_{s_r} \phi(x) \qquad \text{for all } r \in (cR:I) \setminus Pc,$$

or, equivalently, that

$$r(s_r^2 + s_r - p^2) x = 0 \qquad \text{for all } r \in (cR:I) \setminus Pc.$$

Clearly, this is equivalent to

$$s_r^2 + s_r - p^2 \in r^{-1}cR \qquad \text{for all } r \in (cR:I) \setminus Pc$$

or to

$$u_r^2 + u_r - p^2 \in rR \qquad \text{for all } r \in R \setminus I.$$

But these relations are surely true, because $u - u_r \in rS$ for all $r \in R \setminus I$ and $u^2 + u - p^2 \in IS$. The matrix representing ϕ and the matrix representing $1-\phi$ are both non-invertible; this proves that $\text{End } M$ is not local.

EXERCISES

1. Let M be a finitely generated module over an arbitrary valuation ring. Show that an epic endomorphism of M is an automorphism.

2. (Vasconcelos [1]) Let M be a finitely generated module over a valuation ring with P as unique non-zero prime ideal. Show that a monic endomorphism of M is an automorphism.

5. POLYSERIAL MODULES

§5. POLYSERIAL MODULES

A remarkable class of modules over valuation rings was discovered by Fuchs-Salce [1]. This includes the class of finitely generated as well as the class of uniserial modules.

An R-module M is said to be <u>polyserial</u> if it has a pure-composition series, i.e. if there is a finite chain

$$0 = M_0 < M_1 < \ldots < M_n = M \qquad (1)$$

of pure submodules M_i such that every quotient M_{i+1}/M_i ($i = 0,\ldots,n-1$) is uniserial. M is called <u>standard</u> if all its uniserial factors are standard uniserials. n is the <u>length</u> of M.

First, we settle the (almost) maximal case.

PROPOSITION 5.1. A (torsion) module over an (almost) maximal valuation ring (domain) is polyserial if and only if it is a finite direct sum of uniserial modules.

<u>Proof.</u> Only necessity requires a proof. Hypothesis ensures that all quotients M_{i+1}/M_i are pure-injective; cf. (XI.4.2). Consequently, $M \cong M_1 \oplus M/M_1$, and a straightforward induction completes the proof. □

Returning to the general case, we start off with a uniqueness result.

LEMMA 5.2. Let R be a valuation domain and M a standard polyserial R-module with pure-composition series (1). Then the collection of factors M_{i+1}/M_i ($i = 0,\ldots,n-1$) is uniquely determined by M.

<u>Proof.</u> Let S be a maximal immediate extension of R. As $M_{i+1}/M_i = U_i$ is a uniserial R-module, $U_i \otimes_R S$ is a uniserial S-module. As such, it is pure-injective. Recall that tensoring preserves purity, thus $M \otimes S$ has a pure-composition series

$$0 = M_0 \otimes S < M_1 \otimes S < \ldots < M_n \otimes S = M \otimes S$$

where the factors are pure-injective. Hence $M \otimes S \cong \bigoplus_{i=1}^{n} (U_i \otimes S)$ is a direct sum of uniserial S-modules. In view of (VII.4.2), the summands $U_i \otimes S$ are uniquely determined by $M \otimes S$. By (VII.1.3) we can write $U_i \otimes S \cong J_i S / I_i S$, for R-submodules $I_i < J_i$ of Q. By hypothesis, $U_i \cong J_i'/I_i'$ for R-submodules $I_i' < J_i'$ of Q whence $U_i \otimes S \cong J_i' S / I_i' S$ follows. From I, §1 we conclude $J_i/I_i \cong J_i'/I_i'$, thus the $U_i \cong J_i/I_i$ are uniquely determined by M, indeed. □

The following two lemmas will be required.

LEMMA 5.3. (Fuchs-Salce [1]) Let R be a valuation ring and M a submodule of $U_1 \oplus \ldots \oplus U_n$ where the U_i are uniserial. Then one of $M_i = M \cap U_i$ is pure in M ($i = 1, \ldots, n$).

Proof. This is obvious if either $n = 1$ or one of M_i is 0 or equal to U_i. So we assume $n \geq 2$ and $0 < M_i < U_i$ for every i, and induct on n. If none of M_i is pure in M, then there exist elements $x_i \in M$ and $r_i \in R$ such that

$$r_i x_i = a_i \in M_i, \text{ but } a_i \notin r_i M_i \quad (i = 1, \ldots, n).$$

Write $x_i = u_{i1} + \ldots + u_{in}$ ($u_{ij} \in U_j$). Thus $r_i u_{ii} = a_i$ and $r_i u_{ij} = 0$ for $j \neq i$. Suppose that $Rr_i \geq Rr_n$ for all i. Then $r_n u_{nn} \neq 0$ and $r_n u_{in} = 0$ for $i \leq n-1$ imply that $u_{in} = s_i u_{nn}$ for suitable $s_i \in R$. Note that $r_i s_i u_{nn} = r_i u_{in} = 0$, hence $r_n \mid r_i s_i$ and so $r_i s_i u_{nj} = 0$ for all $j \leq n-1$. Thus $r_i s_i x_n = 0$ for $i \leq n-1$. Consider the elements $y_i = x_i - s_i x_n$ ($i = 1, \ldots, n-1$) of $M' = M \cap (U_1 \oplus \ldots \oplus U_{n-1})$. They satisfy

$$r_i y_i = r_i x_i - r_i s_i x_n = a_i \in M_i = M' \cap U_i \quad (i \leq n-1).$$

Induction hypothesis applied to M' implies that one of $M_1, \ldots, M_{n-1}$ is pure in M', i.e. $a_j \in r_j M_j$ for some $j \leq n-1$, a contradiction. □

Observe that finitely generated R-modules satisfy the hypothesis of the next lemma.

LEMMA 5.4. (Fuchs [5]) Let R be a valuation ring, and M an R-module. Suppose $M = \sum_{i=1}^{n} U_i$ with U_i uniserial. Then one of U_i is pure in M.

Proof. Assume the contrary. Then there exist $u_{ij} \in U_j$, $r_i \in R$ and $a_i \in U_i$ such that

5. POLYSERIAL MODULES

$$r_i(u_{i1} + \ldots + u_{in}) = a_i \notin r_i U_i$$

for every i. Without loss of generality we can assume $u_{ii} = 0$ for each i. We can choose the indices such that $Rr_1 \leq Rr_i$ for all i. Write $r_1 = s_i r_i$ ($s_i \in R$).

Notice that $s_j a_j = 0$ implies $r_1 U_j = 0$ for any j. In fact, $a_j \notin r_j U_j$ implies $r_j U_j < Ra_j$, so $s_j a_j = 0$ implies $r_1 U_j = s_j r_j U_j = 0$. On the other hand, from uniseriality it is clear that if $s_j a_j \neq 0$, then $s_j a_j \notin r_1 U_j$.

Suppose that $b_i = s_i a_i \neq 0$ for $i = 1, \ldots, k$, and $= 0$ for $i = k+1, \ldots, n$. We have then a system

$$r_1(u_{i1} + \ldots + u_{ik}) = b_i \notin r_1 U_i$$

for $i = 1, \ldots, k$ (note that $r_1 u_{ij} = 0$ for $j > k$) where $u_{ii} = 0$. As $b_i \notin r_1 U_i$, but $r_1 u_{ji} \in r_1 U_i$, there are $t_{ji} \in P$ such that

$$r_1 u_{ji} = t_{ji} b_i \quad (i \neq j).$$

We obtain

$$\sum_{\substack{j=1 \\ j \neq i}}^{k} t_{ij} b_j = b_i \quad (i = 1, \ldots, k).$$

View this as a homogeneous linear system in the unknowns $b_1, \ldots, b_k$ whose determinant is

$$\begin{vmatrix} -1 & t_{12} & \cdots & t_{1k} \\ t_{21} & -1 & \cdots & t_{2k} \\ \cdot & \cdot & \cdot & \cdot \\ t_{k1} & \cdot & \cdot & -1 \end{vmatrix} \in (-1)^k + P,$$

so it is a unit in R. Consequently, all b_i have to be 0, a contradiction. □

We are now able to prove:

THEOREM 5.5. [VR] Submodules and factor modules of direct sums of finitely many uniserial modules are polyserial.

Proof. By induction on the number of uniserial modules involved. By (5.3) or (5.4), one of the uniserial modules is pure in the module in question. Then by factoring out this uniserial submodule, the induction hypothesis applies to the factor module, and the claim follows. □

Actually, we have proved a slightly stronger result than stated. In fact, our proof shows that if M is as in (5.3), then after a suitable rearrangement of the U_i, a pure-composition series of M can be selected from the submodules $M \cap (U_1 \oplus \ldots \oplus U_i)$ ($i = 0, \ldots, n$). And if M is as in (5.4), then an analogous statement can be made of the submodules $U_1 + \ldots + U_i$ of M.

The following result seems to be quite useful.

THEOREM 5.6. [VR] (Fuchs-Salce [1]) Pure submodules of direct sums of finitely many uniserial modules are summands, so they are themselves direct sums of uniserials.

Proof. If M is pure in $U_1 \oplus \ldots \oplus U_n = A$ with U_i uniserial, then by (5.5) M is polyserial. An obvious induction shows that it suffices to prove the assertion for a uniserial module M. Consider the projections $\pi_i : M \to U_i$ ($i = 1, \ldots, n$). It is clear that if one of them, say π_j, is an isomorphism, then in the above direct decomposition of A, U_j can be replaced by M. As M is subdirectly irreducible, at least one of π_i has to be monic. Let $\pi_1, \ldots, \pi_k$ be monic, and let $\pi_{k+1}, \ldots, \pi_n$ have non-zero kernels. By way of contradiction, assume that none of π_i ($i \leq k$) is onto. Select $u_i \in U_i \setminus \pi_i M$ for $i \leq k$, and pick any $0 \neq a \in M$ such that it satisfies $a \in \operatorname{Ker} \pi_i$ ($i \geq k+1$) whenever $k < n$. Write $r_i u_i = \pi_i a$ ($r_i \in R$, $i \leq k$), and suppose $Rr_i \leq Rr_k$ for $i < k$, say, $r_i = r_k s_i$ ($s_i \in R$, $i < k$). Then $x = (s_1 u_1, \ldots, s_{k-1} u_{k-1}, u_k, 0, \ldots, 0) \in A$ satisfies $r_k x = a$. But this equation has no solution in M (look at the k-th projection), in contradiction to the purity of M in A. □

It follows at once that a pure cyclic submodule of a direct sum of (possibly infinitely many) uniserial modules over a valuation ring is a summand.

EXERCISES

1. [VR] (Simmons [1]) Prove that in (5.3), $M \cap J_i$ is cyclically pure in M whenever it is pure in M. Derive Matlis' result [1] stating that finitely generated submodules of direct sums of uniserials are direct sums of cyclics.

2. [VR] Let M be polyserial with a pure-composition series of length n. Then $g(M) \leq n$.

3. [VR] Let N be a pure submodule in a polyserial module M. Then N is polyserial. Compare factors in pure-composition series for M and N.

4. [VD] An R-module of finite Fleischer rank M satisfies $g(M) \leq r(M)$.

NOTES

If one plans to start a systematic program of investigating modules over a ring R and a search for tractable classes of R-modules, then a natural point of departure is the discussion of finitely generated R-modules. In fact, we can hardly understand R-modules until we at least know something about the finitely generated ones, and in addition, their study provides some vague idea as to what can be expected from broader classes of R-modules. We have briefly touched upon finitely generated modules over valuation rings in II, §3, but postponed their more delicate study until this chapter.

It was I. Kaplansky [1] who initiated the study of finitely generated modules over valuation domains. Subsequent papers by Matlis [1], Lafon [1] and Gill [1] culminated in the result (1.1). The idea of pure-composition series goes back to Fuchs-Salce [1]. A more extensive study of finitely generated modules over valuation domains is due to Salce-Zanardo [2] and Zanardo [2].

Major points that remain to be settled are:

PROBLEM 9. [VD] Classify the indecomposable finitely generated modules.

PROBLEM 10. [VD] Are decompositions of finitely generated modules into direct sums of indecomposable summands unique up to isomorphism?

PROBLEM 11. [VD] Find the valuation domains R such that all finitely generated indecomposable R-modules have local endomorphism rings.

Zanardo's conjecture is that they are those R for which each R/I (I $\neq$ 0 a prime) is complete.

Fleischer [1] considered the more general class of finite rank modules over almost maximal valuation domains. His proof was extended to the non-archimedean case by Salce-Zanardo [1]. Polyserial modules were introduced by Fuchs-Salce [1] as bona fide generalizations of finite direct sums of uniserial modules. Our knowledge of polyserial modules is very limited.

We believe that this class deserves attention. Here are a few, more or less specific questions on polyserial modules.

PROBLEM 12. [VD] Classify the finite rank torsion modules of projective dimension 1.

PROBLEM 13. [VD] Is the class of polyserial torsion modules closed under taking pure submodules and quotients mod pure submodules?

PROBLEM 14. [VD] Are torsion images of finite rank torsion-free modules necessarily polyserial?

PROBLEM 15. [VD] Prove the uniqueness of factors in pure-composition series of polyserial modules over valuation rings that are not images of valuation domains.

PROBLEM 16. [VD] Is a standard polyserial module always cohesive?

PROBLEM 17. [VD] Is a smooth standard polyserial module a direct sum of uniserial modules?

X. Invariants and Basic Submodules

Guided by the idea that the structure of algebraic objects should be exhibited by invariants, in this chapter we embark on the study of numerical invariants for modules over valuation domains.

The definition of α-invariant, due to the authors [2], is motivated by the utmost useful Ulm-Kaplansky invariants of abelian p-groups as well as the Baer invariants associated with the types of torsion-free abelian groups. They embody a great deal of information about the modules themselves and serve as complete and independent sets of invariants e.g. for direct sums of standard uniserials and for their RD-injective hulls. These α-invariants are intimately related to the α-basic submodules which will also be studied here.

A need for additional invariants arises from the pathology of the existence of large classes of modules with trivial α-invariants. Our future goal is to get an insight into the structure of these modules via suitably defined β-basic submodules. The ultimate goal at this stage is to classify the RD-injectives with trivial α-invariants by means of suitable β-invariants whose proper definition is still unknown to us. Consequently, we have to be satisfied now with a weaker result (see (XI.4.9)).

§1. α-INVARIANTS

In this section, we introduce what we call the α-invariants, for modules over valuation domains. The rings will be throughout valuation domains. M will denote an R-module and Σ the set of heights introduced in VIII.§1.

For every pair (σ, I) with $\sigma \in \Sigma$ and I a proper ideal of R, we consider the factor module

$$\alpha_M(\sigma, I) = \frac{M^\sigma[I]}{M^\sigma[I^+] + M^{\sigma^+}[I]} .$$

The correspondence $M \longmapsto \alpha_M(\sigma, I)$ is functorial in the sense that every homomorphism $\phi : M \to M'$ induces a homomorphism

$$\phi_{(\sigma, I)} : \alpha_M(\sigma, I) \to \alpha_{M'}(\sigma, I),$$

and this correspondence satisfies the usual properties of functors.

It is easy to verify the isomorphisms

$$\alpha_M(\sigma, I) \cong \frac{M^\sigma[I] + M^{\sigma^+}}{M^\sigma[I^+] + M^{\sigma^+}} \cong \frac{M^\sigma[I] + M[I^+]}{M^{\sigma^+}[I] + M[I^+]} . \tag{1}$$

In fact, the first quotient is isomorphic to

$$M^\sigma[I]/(M^\sigma[I] \cap (M^\sigma[I^+] + M^{\sigma^+}))$$

where the denominator is equal to

$$M^\sigma[I^+] + (M^\sigma[I] \cap M^{\sigma^+}) = M^\sigma[I^+] + M^{\sigma^+}[I].$$

The second isomorphism follows similarly.

The following lemma provides us with a most relevant information on $\alpha_M(\sigma, I)$.

LEMMA 1.1. [VD] (Fuchs-Salce [2]) An element $a \in M$ represents a non-zero element of $\alpha_M(\sigma, I)$ if and only if it satisfies
    (i)   Ann $a = I$,
    (ii) $h(a) = \sigma$,
    (iii) $h(ra) = r^{-1}h(a)$ for all $r \in R \setminus I$.

1. α-INVARIANTS

Proof. Suppose $a \in M$ represents a non-zero element of $\alpha_M(\sigma,I)$. Then obviously Ann $a \geq I$ and $h(a) \geq \sigma$. If either of these is a strict inequality, then $a \in M^\sigma(I^+) + M^{\sigma+}[I]$, so (i) and (ii) must hold. If (iii) fails, i.e., if there is an $r \in R \setminus I$ such that $h(ra) > r^{-1}(a)$, then, by (VIII.1.5), $ra = rb$ for some $b \in M^{\sigma+}$. Hence $a-b \in M^\sigma[r] \leq M^\sigma[I^+]$ where b is annihilated by I.

Conversely, let $a \in M$ satisfy (i)-(iii). Manifestly, $a \in M^\sigma[I]$ follows from (i) and (ii). If $a = b+c$ with $b \in M^\sigma[I^+]$ and $c \in M^{\sigma+}[I]$, then there is an $r \in R \setminus I$ such that $rb = 0$. Hence $ra = rc$, which shows that, for this r,

$$h(ra) = h(rc) \geq r^{-1}h(c) > r^{-1}h(a).$$

This contradicts (iii), so $a \notin M^\sigma[I^+] + M^{\sigma+}[I]$. □

If we agree that $Q^\# = 0 = 0^\#$, then we have the following result.

COROLLARY 1.2. [VD] Let $\sigma = U$ or U^- with $U = J/R$, and I a proper ideal of R. If π is the prime ideal $J^\# \cup I^\#$ of R, then $\alpha_M(\sigma,I)$ is a vector space over the field $R_\pi/\pi R_\pi$.

Proof. The preceding lemma shows that multiplication by $r \in R \setminus \pi$ does not annihilate any non-zero element of $\alpha_M(\sigma,I)$, so $\alpha_M(\sigma,I)$ is a torsion-free R/π-module. For each $r \in R \setminus J^\#$, $rx = a$ is solvable in M, thus $\alpha_M(\sigma,I)$ is a vector space over $R_\pi/\pi R_\pi$. □

Another relevant consequence of (1.1) is the following.

COROLLARY 1.3. [VD] Let U be a uniserial module, $\sigma = J/R$ and I a proper ideal of R. Then

$$\dim \alpha_U(\sigma,I) = \begin{cases} 1 & \text{if } U \cong J/I, \\ 0 & \text{otherwise}. \end{cases}$$

Proof. Clearly, $\dim \alpha_U(\sigma,I) \leq 1$. If $U = J/I$, then $a = 1+I$ satisfies conditions (i)-(iii) of (1.1), so $\dim \alpha_U(\sigma,I) \geq 1$. Conversely, if $\dim \alpha_U(\sigma,I) = 1$, then, by (1.1), there exists an $a \in U$ with (i)-(iii). There is a homomorphism $\phi : J \to U$ such that $\phi(1) = a$. Here ϕ is epic and Ker $\phi = I$, so $U \cong J/I$. □

A uniserial module U does not determine uniquely the pair (σ,I) for which $\dim \alpha_U(\sigma,I) = 1$. In order to make the invariants independent of the selection of $a \in U$ (for which $h(a) = \sigma$ and $\text{Ann } a = I$), we introduce an equivalence relation among the pairs (σ,I). We require a preliminary result.

LEMMA 1.4. [VD] For every $r \in R\setminus I$, the map $a \mapsto ra$ induces a monomorphism
$$\mu_r : \alpha_M(\sigma,I) \to \alpha_M(r^{-1}\sigma, r^{-1}I).$$
If σ is a non-limit height, or if M is smooth, then μ_r is an isomorphism.

Proof. If $a \in M$ represents an element $\neq 0$ of $\alpha_M(\sigma,I)$, then (1.1) shows that ra represents an element $\neq 0$ of $\alpha_M(r^{-1}\sigma, r^{-1}I)$, as ra also satisfies (iii). The second part is a direct consequence of (VIII.5.e)). □

Let now σ, σ' be heights and I, I' proper ideals of R. Call the pairs (σ,I) and (σ',I') equivalent if, for some $r \in R\setminus I$, $\sigma' = r^{-1}\sigma$ and $I' = r^{-1}I$, or vice versa. It is readily seen that this is indeed an equivalence relation. The equivalence class of (σ,I) will be denoted by $[\sigma,I]$.

For each equivalence class $[\sigma,I]$, we define the α-invariant of M as
$$\alpha_M[\sigma,I] = \begin{cases} \alpha_M(\sigma,I) & \text{if } \sigma \text{ is a non-limit height,} \\ \varinjlim_{r \in R\setminus I} \alpha_M(r^{-1}\sigma, r^{-1}I) & \text{if } \sigma \text{ is a limit height.} \end{cases}$$
Obviously, the connecting maps in the direct limit are the maps μ_r. From (1.4) it is evident that this definition is independent (up to isomorphism) of the choice of the representative of the equivalence class $[\sigma,I]$.

EXAMPLE 1.5. Let G be a $\mathbb{Z}_p$-module, σ a height, and $I < \mathbb{Z}_p$. Then $\alpha_G(\sigma,I)$ is a vector space over a field with p elements, except when $I = 0$ and $\sigma = \mathbb{Q}/\mathbb{Z}_p$. In the latter case
$$G^\sigma[0] = dG, \quad G^\sigma[0^+] = t(dG), \quad G^{\sigma+} = 0;$$

2. α-INVARIANTS OF SUBMODULES

therefore

$$\alpha_G(\mathbb{Q}/\mathbb{Z}_p, 0) = dG/t(dG)$$

which is a vector space over $\mathbb{Q}$.

An obvious consequence of (1.3) is the following.

LEMMA 1.6. [VD] There is a bijection between the isomorphism classes of standard uniserial modules U and the equivalence classes $[\sigma, I]$ with $\sigma \in \Sigma$ a non-limit height and I a proper ideal of R. Here U corresponds to $[\sigma, I]$ if and only if $\dim \alpha_U[\sigma, I] = 1$. □

The α-invariants are obviously additive. Thus from (1.6) we obtain at once:

THEOREM 1.7. [VD] (Fuchs-Salce [2]) In the class of direct sums M of standard uniserial modules, the invariants $\alpha_M[\sigma, I]$, with σ a non-limit height and I a proper ideal, form a complete and independent system of invariants for M. □

EXERCISES

1. Let $G = \bigoplus_\kappa \mathbb{Z}(p^\infty)$ be a divisible abelian p-group, κ a cardinal. Show that $\dim \alpha_G(\mathbb{Q}/\mathbb{Z}_p, p\mathbb{Z}_p) = \kappa$.
2. [VD] Find the α-invariants of the module ∂ in VI.§3.
3. [VD] For a cohesive module M, $\alpha_M[\sigma, I] = 0$ for all limit heights σ.
4. [VD] For non-standard uniserials U, (1.6) fails.

§2. α-INVARIANTS OF EQUIHEIGHT SUBMODULES

We examine now the relation between the α-invariants of a module and those of a pure or equiheight submodule. More complete information can be derived when the submodule is smooth. The underlying ring is assumed to be a valuation domain.

We will use the concept of pure-essential submodules which will be defined in XI.§2.

LEMMA 2.1. [VD] If N is an equiheight submodule of M, then there is a canonical embedding of $\alpha_N(\sigma, I)$ in $\alpha_M(\sigma, I)$, for all heights σ and proper ideals I of R.

Proof. From (1.1) we get
$$N \cap (M^\sigma[I^+] + M^{\sigma+}[I]) = N^\sigma[I^+] + N^{\sigma+}[I].$$
In fact, whichever of (i)-(iii) in (1.1) fails for $a \in N$ in M, fails in N too. Therefore, we have
$$\alpha_N(\sigma,I) \cong \frac{N^\sigma[I] + M^\sigma[I^+] + M^{\sigma+}[I]}{M^\sigma[I^+] + M^{\sigma+}[I]} \leq \alpha_M(\sigma,I). \quad \square$$

The next result gives a sufficient condition in order that a submodule N of M have the same α-invariants as M.

THEOREM 2.2. [VD] (Fuchs-Salce [2]) Let N be an equiheight and smooth submodule of M, $\sigma \in \Sigma$ and I a proper ideal of R. Then $\alpha_N[\sigma,I] = \alpha_M[\sigma,I]$ whenever one of the following conditions is satisfied:

(A) N is pure-essential in M;

(B) σ is non-limit and, for every standard uniserial submodule U of M with $\dim \alpha_U[\sigma,I] = 1$, either $N \cap U \neq 0$ or $N \oplus U$ is not pure in M.

Proof. From the proof of (2.1) it is evident that it suffices to establish the inclusion
$$M^\sigma[I] \leq N^\sigma[I] + M^\sigma[I^+] + M^{\sigma+}[I]. \qquad (1)$$
In view of (1.1) it is enough to show that every $a \in M$ of height σ and with annihilator I, satisfying $h(ta) = t^{-1}h(a)$ for all $t \in R \setminus I$, belongs to the right member of (1). Notice that for such an $a \in M$, Ra can be embedded in a pure standard uniserial submodule U of M with $\dim \alpha_U[\sigma,I] = 1$ whenever σ is non-limit.

If $Ra \cap N \neq 0$, then for some $r \notin I$ and $b \in N$, $ra = rb$ holds. By (VIII.5.e), $b \in N^\sigma$ can be assumed. Clearly $\operatorname{Ann} b = \operatorname{Ann} a$, thus $b \in N^\sigma[I]$. Furthermore, $c = a - b \in M^\sigma$ is annihilated by $r \notin I$, so $c \in N^\sigma[I^+]$, and we obtain $a \in N^\sigma[I] + M^\sigma[I^+]$.

Let $Ra \cap N = 0$. If condition (A) holds, then $(N \oplus Ra)/Ra$ is not pure in M/Ra. If condition (B) holds, then $N \oplus U$ is not pure in M. In either case, there exist elements $r, s \in R$, $c \in M$ and $b \in N \backslash rN$ such that

$$0 \neq rc = b + sa \qquad (2)$$

where $r \in Ps$ and $sa \notin rM$. If $\text{Ann } x < \text{Ann } a$, then for $t \in \text{Ann } a \backslash \text{Ann } x$, $trs^{-1}c = tx$ shows that $h(tx) \geq t^{-1}r^{-1}sR/R$, thus $i_N(x)^I \geq r^{-1}sR/R$. Because of the smoothness of N, we can write $x = y + z$ with $y \in N[I]$ and $z \in rs^{-1}N$. Set $z = rs^{-1}u$ ($u \in N$), and replace c by $d = c - u$, to obtain

$$0 \neq rd = sy + sa \quad (sy \notin rN, sa \notin rM).$$

Necessarily, $\text{Ann } rd \geq s^{-1}I$. We conclude that in (2) we can assume $b = sx$ with $\text{Ann } x \geq I$ and $\text{Ann } rc \geq s^{-1}I$.

Finally, notice that $h(sa) < h(rc)$, for otherwise $sa \in rM$. Without loss of generality, $h(rs^{-1}c) > \sigma$, and $h(sx) = h(sa) = s^{-1}h(a) = s^{-1}\sigma$. Therefore, $x \in N$ may be assumed to belong to N^σ, and thus to $N^\sigma[I]$. We now see that $rs^{-1}c - x - a$ is annihilated by s, while $rs^{-1}c \in M^{\sigma+}[I]$, $x \in N^\sigma[I]$. Consequently, $a \in N^\sigma[I] + M^\sigma[I^+] + M^{\sigma+}[I]$. □

EXERCISES

1. [VD] Let M and B have the same meaning as in (IX.2.2). Prove that M and B have the same α-invariants.

2. [VD] Show that (2.2) may not be true if the condition of smoothness is dropped, by comparing the α-invariants of divisible R-modules and their injective hulls.

§3. α-BASIC SUBMODULES

Let R be a discrete rank one valuation domain, and M an R-module. A basic submodule B of M is traditionally defined by the properties:

(i) B is a direct sum of cyclic R-modules,
(ii) B is pure in M,
(iii) M/B is divisible.

It is well known that every R-module contains a basic submodule and all basic submodules are isomorphic.

Unfortunately, such basic submodules do not exist over valuation domains in general. To justify this claim, it is enough to choose a valuation domain R and a bounded R-module $M \neq 0$ which fails to be a direct sum of cyclic modules.

Following Fuchs-Salce [2], we shall introduce a generalization of basic submodules, which applies to modules over arbitrary valuation domains (it also works for divisible modules). From now on, in this section R will denote an arbitrary valuation domain.

We begin with a definition: a family of submodules $\{N_i\}_{i \in I}$ of an R-module M is called <u>pure-independent</u> if it satisfies:

(a) the N_i ($i \in I$) are independent (i.e. they generate their direct sum in M); and

(b) $\oplus_{i \in I} N_i$ is pure in M.

Notice that an element $x = \sum_1^k x_{i_j} \in \oplus_{i_j \in I} N_{i_j}$ ($x_{i_j} \in N_{i_j}$) is divisible by $r \in R$ exactly if each x_{i_j} is divisible by r in N_{i_j}. Pure-independence is a property of finite character, thus there exist maximal pure-independent families $\{U_i\}_{i \in I}$ of standard uniserial submodules U_i of M. An <u>α-basic</u> submodule B of M is a submodule generated by such a maximal independent family. Therefore,

1) $B = \oplus_{i \in I} U_i$, where each U_i is a standard uniserial module,
2) B is pure in M, and
3) if V is a standard uniserial submodule of M, then either $B \cap V \neq 0$, or $B \oplus V$ is not pure in M.

Note that B is cohesive and thus equiheight in M. Obviously, α-basic submodules B always exist and are determined, up to isomorphism, by their α-invariants $\alpha_B[\sigma, I]$.

EXAMPLE 3.1. Let R be a discrete rank one valuation domain. It is well known that a pure-independent family of cyclic submodules $\{Rb_i\}_{i \in I}$ of an R-module M is maximal if and only if $B = \oplus_{i \in I} Rb_i$ is a basic submodule of M. Therefore the only difference between basic submodules defined by (i)-(iii) and α-basic submodules defined by 1)-3) is that in the latter case, we also admit

3. α-BASIC SUBMODULES

non-cyclic uniserial summands in B. Thus, for M h-reduced, the basic R-submodules coincide with the α-basic submodules just defined above.

Here is the main result on α-basic submodules.

THEOREM 3.2. [VD] (Fuchs-Salce [2]) Every α-basic submodule B of a module M satisfies

$$\alpha_B[\sigma, I] = \alpha_M[\sigma, I]$$

for each equivalence class $[\sigma, I]$ of non-limit heights σ and proper ideals I of R. Consequently, all α-basic submodules of M are isomorphic.

Proof. We already noted that B was equiheight in M. Moreover, it is evidently smooth, and condition (B) of (2.2) is satisfied for B. A simple appeal to (1.7) establishes the second statement. □

It is of importance to know whether or not an α-basic submodule B of M is pure-essential in M. If this is the case, B is said to be basic in M. Notice that the quotient M/B need not be divisible. In fact, R is a basic R-submodule of any maximal immediate extension S of R. However, S/R is never divisible unless R is almost maximal (see Exercise 3).

There are several interesting classes of modules for which every α-basic submodule is basic. To study this situation, we require a lemma.

LEMMA 3.3. [VD] Every α-basic submodule of the module M is basic whenever the following condition is satisfied: for every $a \neq 0$ in M, there is an $r \in R$ such that $ra \neq 0$ and the indicator of ra is, up to Ann ra, constantly equal to a non-limit height.

Proof. We show that an α-basic submodule $B = \oplus_i U_i$ of M is pure-essential in M. Assume, by way of contradiction, that M contains a submodule $C \neq 0$ such that $B \cap C = 0$ and $(B \oplus C)/C$ is pure in M/C. As every non-zero submodule of C satisfies the same two conditions, C can be assumed cyclic, C = Rc. Moreover, by hypothesis, we may assume that the indicator i(c) is constantly equal to a non-limit height up to Ann c. By (VIII.3.4), C is contained in a pure standard uniserial submodule U of M. As $B \cap U = 0$, $B \oplus U$ is no longer pure in M. Thus there exist an $a \in M$ and an $r \in R$ such that

$$ra = u + u_1 + \ldots + u_k \quad (0 \neq u \in U; u_i \in U_i) \quad (1)$$

where r does not divide any of $u, u_1, \ldots, u_k$. If necessary, we multiply this equation by a ring element so as to have $0 \neq u \in C$. This multiplication does not change divisibility relations in (1) (cf. (II.2.1)), so $u \in C$ may be assumed. In view of the purity of $(B \oplus C)/C$ in M/C,

$$ra = v + rv_1 + \ldots + rv_k + \ldots + rv_\ell$$

where $v \in U$, $v_i \in U_i$. Equating the two expressions for ra, we obtain $u_1 = rv_1, \ldots, u_k = rv_k$, a contradiction. □

The preceding result is used to show

THEOREM 3.4. [VD] Every α-basic submodule of a module M is basic if M has at least one of the following properties:
 a) torsion-free,
 b) h-divisible,
 c) finitely generated.

Proof. Every element of a torsion-free or an h-divisible module has an indicator (up to its annihilator) constantly a non-limit height; therefore the hypothesis of (3.3) is trivially satisfied. The claim for finitely generated modules is an immediate consequence of (IX.2.2). □

The following is an example of an α-basic submodule which is not basic.

EXAMPLE 3.5. Let D be a divisible module such that $0 \neq hD = tD$ and $D/hD \cong Q$ (see (VI.1.4)). Any α-basic submodule B of D is contained in hD, because hD contains all the elements of non-limit heights. B is h-divisible. It is not essential in D, so

4. MODULES WITH TRIVIAL α-INVARIANTS   205

it is not basic in D.

EXERCISES

1. Let G be a reduced abelian p-group. Show that a basic subgroup B, defined as in (i)-(iii), is pure-essential if and only if $G^1 = p^\omega G = 0$.

2. Consider the module M of (IX.1.1) which can not be smooth (IX.§2.Ex.4). Show that the invariants of M are different from those of its pure-injective hull $\hat{M}$. (Hint: see the structure of the pure-injective hull of finitely generated modules to be given in (XI.5.9). Notice that M is equiheight and pure-essential in $\hat{M}$; thus compare with (2.3).)

3. Let S be a maximal immediate extension of a not almost maximal valuation domain R. Show that
   (a) R is a basic submodule of S (use XI.5.9);
   (b) S/R is not divisible.

4. [VD] In a divisible module D, an α-basic submodule is basic if and only if D is h-divisible.

5. [VD] Find the α-basic submodules of the module ∂ defined in VI.§3.

§4. MODULES WITH TRIVIAL α-INVARIANTS

So far we have not seen any examples of cohesive modules $M \neq 0$ over valuation domains R such that $\alpha_M[\sigma,I] = 0$ for all non-limit heights σ and proper ideals I. (For the module ∂ of VI.§3, all these α-invariants are trivial whenever $p.d.Q \geq 2$.) The only aim of this section is to exhibit such examples of cohesive modules. Later on (see XIII.§6), we will investigate these modules and characterize the valuation domains which fail to admit this kind of modules.

It is an easy consequence of a well-known result that every non-zero module over a discrete rank one valuation domain has a non-zero α-invariant. Henceforth, in order to construct examples, let us start with a valuation domain R such that its maximal ideal P is not principal. We will concentrate on torsion R-modules, because torsion-free modules $\neq 0$ necessarily have, in view of (3.4), non-zero α-invariants. The following example is due to Fuchs-Salce [1].

Fix an element $\theta > 0$ in $\Gamma = \Gamma(R)$, and let
$$X = \{\gamma \in \Gamma \mid 0 \leq \gamma \leq \theta\}.$$
For each $\gamma \in X$, fix an $r_\gamma \in R$ such that $v(r_\gamma) = \gamma$, and let $C_\gamma = R/Rr_{2\gamma}$ be the cyclic R-module generated by $c_\gamma = 1 + Rr_{2\gamma}$. Clearly, Ann $c_\gamma = Rr_{2\gamma}$. We form the direct product
$$T = \prod_{\gamma \in X} C_\gamma$$
which is a torsion module annihilated by $r_{2\theta}$.

For all pairs $\alpha < \beta$ in X, let $g_{\alpha\beta} \in T$ be the vector whose coordinate in C_γ is either $r_\gamma r_\alpha^{-1} c_\gamma$ or 0, according as γ satisfies $\alpha < \gamma \leq \beta$ or not. It is straightforward to verify that the following rules hold:

$$g_{\alpha\beta} = g_{\alpha\delta} + r_\delta r_\alpha^{-1} g_{\delta\beta} \quad \text{for} \quad \alpha < \delta < \beta, \tag{1}$$

$$r_\gamma g_{\alpha\beta} = u_{\alpha\gamma} r_{2\delta} g_{\delta\beta} \quad \text{if} \quad \gamma > 2\alpha \quad \text{and} \quad \delta = \gamma - \alpha, \tag{2}$$

where $u_{\alpha\gamma}$ is an appropriate unit in R. In particular, $r_{\alpha+\beta} g_{\alpha\beta} = 0$, but clearly $r_\delta g_{\alpha\beta} \neq 0$ if $\delta < \alpha + \beta$.

Define A as the submodule of T generated by all the elements $g_{\alpha\beta}$, for $0 \leq \alpha < \beta \leq \theta$. By virtue of (1), we can write every $a \in A$ in the form

$$a = s_1 g_{\alpha_0 \alpha_1} + \ldots + s_n g_{\alpha_{n-1} \alpha_n} \quad (s_i \in R) \tag{3}$$

where $0 = \alpha_0 < \alpha_1 < \ldots < \alpha_{n-1} < \alpha_n = \theta$ is a partition of X with $\alpha_i \in X$. Rule (2) shows that we can assume that, for every i, either $s_i = 0$ or $v(s_i) \leq 2\alpha_{i-1}$. Under these assumptions, it is clear that $a \in A$ belongs to rA if and only if $v(r) \leq v(s_i)$ for all $s_i \neq 0$. It follows at once that $a \in A$ belongs to rA ($r \in R$)

4. MODULES WITH TRIVIAL α-INVARIANTS

if and only if it belongs to rT. Consequently, A is a pure submodule in T.

In order to show that $\alpha_A[\sigma,I] = 0$ for all non-limit heights σ and proper ideals I, it is enough to prove, by (VIII.3.4), that every non-zero element $a \in A$ has cyclic height and $i_A(a)$ is never constant.

Write a as in (3), where $s_i \neq 0$ implies $v(s_i) \leq 2\alpha_{i-1}$. Let j be the largest index such that $v(s_j) = \inf\{v(s_1),\ldots,v(s_n)\}$. Then $a \in s_j A \setminus s_j PA$; therefore $h_A(a) = s_j^{-1} R/R$. By hypothesis, P is not principal, so we can choose and fix a $p \in P$ such that

$$v(p) < \inf\{\alpha_j - \alpha_{j-1};\ v(s_i) - v(s_j) \mid i > j\}.$$

Let now $r = s_j p$ and $s = r^2_{\alpha_{j-1}} s_j^{-1} p$; s belongs to R, as $v(s_j) \leq 2\alpha_{j-1}$ implies $v(s) > 0$. Consider the terms of sa.

For $i < j$, $ss_i g_{\alpha_{i-1} \alpha_i} = 0$ because $v(ss_i) > v(r^2_{\alpha_{j-1}}) = 2\alpha_{j-1} \geq v(r_{2\alpha_{j-1}})$. Therefore, for some unit u of R,

$$ss_j g_{\alpha_{j-1} \alpha_j} = r^2_{\alpha_{j-1}} p g_{\alpha_{j-1} \alpha_j} = u r_{2\delta} g_{\delta \alpha_j}$$

with $\delta = \alpha_{j-1} + v(p) < \alpha_j$. Thus $sa \neq 0$ and $ss_j g_{\alpha_{j-1} \alpha_j}$ belongs to srA; note that $v(sr) = 2\alpha_{j-1} + 2v(p)$.

For $i > j$, $ss_i g_{\alpha_{i-1} \alpha_i} = r^2_{\alpha_{j-1}} s_j^{-1} s_i p g_{\alpha_{i-1} \alpha_i}$ again belongs to srA, because

$$v(r^2_{\alpha_{j-1}} s_j^{-1} s_i p) = 2\alpha_{j-1} - v(s_j) + v(s_i) + v(p) \geq$$
$$\geq 2\alpha_{j-1} + 2v(p) = v(sr).$$

It follows that $sa \in srA$, therefore $h_A(sa) > s^{-1} s_j^{-1} R/R$. Thus the indicator $i_A(a)$ strictly increases between 0 and $v(s)$, as we wished to prove.

EXERCISES

1. [VD] Consider the module A constructed in this section and assume that $\Gamma(R)$ is a dense subgroup of the reals $\mathbb{R}$. Show that A is smooth if and only if $\Gamma(R) = \mathbb{R}$.

2. [VD] Prove that the module A constructed in this section is not pure-injective. (Hint: compare the socles of T and T/A.)

3. If $\Gamma \cong \mathbb{R}$, then the pure-injective hull $\hat{A}$ of our A (see XI) has zero α-invariants for all non-limit heights. (Hint: use (2.2) and Ex. 1.)

NOTES

The α-invariants and the α-basic submodules discussed in this chapter are indispensable tools in the study of modules over valuation domains. Their roles are reminiscent of those played by the Ulm-Kaplansky invariants and the basic subgroups in abelian group theory. Because of the complexity in the present setting, they do not reveal as satisfactory information as their counterparts in abelian groups. It is clear from §4 that additional information is required to handle the general case.

The present authors have made a good progress in the search of suitable new invariants. The leading principle in this search is that the new invariants have to provide satisfactory (numerical) invariants for superdecomposable pure-injectives over valuation domains (see (XI.4.9)). At the time of the publication of this volume, no satisfactory results are available, and since the fragmentary results might undergo drastic changes before publication, we chose not to include them here.

The material in Chapter X is based on the authors' paper [2]. The α-invariants were introduced earlier in a special case by Salce [1].

PROBLEM 18. [VD] Are all α-basic submodules basic whenever one of them is?

PROBLEM 19. [VD] Is an R-module with zero α-invariants necessarily superdecomposable (i.e. all summands ≠ 0 are decomposable)?

XI. RD-Injectivity and Pure-Injectivity

The pure-injective modules constitute an extremely attractive class. They share a number of remarkable properties, both algebraic, homological and topological. This class will be discussed here over arbitrary domains along with its twin sister: the class of RD-injectives. A more detailed exposition will be given, since these classes are usually ignored in texts on modules.

Specialization of the domains to Prüfer domains leads to a characterization of pure-injective modules as modules ultracomplete in a certain filtration. Further specialization to valuation domains makes it possible to establish additional properties relevant in our endeavor to classify the pure-injectives in terms of suitable invariants. Our classification is however incomplete inasmuch as the pure-injective modules with trivial α-invariants require further study.

We include a brief discussion of pure-injective hulls of polyserial modules over valuation domains. This yields more explicit examples of the formation of pure-injective hulls.

§1. RD-INJECTIVE MODULES

In this section, R denotes a commutative domain.

An R-module M is <u>RD-injective</u> if it has the injective property relative to all RD-exact sequences, i.e. if each diagram

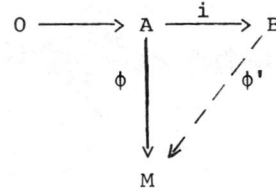

where iA is an RD-submodule of B can be completed by a map $\phi' : B \to M$ such that $\phi' i = \phi$.

The class of RD-injective modules is obviously closed under taking summands and direct products, and contains all injective modules. More examples of RD-injective modules are given by

LEMMA 1.1. [D] (Warfield [4]) Let E be an injective R-module, and $r \in R$. Then $E[r]$ is RD-injective.

<u>Proof</u>. Let A be an RD-submodule of B and $\phi : A \to E[r]$ a homomorphism. ϕ induces a map $\bar{\phi} : A/rA \to E[r]$. This extends to a homomorphism $\bar{\psi} : B/rB \to E$, because there is an embedding of A/rA as $(A+rB)/rB$ in B/rB, and E is injective. Since Im $\bar{\psi} \leq E[r]$, the composition of the canonical surjection of B onto B/rB with $\bar{\psi}$ extends ϕ. □

Recall that a <u>cocyclic</u> module is an essential extension of a simple module.

PROPOSITION 1.2. [D] Every R-module is embeddable as an RD-submodule in a product of cocyclic RD-injective modules.

<u>Proof</u>. Given $0 \neq a \in M$, set $\Lambda_a = \{r \in R \mid a \notin rM\}$. For each $r \in \Lambda_a$, let $K(a,r)$ be a submodule of M maximal with respect to the properties: $rM \leq K(a,r)$ and $a \notin K(a,r)$. Clearly, $M/K(a,r)$ is a cocyclic module, so $E(M/K(a,r))$ is indecomposable injective. By (1.1), $E(a,r) = E(M/K(a,r))[r]$ is a

cocyclic RD-injective module. The canonical surjection of M onto $M/K(a,r)$ can be regarded as a homomorphism $\phi_{a,r} : M \to E(a,r)$. Consider the diagonal map of the $\phi_{a,r}$:

$$\phi : M \longrightarrow \prod_{\substack{0 \neq a \in M \\ r \in \Lambda_a}} E(a,r) = A \, .$$

Evidently, ϕ is monic and A is RD-injective. To show that ϕM is an RD-submodule of A, let $x \in M$ be such that $\phi x \in \phi M \cap sA$ ($s \in R$), and assume, by way of contradiction, that $\phi x \notin s\phi M$. Then $x \notin sM$, so $s \in \Lambda_x$. By hypothesis, $\phi_{x,s}(x) = sy + K(x,s)$ for some $y + K(x,s) \in E(x,s)$. Therefore, $\phi_{x,s}(x) = 0$, which is absurd. □

We hence deduce the following characterization of RD-injectivity.

COROLLARY 1.3. [D] A module is RD-injective if and only if it is a summand of a product of cocyclic RD-injective modules. □

An RD-submodule A of a module B is said to be <u>RD-essential</u> in B, and B an <u>RD-essential extension</u> of A, if every homomorphism $\phi : B \to M$, with $\phi|A$ monic and ϕA an RD-submodule of M, is monic. Equivalently, given a submodule $K \neq 0$ of B, either $K \cap A \neq 0$, or $(K + A)/K$ fails to be an RD-submodule of B/K. An RD-essential extension B of a module A is <u>maximal</u> if there are no RD-essential extensions of A properly containing B.

An <u>RD-injective hull</u> of an R-module A is an RD-injective R-module which is an RD-essential extension of A. To establish the existence of such modules, we require the following two lemmas.

LEMMA 1.4. [D] Let B be an RD-essential extension of the R-module A, and let $\phi : A \to M$ be an RD-embedding in an RD-injective module M. Then ϕ extends to a monomorphism $\psi : B \to M$.

<u>Proof.</u> Evidently, there exists a homomorphism $\psi : B \to M$ extending ϕ. Now $K = \text{Ker } \psi$ satisfies: $A \cap K = 0$ and $\psi(A \oplus K)$ is an RD-submodule of ψB. Consequently, $(A \oplus K)/K$ is an RD-submodule of B/K. As A is RD-essential in B, $K = 0$ and ψ is monic. □

LEMMA 1.5. [D] Suppose $A \leq B \leq M$ where A is an RD-submodule of M and B is a maximal RD-essential extension of A. Then B is a summand of M.

Proof. The set of all submodules K of M such that $K \cap A = 0$ and $(A + K)/K$ is an RD-submodule in M/K is inductive. If H is a maximal member in this set, then $(A + H)/H$ is RD-essential in M/H, as is readily checked. Recall that $(B + H)/H \cong B$ is a maximal RD-essential extension of $(A + H)/H \cong A$. Therefore, $B + H = M$ and $B \cap H = 0$, whence $M = B \oplus H$. □

We can now prove the existence of RD-injective hulls.

THEOREM 1.6. [D] (Stenström [1], Warfield [4]) Every R-module M has a maximal RD-essential extension. This is an RD-injective hull of M. Any two RD-injective hulls of M are isomorphic over M.

Proof. By (1.2), M is an RD-submodule of some RD-injective R-module A. The family

$$\mathcal{F} = \{N \leq A \mid N \text{ is an RD-essential extension of } M\}$$

is inductive, as is easily verified. We claim: a maximal member B in $\mathcal{F}$ is a maximal RD-essential extension of M. For, let $B \leq B'$ and B' an RD-essential extension of M. In view of (1.4), there exists a monomorphism of B' into A. The maximality of B in ensures that $B = B'$. By (1.5), B is a summand of A; thus it is RD-injective.

The uniqueness of B up to isomorphism over M is an immediate consequence of (1.4). □

We shall use the notation $\hat{M}$ for the RD-injective hull of M.

The next result supplements (1.6).

LEMMA 1.7. [D] An RD-injective hull $\hat{A}$ of a module A is a maximal RD-essential extension of A.

Proof. Assume that $A \leq \hat{A} \leq B$ with A RD-essential in B. There exists a homomorphism $\phi : B \to \hat{A}$ extending the inclusion map $A \to \hat{A}$. By (1.4), ϕ is monic, thus ϕB is still a maximal RD-essential extension of A. Therefore, $\phi B = B$ and $\hat{A} = B$. □

1. RD-INJECTIVE MODULES 213

The following result is an easy consequence of the preceding ones.

THEOREM 1.8. [D] (Warfield [4]) An RD-monomorphism $\phi : A \to B$ extends to a splitting monomorphism $\hat{\phi} : \hat{A} \to \hat{B}$.

Proof. The composition of ϕ and the embedding of B in $\hat{B}$ extends to a homomorphism $\hat{\phi} : \hat{A} \to \hat{B}$, which is monic by (1.4). Apply now (1.5) to the inclusions $\phi A \leq \hat{\phi} \hat{A} \leq \hat{B}$, noting that, in view of (1.7), $\hat{\phi}\hat{A}$ is a maximal RD-essential extension of ϕA. □

We wish to obtain more information about RD-injectives in general. To this end, we prove (1.10). First, a lemma:

LEMMA 1.9. [D] For a divisible module M, we have $\hat{M} = E(M)$.

Proof. Clearly, M is both RD and essential in $E(M)$. Hence M is an RD-essential submodule of the RD-injective R-module $E(M)$. □

THEOREM 1.10. [D] An RD-injective R-module M has a decomposition

$$M = E \oplus N \qquad (1)$$

where E is injective and $N^1 = 0$. Both E and N are unique up to isomorphism.

Proof. It suffices to verify that M^1 is injective. In fact, it is then a summand and (1) follows, while uniqueness is readily checked.

Consider the R-module $X = \langle x, x_r \ (0 \neq r \in R) \rangle$ where the generators are subject to the defining relations $rx_r = x$ (for all $0 \neq r \in R$). Define the R-module Y via the pushout diagram

$$\begin{array}{ccc} R & \xrightarrow{\alpha} & X \\ \text{incl.} \downarrow & & \downarrow \\ Q & \longrightarrow & Y \end{array}$$

where $\alpha(1) = x$. It is easy to see that Y is a direct sum of Q and a direct sum of cyclically presented modules of annihilators

Rr, and X is RD in Y. Hence every homomorphism $\eta : X \to M$ extends to a homomorphism $Y \to M$. As every $a \in M^1$ is the image of x under a suitable η, we obtain $a \in hM$. Consequently, $M^1 = hM$. It follows that M^1 is RD in M, so by (1.8) and (1.9), $E(hM)$ is contained in M. This being h-divisible, $E(hM) = hM = M^1$. □

EXERCISES

1. A cocyclic module over an almost maximal valuation domain is standard uniserial.

2. [D] (Warfield [4]) Show that if M is a torsion-free R-module, then
$$\hat{M} \cong \operatorname{Hom}_R(K, E(K \otimes M))$$
where $K = Q/R$.

3. [D] (Stenström [1]) For a divisible abelian group D, $\operatorname{Hom}_{\mathbb{Z}}(F,D)$ is an RD-injective R-module whenever F is an RD-projective R-module.

4. [D] If $0 \to A \to B \to C \to 0$ is an RD-exact sequence, then so is $0 \to C^* \to B^* \to A^* \to 0$ where $M^* = \operatorname{Hom}_{\mathbb{Z}}(M, \mathbb{Q}/\mathbb{Z})$.

5. [D] Using the notation of Ex.4, show that the evaluation map $M \to (M^*)^*$ is an RD-embedding.

6. [D] (Stenström [1]) Give an alternate proof of the existence of enough RD-injectives by applying Ex. 3-5 to an RD-projective resolution of M^*.

7. [D] From (1.2) derive that $rM = 0$ implies $r\hat{M} = 0$.

§2. PURE-INJECTIVE MODULES

A module M is <u>pure-injective</u> if it has the injective property relative to all pure-exact sequences. The class of pure-injective modules is obviously closed under taking summands and direct products. Manifestly, it contains all RD-injective modules.

2. PURE-INJECTIVE MODULES   215

If A is a pure submodule of the module B over any commutative ring R, then $IB \cap A = IA$ for every ideal I of R. An argument similar to the one in (1.1) leads to

LEMMA 2.1. *Let E be an injective module over a ring R, and let $I \leq R$. Then $E[I]$ is pure-injective.* □

A homomorphism ϕ of a submodule A of an R-module B into an R-module M will be called a <u>partial homomorphism</u> of B into M if, for every finite system of equations,

$$\sum_{j=1}^{n} r_{ij} x_j = a_i \quad (r_{ij} \in R, \ i = 1, \ldots, m)$$

with constants a_i in A, that is solvable in B, the system $\sum_{j=1}^{n} r_{ij} x_j = \phi a_i$ $(i = 1, \ldots, m)$ has a solution in M. Manifestly, if A is pure in B, then every homomorphism $A \to M$ is a partial homomorphism.

We now come to a main characterization theorem.

THEOREM 2.2. *For an R-module M, the following conditions are equivalent:*

(i) M *is pure-injective;*

(ii) M *is a summand in every R-module that contains M as a pure submodule;*

(iii) M *is <u>algebraically compact</u>, i.e. if a system of linear equations over M is finitely solvable in M, then it has a global solution in M;*

(iv) *every partial homomorphism from an R-module B into M extends to a homomorphism of B into M.*

<u>Proof.</u> (i) ⟹ (ii) is trivial.

(ii) ⟹ (iii) Suppose (ii) for M, and select a system of equations

$$\sum_{j \in J} r_{ij} x_j = a_i \in M \quad (i \in I) \qquad (1)$$

whose finite subsystems are solvable in M; here $r_{ij} \in R$ and for each fixed $i \in I$, almost all $r_{ij} = 0$. The index sets I and J may have arbitrary cardinalities. Define an R-module A to be generated by M and generators b_j ($j \in J$) subject to the relations $\sum_j r_{ij} b_j = a_i$ for all $i \in I$. The hypothesis guarantees that the new relations do not cause any collapse of elements of M; in other words, M is a genuine submodule in A. It is straightforward to verify the purity of M in A. By (ii), M is a summand of A. The coordinates of b_j in M yield a solution of the entire system (1) in M.

(iii) $\implies$ (iv) We now assume that M satisfies (iii), and let $\phi: A \to M$ be a partial homomorphism of B into M where A is a submodule of the R-module B. Let $\{b_j \mid j \in J\}$ be generators of B mod A, and $\sum_{j \in J} r_{ij} b_j = a_i \in A$ ($i \in I$) the list of all relations between these generators and elements of A. Owing to hypothesis, $\sum_{j \in J} r_{ij} x_j = \phi a_i \in M$ is a finitely solvable system in M. By (iii), there is a solution $x_j = v_j \in M$ ($j \in J$). The correspondence $b_j \mapsto v_j$ gives rise to an extension $\psi: B \to M$ of ϕ. Thus M satisfies (iv).

(iv) $\implies$ (i) is clear. $\square$

The following sufficient conditions for pure-injectivity are of utmost importance. (Recall that the rings are supposed to carry the discrete topology.)

PROPOSITION 2.3. A module M that is compact or linearly compact in a Hausdorff topology is pure-injective.

Proof. We show that such a module M satisfies (2.2)(iii). Assume (1) is a finitely solvable system. Consider the homomorphism $\phi_i: M^J \to M$ ($i \in I$) defined via

$$\phi_i(,\ldots,b_j,\ldots) = \sum_{j \in J} r_{ij} b_j \qquad (b_j \in M).$$

If a solution $x_j = c_j \in M$ ($j \in J$) of the ith equation is thought of as an element $y = (\ldots, c_j, \ldots)$ of M^J (any value can be assigned to unknowns not occurring explicitly in the ith equation), then the set S_i of all solutions to the ith equation is $S_i = y + \text{Ker } \phi_i$. As ϕ_i is a continuous homomorphism, S_i is a closed subspace in

2. PURE-INJECTIVES MODULES

the (linearly) compact space M^J. The finite solvability of (1) amounts to the finite intersection property of the family $\{S_i \mid i \in I\}$. Hence $\cap S_i$ is not empty. Every $y \in M^J$ in this intersection yields a global solution to (1). □

We now proceed to the proofs of three preliminary lemmas.

LEMMA 2.4. If M is any R-module and E is a pure-injective R-module, then $\text{Hom}_R(M,E)$ is a pure-injective R-module.

Proof. Let $0 \to A \to B \to C \to 0$ be a pure-exact sequence. Because of (II.4.6), $0 \to A \otimes_R M \to B \otimes_R M \to C \otimes_R M \to 0$ is likewise pure-exact. Thus the induced map $\text{Hom}_R(B \otimes M, E) \to \text{Hom}_R(A \otimes M, E)$ is surjective. Equivalently, $\text{Hom}_R(B, \text{Hom}_R(M,E)) \to \text{Hom}_R(A, \text{Hom}(M,E))$ is an epimorphism. □

LEMMA 2.5. If $0 \to A \to B \to C \to 0$ is a pure-exact sequence and E is pure-injective, then the sequence

$$0 \to \text{Hom}_R(C,E) \to \text{Hom}_R(B,E) \to \text{Hom}_R(A,E) \to 0$$

is splitting exact.

Proof. It is exact, since E is pure-injective. As in the proof of (2.4), we can argue that, for any R-module M, the map $\text{Hom}_R(M, \text{Hom}_R(B,E)) \to \text{Hom}_R(M, \text{Hom}_R(A,E))$ is surjective. This implies splitting. □

LEMMA 2.6. If E is an injective cogenerator of R-Mod, then there is a natural pure embedding

$$M \to \text{Hom}_R(\text{Hom}_R(M,E),E)$$

for each R-module M.

Proof. For simplicity, write $M^* = \text{Hom}_R(M,E)$. Consider the evaluation map $\phi : M \to M^{**}$ given by $(\phi x)\alpha = \alpha x$ where $x \in M$ and $\alpha \in M^*$. Owing to the hypothesis on E, ϕ is monic. Let

$$\sum_{j=1}^{n} r_{ij} x_j = a_i \in M \quad (i = 1, \ldots, m) \tag{2}$$

be a finite set of equations over M ($r_{ij} \in R$) which is solvable in M^{**}. Let $\psi : (M^{**})^n \to (M^{**})^m$ denote the homomorphism

$$\psi : (\eta_1,\ldots,\eta_n) \longmapsto (\sum_{j=1}^{n} r_{1j}\eta_j, \ldots, \sum_{j=1}^{n} r_{mj}\eta_j)$$

where $\eta_k \in M^{**}$. Here $\psi(M^{**})^n = (\phi M^n)^{**}$, since $\psi : (M^{**})^n \to (\phi M^n)^{**}$ is surjective, being induced by the epimorphism $\psi : M^n \to \phi M^n$.

Now, for any submodule N of M, there is a canonical isomorphism $(M/N)^{**} \cong M^{**}/N^{**}$. Therefore, ϕ induces an embedding $M/N \to M^{**}/N^{**}$ which leads to $\phi M \cap N^{**} = \phi N$. Applying this to the module M^m and its submodule ϕM^n, we obtain

$$\phi M^m \cap \psi(M^{**})^n = \phi M^m \cap (\phi M^n)^{**} = \phi M^n.$$

(2) is solvable in M^{**} means $(\phi a_1,\ldots,\phi a_m) \in \psi(M^{**})^n$, thus $(\phi a_1,\ldots,\phi a_m) \in \phi M^n$, proving the solvability of (2) in M. □

A module M is called <u>finitely cogenerated</u> if $E(M)$ is a finite direct sum of injective hulls of simple modules; or, equivalently, if it has a finitely generated essential socle. Cocyclic modules are finitely cogenerated. We shall show that every module is purely embeddable in a direct product of finitely cogenerated pure-injective modules.

We need the following easy result.

LEMMA 2.7. Let M be a finitely generated module and P a maximal ideal of the ring R. Then $\text{Hom}_R(M, E(R/P))$ is a finitely cogenerated R-module.

<u>Proof</u>. It is straightforward to show that $S = \{\phi \mid \phi M \leq R/P\}$ is the socle of $\text{Hom}_R(M, E(R/P))$, is finitely generated and essential in $\text{Hom}_R(M, E(R/P))$. □

We are now in a position to prove the main embedding theorem.

THEOREM 2.8. Every R-module M can be embedded as a pure submodule in a direct product of finitely cogenerated pure-injective R-modules.

2. PURE-INJECTIVE MODULES

Proof. Apparently, $E = \prod_P E(R/P)$ is an injective cogenerator for R-Mod if P ranges over the maximal ideals of R. Consider a pure-projective resolution (cf. (II.4.2))

$$0 \to K \to \bigoplus_{i \in I} F_i \to \mathrm{Hom}_R(M,E) \to 0$$

where the F_i are finitely presented. (2.5) yields a splitting exact sequence

$$0 \to \mathrm{Hom}_R(\mathrm{Hom}_R(M,E),E) \to \mathrm{Hom}_R(\bigoplus_{i \in I} F_i, E) \cong \prod_P \prod_{i \in I} \mathrm{Hom}_R(F_i, E(R/P)).$$

Owing to (2.4) and (2.7), this is a product of finitely cogenerated pure-injective modules. By (2.6), M is purely embeddable in the product. □

Following the pattern of §1, one can define **pure-essential extensions** and, among these, the maximal ones, just replace, in the definitions, "relative divisibility" by "purity". A **pure-injective hull** PE(M) of a module M is a pure-essential extension of M which is pure-injective. Since (1.4), (1.5) and (1.7) continue to hold if RD-property is replaced by purity, the arguments used in (1.6) and (1.8) lead to the following result.

THEOREM 2.9. (Warfield [1]) Every R-module M has a pure-injective hull; it is unique up to isomorphism over M.

A pure embedding $A \to B$ extends to a splitting monomorphism PE(A) → PE(B). □

A result of considerable importance is that pure-injective modules have the finite exchange property. The proof is based on a delicate analysis of their endomorphism rings. This being beyond the scope of this volume, we content ourselves with stating without proof the relevant result:

THEOREM 2.10. (Zimmermann-Huisgen, Zimmermann [1]) Pure-injective modules have the finite exchange property.

A pure-injective module is indecomposable if and only if its endomorphism ring is local.

EXERCISES

1. A module is pure-injective if and only if it is a summand of a direct product of finitely cogenerated pure-injective modules.

2. The property of being a pure-essential submodule is not transitive.

3. $\text{Hom}_{\mathbb{Z}}(M,C)$ is a compact R-module if M is an R-module and C is a compact abelian group.

4. (Warfield [1]) For any R-module M, the canonical map

$$\mu : M \to \text{Hom}_{\mathbb{Z}}(\text{Hom}_{\mathbb{Z}}(M, \mathbb{R}/\mathbb{Z}), \mathbb{R}/\mathbb{Z})$$

given by $\mu(a)(\chi) = \chi(a)$ for $a \in M$ and $\chi \in \text{Hom}(M, \mathbb{R}/\mathbb{Z})$ is a pure embedding; here $\mathbb{R}/\mathbb{Z}$ denotes the circle group. Conclude that every R-module can be embedded as a pure submodule in a compact R-module.

§3. PURE-INJECTIVE MODULES OVER PRÜFER DOMAINS

As purity and RD-property coincide for modules over Prüfer domains, one can expect a simplification in the theory of pure- and RD-injectives in this case. As a matter of fact, several interesting results hold for Prüfer domains.

We shall consider the filtration of the module M given by the submodules

$$rM : s \qquad (r, s \in R). \tag{1}$$

The main result is:

THEOREM 3.1. (Warfield [1]) A module M over a Prüfer domain R is pure-injective if and only if it is ultracomplete in the filtration (1).

Proof. Let M be pure-injective, and suppose that $\{a_{r,s} + rM:s\}_{(r,s) \in Y}$ ($a_{r,s} \in M$) is a family of cosets with the finite intersection property. Notice that $x \in a_{r,s} + rM:s$ means that the equation $s(x - a_{r,s}) = ry$ is solvable for y in M. Thus

3. PURE-INJECTIVES OVER PRÜFER DOMAINS

the finite intersection property of the given family is equivalent to the finite solvability of the system

$$sx - ry_{r,s} = sa_{r,s} \quad ((r,s) \in Y)$$

with unknowns x and $y_{r,s}$ $((r,s) \in Y)$. Because of (2.2), there is a global solution, and for a solution $x = a$, a belongs to each of the given cosets.

Conversely, let M be ultracomplete in the filtration (1), and suppose that there exists an $a \in \hat{M} \setminus M$. Consider the system consisting of the equations

$$rx = ra \qquad (2)$$

for all $r \in R$ with $ra \in M$, and of the equations

$$ty_{t,s,b} = sx + b \qquad (3)$$

for all $t, s \in R$ and $b \in M$ such that $t | sa + b$ in $\hat{M}$.

This system has a solution in $\hat{M}$. Since M is pure in $\hat{M}$, we infer that the system of equations (2) and (3) is finitely solvable in M. Let $a_r \in M$ be such that $ra_r = ra$; then $x = c_r \in M$ is a solution to (2) exactly if $c_r \in a_r + M[r]$. Similarly, if $x = a_{t,s,b} \in M$ is a solution (for suitable $y_{t,s,b}$) of (3), then the set of all solutions for x coincides with the coset $a_{t,s,b} + tM:s$. We see that $\{a_r + M[r], c_{t,s,b} + tM:s\}$ is a family of cosets with the finite intersection property. As $M[r] = 0M:r$, all these cosets are of the form as required by the filtration (1). Therefore, by hypothesis, there is a $c \in M$ that belongs to all the cosets, i.e. $x = c$ is a solution to (2) and (3). Thus $rc = ra$ for all $r \in R$ with $ra \in M$ and $t | sa + b$ implies $t | sc + b$.

Set $C = R(a - c) \neq 0$. Then $M \cap C = 0$; in fact, $r(a - c) \in M$ implies $ra \in M$, thus $rc = ra$ and $r(a - c) = 0$. Furthermore, $(M + C)/C$ is RD in $\hat{M}/C$, since if $g \in \hat{M}$ satisfies $tg = b + s(a - c)$ for some $t, s \in R$, $b \in M$, then $t | sa + (b - sc)$ which implies $t | sc + (b - sc) = b$. This means that M is not RD-essential in $\hat{M}$, a contradiction. Thus $\hat{M} = M$. □

If M is divisible, then (1) is the annihilator filtration, thus (3.1) specializes to (VI.4.3).

A careful analysis of the proof above reveals that in the torsion-free case a stronger assertion holds true.

THEOREM 3.2. Over any domain R, for a torsion-free R-module M, the following are equivalent:

(i) M is RD-injective;

(ii) M is pure-injective;

(iii) M is ultracomplete in the filtration (1).

Proof. (i) $\Longrightarrow$ (ii) is evident.

(ii) $\Longrightarrow$ (iii) The proof given in (3.1) is valid without any restriction on R or M.

(iii) $\Longrightarrow$ (i) We proceed as in the second part of the proof of (3.1). However, we ignore equations (2), since because of the torsion-freeness of $\hat{M}/M$, only the case $r = 0$ can occur. At this stage we reach back to (II.4.4) to argue that, due to M being RD in $\hat{M}$, the system (3) is finitely solvable in M. Having passed this hurdle, we can proceed as in the proof above to infer that $\hat{M} = M$. □

In order to obtain more information on the pure-injectives over Prüfer domains, we may confine our attention, owing to (1.10), to modules without elements of infinite heights. In the h-local case, the discussion can further be reduced to the local (i.e. valuation domain) case which will be dealt with in the subsequent sections.

THEOREM 3.3. (Warfield [1]) Let R be an h-local Prüfer domain and M a pure-injective R-module without elements of infinite height. Then M is the product of pure-injective R_P-modules with P ranging over the set of maximal ideals of R.

Proof. Because of $M^1 = 0$, M is Hausdorff in the R-topology. From (3.1) we infer that M is R-complete (indeed, $rM = rM:1$ for $r \in R$). Consequently,

$$M \cong \varprojlim M/rM$$

where r ranges over the non-zero elements of R. By (I.6.5), M/rM is a direct sum of torsion modules over the valuation domains R_P; here $(M/rM)_P = 0$ for almost all P, since r is contained in but a finite number of maximal ideals. Clearly, $M \cong \prod_P \varprojlim (M/rM)_P$. Thus M is a product of R_P-modules which are necessarily pure-injective. □

EXERCISE

1. Prove that pure-injective modules over any domain are ultracomplete in the filtration (1).

§4. PURE-INJECTIVITY OVER VALUATION DOMAINS

Throughout this section R will denote a valuation domain. In this case, for any R-module M and any $r, s \in R$, the submodule $rM:s$ either coincides with M or equals $M[s] + rs^{-1}M$ according as $r|s$ or $s|r$. Therefore, (3.2) becomes:

THEOREM 4.1. [VD] An R-module M is pure-injective exactly if it is ultracomplete in the filtration

$$\{M[s] + rM \mid r, s \in R\}.$$

In particular, a torsion-free R-module is pure-injective if and only if it is R-ultracomplete. □

It is rather obvious that a uniserial module is linearly compact (in the discrete topology) if its cyclic submodules are linearly compact. A simple appeal to (2.3) thus gives:

THEOREM 4.2. [VR] (Warfield [1]) Uniserial modules over maximal valuation rings and torsion uniserial modules over almost maximal valuation domains are pure-injective. □

Next we examine some general properties of pure-injective R-modules. The first result implies that all of their α-invariants corresponding to limit heights vanish.

PROPOSITION 4.3. [VD] Pure-injective R-modules are cohesive.

Proof. Let $z \in M$ where M is pure-injective. We show that $h_M(z) \geq (J/R)^-$ implies $h_M(z) \geq J/R$. For all $a \in J\setminus R$, pick a cyclic module $Rx_a \cong Ra/R$ and form

$$K = \bigoplus_{a \in J\setminus R} Rx_a \oplus J.$$

The submodule $H = \langle x_a - a \mid a \in J\setminus R \rangle$ is easily seen to be pure in K. There clearly exists a homomorphism $\phi : H \to M$ such that $\phi(1) = z$. By pure-injectivity, ϕ can be extended to a $\psi : K \to M$. This ψ, restricted to J, yields a homomorphism $J \to M$ sending 1 to z. □

The following result tells us about their smoothness.

PROPOSITION 4.4. [VD] Pure-injective modules are smooth.

Proof. Let M be pure-injective, $a \in M$ and $0 \neq L$ an ideal of R. We wish to decompose $a = b + c$ with $L \leq \text{Ann } b$ and $h(c) \geq \sigma = i(a)^L$.

Consider the following system of equations:

$$\left.\begin{array}{l} rx = 0 \quad \text{for all } r \in L; \\ sz_s = y \quad \text{for all } s \in R \text{ such that } sr\mid ra \\ \qquad\qquad \text{for all } r \in L; \\ x + y = a \end{array}\right\} \quad (1)$$

with unknowns x, y and z_s. Manifestly, the solvability of a finite subsystem of (1) is tantamount to the solvability of a finite system consisting of at most three equations: $rx = 0$, $sz_s = y$ and $x + y = a$. To solve this, choose $w \in M$ satisfying $rws = ra$ and set $z_s = w$, $y = sw$, $x = a - y$. By algebraic compactness, (1) has a global solution in M, say, $x = b$, $y = c$ and $z_s = d_s$. Then $a = b + c$ with $L \leq \text{Ann } b$. Furthermore, c has a non-limit height by (4.3), thus $s\mid c$ for all $s^{-1} \in rH_M(ra)$ with $0 \neq r \in L$ implies $h(c) \geq \sigma$. □

In order to identify the indecomposable pure-injectives, we prove a lemma on indicators.

LEMMA 4.5. [VD] (Monari-Martinez [1]) In an indecomposable pure-injective R-module, the indicators of elements are constant (up to

4. PURE-INJECTIVITY OVER VALUATION DOMAINS

the annihilators).

  Proof. Assume, by way of contradiction, that M is indecomposable pure-injective and there exist $a \in M$, $r \in R \setminus \mathrm{Ann}\, a$ such that $rh(ra) > h(a)$. Because of (VIII.1.5), there is a $b \in M$ such that $rb = ra$ and $h(b) > h(a)$. The homomorphism $\phi : Ra \to M$ defined by $\phi a = b$ is readily seen to be a partial homomorphism of M into M (as $h(sb) \geq h(sa)$ for every $s \in R$). (2.2) guarantees that it can be extended to an endomorphism ψ of M. Certainly, ψ is not an automorphism, as it fails to preserve the height of a. Hence by (2.10), $1 - \psi$ is an automorphism. But this is absurd, since $ra \neq 0$ and $(1 - \psi)ra = ra - rb = 0$. □

  We can now verify:

THEOREM 4.6. [VD] (Ziegler [1], Monari-Martinez [1]) A pure-injective R-module is indecomposable if and only if it is the pure-injective hull of a standard uniserial R-module.

  Proof. Let M be indecomposable pure-injective, and $0 \neq a \in M$. If $\mathrm{Ann}\, a = I$ and $h_M(a) = J/R$ (cf. (4.3)), then there is a monomorphism $\phi : J/I \to M$ with $\phi(1 + I) = a$. Now (4.5) implies $\mathrm{Im}\, \phi = \hat{U}$ is pure in M. Thus $\hat{U}$ is a summand of M. Consequently, $M = \hat{U}$.

  Conversely, let $M = \hat{U}$ where $U = J/I$ is standard uniserial, $I < R < J \leq Q$. It is readily checked that if $M = A \oplus B$, then the projection of U either to A or to B is a (pure) submodule isomorphic to U. By the definition of pure-essential extensions, the projection of M either to A or to B is monic. Hence M is indecomposable. □

  We proceed to the discussion of pure-injective hulls of direct sums of standard uniserial modules.

LEMMA 4.7. [VD] Let $\hat{B}$ be the pure-injective hull of a direct sum B of standard uniserial modules. Then $\hat{B}$ contains no elements whose indicators increase on the left at all gaps.

  Proof. First, let $0 \neq a \in \hat{B}$ satisfy $B \cap Ra \neq 0$. In this case, the claim is clear as the elements of B have piecewise constant indicators. So assume $B \cap Ra = 0$. Since $(B \oplus Ra)/Ra$ is not pure in $\hat{B}/Ra$, there exist $b \in B$, $x \in \hat{B}$, $r,s \in R$ such that $b + sa = rx$ with r not dividing b and sa. We can assume that none of the coordinates of b (in a decomposition of B) is divisible by r. Then multiplying the equation by $t \in R$ with $tb \neq 0$ if necessary, we can also assume that b has a constant indicator up to $\operatorname{Ann} b = L$. Evidently, b and sa have the same constant indicator up to L and $i(sa)$ has a gap at L. □

PROPOSITION 4.8. [VD] The pure-injective hull of a direct sum of standard uniserial modules has no non-zero summand with trivial α-invariants.

  Proof. Owing to (VIII.3.4), it suffices to show that every summand contains an element x with $i(x)$ constant non-limit up to $\operatorname{Ann} x$. By (4.7), for ever non-zero element a, there is an ideal L such that $i(a)$ has a gap at L and does not increase on the left at L. By (4.4), we can write $a = x + y$ with $\operatorname{Ann} x \geq L$ and $h(y) \geq i(a)^L$. As $i(a) = i(x)$ up to L (see (VIII.5.1)), this x is a desired element. □

  We are ready for the main decomposition theorem on pure-injectives. (Parts (i) and (ii) are due to Warfield [1].) Recall that a module which has no indecomposable summand $\neq 0$ is called superdecomposable.

THEOREM 4.9. [VD] A pure-injective R-module M decomposes as

$$M = E \oplus F \oplus T \oplus N$$

where

  (i) E is an injective R-module;

  (ii) F is the pure-injective hull of a direct sum of R-modules isomorphic to ideals of R;

  (iii) T is the pure-injective hull of a direct sum of standard uniserial torsion R-modules;

  (iv) N is a superdecomposable pure-injective R-module with trivial α-invariants.

4. PURE-INJECTIVITY OVER VALUATION DOMAINS

The summands E, F, T are uniquely determined, up to isomorphism, by the α-invariants of M. The summand N is likewise unique up to isomorphism.

Proof. Let B be an α-basic submodule of M. By (X.3.2), $\alpha_B[\sigma,I] = \alpha_M[\sigma,I]$ for all non-limit heights σ and proper ideals I; because of (4.3), the α-invariants of M for limit heights vanish. Write

$$B = H \oplus B_0 \oplus B_1$$

where H is the h-divisible part of B, and B_0, B_1 are direct sums of torsion-free and torsion, non-divisible uniserial modules. Evidently, $\hat{B} = E \oplus F \oplus T$ with $E = \hat{H}$, $F = \hat{B}_0$, $T = \hat{B}_1$ is a direct summand of M. In view of the definition of B, a complementary summand N must have trivial α-invariants. (4.6) implies that N can not have a non-zero indecomposable summand.

As the direct sum of α-basic submodules of E, F and T is α-basic in M, the uniqueness statement on E, F, T is a simple consequence of the uniqueness, up to isomorphism, of α-basic submodules in general. To verify the uniqueness of N, let B, B' be α-basic submodules of M, and $M = \hat{B} \oplus N = \hat{B}' \oplus N'$ corresponding decompositions. (2.10) applied to N' yields $M = C_0 \oplus N_0 \oplus N'$ where C_0, N_0 are summands of $\hat{B}$ and N, respectively, such that $\hat{B}' \cong C_0 \oplus N_0$. Hence N_0 has trivial α-invariants, so by (4.8), $N_0 = 0$. We conclude that $C_0 = \hat{B}$ and $N \cong M/\hat{B} \cong N'$. □

The uniqueness of the direct sum in (VI.4.4) follows at once from the last theorem.

EXERCISES

1. [VD] (Monari-Martinez [1]) Verify (4.3) by setting up a finitely solvable system of equations expressing what the height of z is.

2. [VD] The pure-injective hull of a non-standard uniserial module coincides with the pure-injective hull of a standard uniserial module.

3. [VD] A torsion uniserial module U is pure-injective if and only if Ann U $\geq$ I_F as defined in I, §5.

4. [VD] If M is a pure-injective module which has no super-decomposable summand, then every α-basic submodule of M is basic.

5. [VD] (Fuchs-Salce [2]) Use (X.2.2) to prove that the α-invariants of a smooth, cohesive R-module and of its pure-injective hull are the same.

6. [VD] Give an example to show that smoothness is relevant in Exercise 5.

§5. PURE-INJECTIVE HULLS OF POLYSERIAL MODULES

Having established general results on pure-injectives, it is instructive to analyze the pure-injective hulls of polyserial modules over arbitrary valuation domains R. Our discussion leans heavily on a maximal immediate extension S of R and is based on a significant relationship between the pure-injective hull $(J/I)^\wedge$ of a standard uniserial R-module J/I and JS/IS. Unless stated otherwise, JS/IS will be regarded as an R-module.

Recall that every element $x \in S$ can be written uniquely, up to unit factors of R, in the form $x = r\varepsilon$ with $r \in R$ and ε a unit of S.

LEMMA 5.1. [VD] Let $0 \leq I < R \leq J \leq Q$, and $a = r\varepsilon + IS \in JS/IS$ ($r \in R\setminus I$, $\varepsilon \in U(S)$). Then $h(a) = r^{-1}J/R$.

Proof. Clearly, $h(a) \geq r^{-1}J/R$. Assume, by way of contradiction, that $h(a) > r^{-1}J/R$. Then there exist $t \in Q\setminus J$ and $s \in JS$ such that $r\varepsilon + IS = rt^{-1}s + IS$. It follows that $r(\varepsilon - t^{-1}s) \in IS$ where $\varepsilon - t^{-1}s \in U(S)$. Therefore $r \in IS \cap R = I$, which is absurd. □

An immediate consequence of (5.1) is as follows.

COROLLARY 5.2. [VD] The indicators of the elements of JS/IS are constant and non-limit. If J is principal, they are likewise principal. □

5. PURE-INJECTIVE HULLS OF POLYSERIAL MODULES

Note that (5.2) reduces to (4.5) whenever JS/IS is an indecomposable pure-injective R-module. We will see, however, that this is not the case in general.

If π is a prime ideal of R, we define the __defect__ of R at π to be the rank of $S/\pi S$ as an R/π-module, and denote it by $d(\pi)$. Evidently, the defect at P is 1, since $S/PS \cong R/P$.

LEMMA 5.3. [VD] Let $A = JS/IS$ ($I < R \leq J \leq Q$) and let $\pi = J^\# \cup I^\#$. Then $\dim \alpha_A[\sigma,L] = d(\pi)$ if $[\sigma,L] = [J/R, I]$, and $\alpha_A[\sigma,L] = 0$ otherwise.

__Proof.__ From (5.1) it follows that $\alpha_A[\sigma,L] = 0$ if $[\sigma,L] \neq [J/R, I]$. From (5.1) we deduce, by easy calculation, that

$$A^{J/R} = S_{J^\#S}/IS \;\; ; \;\; A^{(J/R)^+} = J^\# S/IS,$$

$$A[I] = S_{I^\#S}/IS \;\; ; \;\; A[I^+] = I^\# S/IS.$$

It follows that $\alpha_A[J/R,I] \cong S_{\pi S}/\pi S_{\pi S}$. Clearly, the dimension of $S_{\pi S}/\pi S_{\pi S}$ over $R_\pi/\pi R_\pi$ is equal to the rank of the torsion-free module $S/\pi S$ over the valuation domain R/π, i.e. to $d(\pi)$. □

We can easily determine what the structure of JS/IS is like.

PROPOSITION 5.4. [VD] Let the notation be as in (5.3). Then

$$JS/IS \cong (\bigoplus_{d(\pi)} (J/I))^\wedge .$$

__Proof.__ $A = JS/IS$ is a linearly compact S-module in the discrete topology. Thus it is a linearly compact R-module in a Hausdorff topology. By (2.3), it is a pure-injective R-module. From (5.1) and (5.3) it follows that $\bigoplus_{d(\pi)} (J/I)$ is an α-basic submodule of A, while (5.2) along with (X.3.3) shows that it is pure-essential in A. Hence the claim follows. □

An immediate consequence of (5.4) is the following

COROLLARY 5.5. [VD] Let the notation be as in (5.3). Then $JS/IS = (J/I)^\wedge$ if either J or I is archimedean. In particular, $(R/I)^\wedge \cong S/IS$.

__Proof.__ By (5.4), $JS/IS \cong (J/I)^\wedge$ if and only if $d(\pi) = 1$. If J or I is archimedean, then $\pi = P$ whose defect is 1. □

We need the following general result.

LEMMA 5.6. [VD] The canonical embedding of an R-module M in $M \otimes_R S$ is pure.

Proof. As R is pure in S, the claim follows from (II.4.6). □

We can now easily establish a structure theorem for the pure-injective hulls of polyserial modules.

THEOREM 5.7. [VD] Let M be a polyserial module with pure-composition series:
$$0 = M_0 < M_1 < \ldots < M_{n-1} < M_n = M.$$
Then $\hat{M} \cong \bigoplus_1^n (M_i/M_{i-1})^{\wedge}$, where $(M_i/M_{i-1})^{\wedge} \cong (J_i/I_i)^{\wedge}$ for suitable $0 \leq I_i < R \leq J_i \leq Q$.

Proof. A repeated application of (2.9) yields the first assertion, while the second follows from §4, Ex.2. □

COROLLARY 5.8. [VD] Keeping the notation of (5.7), let $\pi_i = J_i^{\#} \cup I_i^{\#}$. Then $\hat{M} \cong M \otimes_R S$ if and only if $d(\pi_i) = 1$ for every i.

Proof. The exact sequence $0 \to I \to J \to J/I \to 0$ and the isomorphism $I \otimes S \cong IS$ imply $JS/IS \cong J/I \otimes S$. The last R-module is pure-injective, thus $M \otimes S \cong \bigoplus_1^n (M_i/M_{i-1} \otimes S) \cong \bigoplus_1^n (J_iS/I_iS)$. The claim follows from (5.4) and (5.7). □

In particular, we are led to

COROLLARY 5.9. [VD] (Warfield [1]) The pure-injective hull $\hat{M}$ of a finitely generated R-module M is isomorphic to $\bigoplus_1^n S/I_i S$ where $\{I_i \mid 1 \leq i \leq n\}$ is the annihilator sequence of M. In particular, S is the pure-injective hull of R. □

EXERCISES

1. [VD] Using IX.§1, give examples of finitely generated R-modules M whose α-invariants differ from those of $\hat{M}$. Find conditions for the equality of these invariants.

2. [VD] (Warfield [1]) Verify $(R/I)^\wedge \cong S/IS$ for an ideal I of R by showing that (a) the first module is a summand in the second; (b) a complement C satisfies $PC = C$; (c) the elements of C have cyclic heights.

NOTES

The study of pure-injectivity has its origin in abelian group theory where a large amount of literature is available on various aspects of pure-injective (= algebraically compact) abelian groups. They form a remarkable class which plays a relevant role in the theory. In fact, they admit several interesting characterizations, they can be classified in terms of numerical invariants, and above all, they are instrumental in various considerations (cf. Fuchs [2]).

In the module case, the first important development was Warfield's paper [1]. He established characterizations of pure-injectives for general rings, modelled after the abelian case, and went on to derive special results for Prüfer and valuation domains. Stenström [1] focussed his attention on abstract purity in a categorical setting and studied the corresponding injective objects. RD-injectives were considered by Warfield in an unpublished preprint [4] (which we used with his kind permission).

Pure-injectives turn out to be injectives in a suitable category. This aspect was studied by various authors. The idea goes back to Gruson-Jensen [1]. With a left R-module M, they associate the functor $* \otimes M$ from the category of finitely presented right R-modules to the category of abelian groups. Now, M is a pure-injective R-module exactly if $* \otimes M$ is an injective object in the functor category. Cf. also Facchini [1].

A new development was the description of endomorphism rings of pure-injectives by Zimmermann-Huisgen and Zimmermann [1].

Recently, a model theoretical approach led to the discovery of new features of pure-injective modules; see Ziegler [1]. In particular, the existence of minimal pure-injective summands containing a given element or sets of elements is of great importance. Some of these results were discussed from a algebraic point of view by Monari-Martinez [1].

A satisfactory classification of pure-injectives over valuation domains is, in the authors' opinion, a feasible project. In the decomposition of (4.9), the first two summands were characterized by invariants by Warfield [1]. The idea of using α-basic submodules to characterize the first three summands is new. The superdecomposable summands are awfully difficult to handle. The present authors have encouraging preliminary results which they plan to develop further (in particular, they have a rather explicit description of a minimal pure submodule containing a given element).

Pure-injective hulls of finitely generated modules over valuation domains were described by Warfield [1]. The more general results in §5 are new.

Several problems on pure-injective and RD-injective modules are still open; of these we point out the following.

PROBLEM 20. [VD]   Let M be an R-module. Under what conditions is $\hat{M}$ a module over a maximal immediate extension of R?

PROBLEM 21. For which valuation rings R can the ring structure of R be extended to a ring structure of $\hat{R}$?

PROBLEM 22. [D]   Give an example of a pure-injective module that fails to be RD-injective.

PROBLEM 23. Describe $\hat{M}$ in terms of M if M is polyserial and R is a valuation ring that is not an epic image of a valuation domain.

XII. Torsion-Complete and Cotorsion Modules

When we search for classes of modules where the ideas and machinery that have been developed so far could successfully be applied, it is natural to turn to classes close to RD-injective modules. This is the motivation that led us to the topics of this chapter.

§1. TORSION-COMPLETE MODULES

R-completeness is a straightforward generalization of $\mathbb{Z}$-complete abelian groups. One wonders if the analogue of a torsion-complete abelian group makes sense for modules over domains. The answer is a firm 'yes'. Actually, there are various immediate generalizations. We discuss two of these, one in this section and another in the next one. The most striking fact is that no restriction ought to be made on the domains in most results.

Assume R is a domain and T a (not necessarily torsion) R-module. We shall say that T is <u>torsion-complete</u> if it is a direct summand in every R-module M such that

1) T is an RD-submodule in M; and

2) M/T is divisible torsion.

An equivalent definition is:

$$\text{RDext}^1_R(D,T) = 0$$

for all divisible torsion R-modules D.

We can easily derive:

(a) torsion-completeness is preserved under formations of direct summands and finite direct sums;

(b) RD-injective modules are torsion-complete;

(c) torsion-free modules are torsion-complete (cf. (III.1.1)).

Characterizations of torsion-completeness resemble those of RD-injectives.

PROPOSITION 1.1. [D] (Schoeman [1]) For a module T, the following are equivalent:

(i) T is torsion-complete;

(ii) T has the injective property relative to all RD-exact sequences $0 \to A \to B \to C \to 0$ of R-modules with C divisible torsion;

(iii) $\hat{T}/T$ contains no divisible torsion submodule $\neq 0$.

Proof. (ii) $\Longrightarrow$ (i) is trivial.

(iii) $\Longrightarrow$ (ii) Assume $\hat{T}/T$ has no divisible torsion submodule $\neq 0$ and ϕ is a map $A \to T$. In view of the RD-injectivity of $\hat{T}$, ϕ extends to a map $\psi : B \to \hat{T}$. Since B/A is divisible torsion and $\hat{T}/T$ contains no such submodule $\neq 0$, the image of ψ has to be contained in T. This is exactly what we wished to prove.

Finally, (i) $\Longrightarrow$ (iii). Let T be torsion-complete and let M/T denote the divisible part of the torsion submodule of $\hat{T}/T$, i.e. $M/T = dt(\hat{T}/T)$. Then T is RD in M, so T is a summand of M. But T is RD-essential in $\hat{T}$ and M is RD in $\hat{T}$, thus T is RD-essential in M too. The only possibility that remains is $M = T$. □

The discussion of torsion-completeness can be reduced to the case where the first Ulm submodules vanish. In fact, we have:

1. TORSION-COMPLETE MODULES

LEMMA 1.2. [D] If T is torsion-complete, then T^1 is an injective summand of T.

Proof. From (XI.1.10), we know that $\hat{T} = E \oplus M$ where E is injective and $M^1 = 0$. If E were not contained in T, then $\hat{T}/T$ would contain the non-zero submodule $(E+T)/T \cong E/(E \cap T)$ which is torsion ($E \cap T$ must be essential in E) and divisible, in contradiction to (1.1). Hence $T = E \oplus (T \cap M)$ where $T^1 = E$ and $(T \cap M)^1 = 0$. □

Another characterization involves the module ∂^0 introduced in VI.§3.

PROPOSITION 1.3. [D] A reduced R-module T is torsion-complete if and only if

$$\text{RDext}^1_R(\partial^0, T) = 0.$$

Proof. Assuming this equality, we prove torsion-completeness. In view of (VI.3.7), for every divisible torsion R-module D there is an exact sequence $0 \to A \to \oplus \partial^0 \to D \to 0$ with A divisible. This being an RD-exact sequence, we are led to the exact sequence

$$0 = \text{Hom}(A,T) \to \text{RDext}^1(D,T) \to \text{RDext}^1(\oplus \partial^0, T)$$

where the last term is 0 by hypothesis. Hence T is torsion-complete. □

Guided by the kinship between R-completeness and torsion-completeness, we wish to explore the analogue of R-completions. The definition is obvious:

By a <u>torsion-completion</u> of an R-module M we mean an R-module $\overline{M}$ such that

(i) M is RD-essential in $\overline{M}$;
(ii) $\overline{M}/M$ is torsion divisible;
(iii) $\overline{M}$ is torsion-complete.

The next goal is to establish the existence and uniqueness of torsion-completions.

THEOREM 1.4. [D] (Schoeman [1]) Every R-module M has a torsion-completion which is unique up to isomorphism over M.

Proof. Given M, embed it in its RD-injective hull $\hat{M}$ and form the divisible part of the torsion submodule of $\hat{M}/M$. Call this $\overline{M}/M$. Then $\overline{M}$ trivially satisfies (i) and (ii). To verify (iii) for $\overline{M}$, note that it is RD in $\hat{M}$, so it is RD-essential in $\hat{M}$. We infer that $\hat{M}$ is the RD-injective hull of $\overline{M}$. As $\hat{M}/\overline{M}$ contains no non-zero divisible torsion modules, (1.1) implies that $\overline{M}$ is torsion-complete.

Uniqueness follows via standard arguments by making use of property (ii) in (1.1). □

EXAMPLE 1.5. If T is divisible torsion, then its torsion-completion is its injective hull.

EXAMPLE 1.6. If T is a bounded module, then $\overline{T} = T$. In fact, XI.§1.Ex.7 shows that $\hat{T}$ is likewise bounded. Hence $\hat{T}/T$ has trivial divisible part.

We now proceed to a topological characterization of torsion-completeness. This is similar to R-completeness which - as we know from V.§§1-2 - works nicely provided that the modules are torsion-free or Q is a countably generated R-module. In view of this, in the rest of this section we assume Q is countably generated. Recall that this condition implies that all divisible R-modules are h-divisible; see (II.2.3).

LEMMA 1.7. [D] (Schoeman [1]) Let Q be countably generated and M an R-module with $M^1 = 0$. Then $\overline{M} \cong N$ where $N/M = t(\check{M}/M)$.

Proof. By (V.2.2), $\check{M}/M$ is h-divisible, thus only the torsion-completeness of N needs a verification. In view of (1.1), it is sufficient to prove that $\hat{M}/N$ contains no divisible torsion submodules $\neq 0$.

If Q is countably generated, then because of (II.2.3), $\hat{M}/\check{M}$ is h-reduced. Thus $\hat{M}/N \cong \check{M}/N \oplus \hat{M}/\check{M}$ does not contain divisible torsion submodules, indeed. □

Let $\{t_n M \mid n < \omega\}$ be a countable subbase of neighborhoods of 0 in the R-topology of M, $0 \neq t_n \in R$ where $t_{n+1} = s_n t_n$ ($s_n \in R$)

1. TORSION-COMPLETE MODULES

can be assumed (see V, §2). A <u>Cauchy sequence</u> in M is a sequence $\{a_n\}_{n<\omega}$ of elements of M such that

$$a_n - a_{n+1} \in t_n M \text{ for every } n.$$

This sequence is <u>bounded</u> if $ra_n = 0$ for some $0 \neq r \in R$ and all n.

THEOREM 1.8. [D] (Schoeman [1]) Suppose Q is countably generated and M is an R-module with $M^1 = 0$. M is torsion-complete if and only if every bounded Cauchy sequence in M converges in M.

<u>Proof</u>. Assume M is torsion-complete and $\{a_n\}$ is a bounded Cauchy sequence in M, say $ra_n = 0$ ($0 \neq r \in R$). By the R-completeness of $\widetilde{M}$, there is an $a \in \widetilde{M}$ such that $a - a_n \in t_n \widetilde{M}$ for all n. Thus $ra - ra_n = ra \in \cap rt_n \widetilde{M} = 0$. As $\widetilde{M}/M$ is torsion-free, $a \in M$ follows.

Conversely, we suppose that every bounded Cauchy sequence has a limit in M and show that $\widetilde{M}/M$ is torsion-free. Let $a \in \widetilde{M}$ with $ra \in M$; as M is an RD-submodule of $\widetilde{M}$, without loss of generality, $ra = 0$ can be assumed. There is a Cauchy sequence $\{a_n\}$ in M that converges to a, $a - a_n \in t_n \widetilde{M}$ for each n. As $ra_n \in rt_n \widetilde{M} \cap M = rt_n M$, we can find $b_n \in M$ such that $ra_n = rt_n b_n$. We see that $\{a_n - t_n b_n\}$ is a bounded Cauchy sequence in M with limit a, whence $a \in M$. □

Another topological characterization of torsion-completeness is obtained if we furnish the submodule $M[r]$ with the topology induced by the R-topology of M. That is, $\{t_n M[r]\}_{n<\omega}$ is a sub-base of neighborhoods about 0.

THEOREM 1.9. [D] (Schoeman [1]) Let Q be countably generated and M an R-module satisfying $M^1 = 0$. M is torsion-complete exactly if $M[r]$ is complete for each $0 \neq r \in R$.

<u>Proof</u>. Note that a Cauchy sequence in M which is bounded by r is nothing else than a Cauchy sequence in $M[r]$. Hence the claim is evident. □

EXERCISES

1. [D] (Schoeman [1]) (a) Let H be a submodule of an R-module M such that M/H is torsion-free. Show that M is torsion-complete if and only if H is.

(b) M is torsion-complete exactly if tM is.

2. [D] (Schoeman [1]) Let M and N have the same torsion-completion T where $T^1 = 0$. Every isomorphism from M onto N extends uniquely to an automorphism of T.

3. [D] (Schoeman [1]) If Q is countably generated and M is reduced, then M is torsion-complete if and only if

$$RDext^1_R(K,M) = 0 \qquad \text{where } K = Q/R.$$

4. [D] (Schoeman [1]) If Q is countably generated and M is an h-reduced torsion-complete module, then a submodule N with M/N h-reduced is itself torsion-complete.

5. [D] Let Q be countably generated, and $r,s \in R$. For an R-module M, $M[rs]$ is complete if and only if both $M[r]$ and $M[s]$ are complete (topologies as above).

§2. TORSION-ULTRACOMPLETE MODULES

Our objective in this section is to study another version of torsion-completeness and to verify results typical for this version. This relates to torsion-completeness studied in §1 as RD-injectivity to R-completeness.

A module T over a domain R is said to be <u>torsion-ultracomplete</u> (Fuchs [5]) if it is a summand in every R-module M satisfying

1) T is RD in M; and
2) M/T is torsion.

Equivalently,

$$RDext^1_R(A,T) = 0$$

for all torsion R-modules A. Using a (torsion) RD-projective

2. TORSION-ULTRACOMPLETE MODULES

resolution for A, from the induced long exact sequence we infer that also

$$RDext_R^n(A,T) = 0 \qquad \text{for } n \geq 1$$

and for all torsion modules A.

It is immediate that torsion-free R-modules as well as RD-injective R-modules are torsion-ultracomplete. On the other hand, torsion-ultracomplete R-modules are necessarily torsion-complete, hence by (1.2) their first Ulm submodules are injective.

The obvious analogue of (1.1) holds for torsion-ultracompleteness; the proof is similar and can be left to the reader.

PROPOSITION 2.1. [D] The following are equivalent for an R-module T.

   (i)   T is torsion-ultracomplete;

   (ii)   T enjoys the injective property relative to all RD-exact sequences $0 \to A \to B \to C \to 0$ of R-modules where C is torsion;

   (iii) $\hat{T}/T$ is torsion-free. □

We now restrict ourselves to Prüfer domains and turn our attention to a characterization in terms of finitely solvable systems of equations. Assume

$$\sum_{j \in J} r_{ij} x_j = a_i \in T \qquad (i \in I) \qquad (1)$$

with $r_{ij} \in R$ and arbitrary index sets J, I (for each fixed i, almost all $r_{ij} = 0$) is a system of equations over T. Let F denote the free R-module with $X = \{x_j\}_{j \in J}$ as free generators and G the submodule of F generated by the left members of the system (1).

THEOREM 2.2. An R-module T over a Prüfer domain R is torsion-ultracomplete if and only if every system (1) of equations over T with F/G torsion which is finitely solvable in T has a global solution in T.

   Proof. Suppose T is torsion-ultracomplete and (1) is a finitely solvable system with F/G torsion. Finite solvability implies that the system is solvable in the pure-injective hull $\hat{T}$

of T, say $x_j = b_j \in \hat{T}$ is a solution. As F/G is torsion, it follows that the $b_j + T$ are torsion elements in $\hat{T}/T$. Therefore, $b_j \in T$ by (2.1), and (1) is solvable in T.

Conversely, assume that the R-module T has the stated property and is contained as an RD-submodule in some R-module M such that M/T is torsion. Let $\{c_j\}_{j \in J}$ be a set of generators of M mod T and $\sum_j r_{ij} c_j = a_i \in T$ ($i \in I$) the list of all relations between the generators c_j and elements of T. Since T is pure in M, the system

$$\sum_j r_{ij} x_j = a_i \in T \qquad (i \in I) \qquad (2)$$

is finitely solvable in T. If F and G have the same meaning as above, then clearly $F/G \cong M/T$. Hence, by hypothesis, the system (2) has a global solution $b_j \in T$. Now M is the direct sum of T and the submodule generated by $\{c_j - b_j\}_{j \in J}$. □

This result is instrumental in establishing the ultra-completeness of our modules in a suitable filtration.

Consider all submodules of an R-module M which are of the form

$$M_{rst} = (rM:s) \cap M[t] \qquad (3)$$

for non-zero elements $r, s, t \in R$. We are going to prove:

THEOREM 2.3. *A module over a Prüfer domain is torsion-ultracomplete if and only if it is ultracomplete in the filtration (3).*

Proof. Let T be torsion-ultracomplete and $\{a_{rst} + T_{rst}\}$ a family of cosets in T with the finite intersection property. $x \in a_{rst} + T_{rst}$ exactly if both

$$s(x - a_{rst}) \in rT \quad \text{and} \quad t(x - a_{rst}) = 0,$$

i.e. if the system consisting of the following two equations (with unknowns x and y_{rst}) is solvable in T:

$$\begin{cases} sx - ry_{rst} = sa_{rst}, \\ tx = ta_{rst}. \end{cases} \qquad (4)$$

2. TORSION-ULTRACOMPLETE MODULES

Consequently, the intersection of the given family of cosets is not empty if and only if there is a global solution to the system (4) with unknowns x and y_{rst} (one y for each coset). By the finite intersection property, (4) is finitely solvable, and moreover, it is of the kind stated in (2.2), viz. F/G is torsion. Thus (2.2) implies that (4) has a global solution, and T is ultracomplete.

Conversely, let T be ultracomplete in the filtration (3). Just as in (XI.3.1), one can verify that then T coincides with any module M in which it is pure-essential with M/T torsion. □

By a <u>torsion-ultracompletion</u> of an R-module M is meant an R-module $\overline{M}$ satisfying:

(i) M is RD-essential in $\overline{M}$;
(ii) $\overline{M}/M$ is torsion;
(iii) $\overline{M}$ is torsion-ultracomplete.

Our next goal is to establish the existence and uniqueness of torsion-ultracompletions:

THEOREM 2.4. [D] Every R-module M has a torsion-ultracompletion. It is unique up to isomorphism over M.

<u>Proof</u>. Embed M in its RD-injective hull $\hat{M}$ and form the torsion submodule $\overline{M}/M$ of $\hat{M}/M$. This $\overline{M}$ trivially satisfies (i) and (ii). In order to check property (iii) for $\overline{M}$, note that it is RD in $\hat{M}$, so it is RD-essential in $\hat{M}$. We infer that $\hat{M}$ is its RD-injective hull. As $\hat{M}/\overline{M}$ is torsion-free, (2.1) implies $\overline{M}$ torsion-ultracomplete.

Uniqueness can be proved by standard arguments by making use of property (ii) of (2.1). □

EXAMPLE 2.5. Let T be a divisible torsion module. Then its torsion-ultracompletion is its injective hull $E(T)$.

EXAMPLE 2.6. Let T be a bounded module. Then $\hat{T}$ is likewise bounded. Thus the torsion-ultracompletion of T is its RD-injective hull.

The classification problem for torsion-ultracomplete torsion modules is equivalent to the classification of RD-injective hulls of torsion R-modules. In fact, by property (iii) of (2.1), torsion-ultracomplete torsion modules are nothing else than torsion submodules of RD-injectives. For their classification over valuation domains, we refer to XI.§4.

EXERCISES

1. [D] A product of R-modules is torsion-ultracomplete if and only if each component is torsion-ultracomplete.

2. [D] The torsion part of a product of torsion R-modules is torsion-ultracomplete exactly if each component is torsion-ultracomplete.

3. [D] Let $0 \to A \to B \to C \to 0$ be an RD-exact sequence of torsion R-modules. Prove that there is an induced RD-exact sequence $0 \to \overline{B} \to \overline{A} \to \overline{C}$ for the torsion-ultracompletions.

4. For principal ideal domains, torsion-completeness and torsion-ultracompleteness are equivalent.

5. Over a valuation domain, a reduced torsion-ultracomplete torsion module is the direct sum of a module with a basic submodule and a superdecomposable module.

§3. COTORSION MODULES

We now embark on the discussion of a concept which is, in a certain sense, similar to torsion-ultracompleteness, but generalizes RD-injectivity in the opposite direction. In fact, while torsion-ultracompleteness means splitting of all RD-extensions by torsion modules, the concept to be studied here will mean splitting of all (necessarily RD-) extensions by torsion-free modules.

A word of warning: 'cotorsion' modules in a wider sense have been studied by Matlis [3]; our definition based on Warfield [5] is more restrictive than his. For Dedekind domains, however, the two definitions are equivalent.

In this section and in the next one, R can be any domain.

3. COTORSION MODULES

An R-module C will be called <u>cotorsion</u> if, for every torsion-free R-module F,

$$\operatorname{Ext}_R^1(F,C) = 0.$$

It is an easy exercise in homological algebra to show that this is equivalent to the injective property of C relative to all exact sequences $0 \to A \to B \to F \to 0$ with torsion-free F.

From the definition it is clear that

(A) RD-injective modules are cotorsion;

(B) a direct product of R-modules C_i is cotorsion if and only if all C_i are cotorsion;

(C) the class of cotorsion R-modules is closed under extensions.

The following result is crucial.

THEOREM 3.1. [D] (Warfield [5]) An R-module C is cotorsion if and only if it satisfies the following two conditions:

(i) $\operatorname{Ext}_R^1(Q,C) = 0$,

(ii) $\operatorname{i.d.}_R C \leq 1$.

<u>Proof.</u> Suppose first C cotorsion. Then (i) is obvious. Let $0 \to A \to B \to M \to 0$ be an exact sequence with B free and M any R-module. Consider the induced sequence $0 = \operatorname{Ext}_R^1(A,C) \to \operatorname{Ext}_R^2(M,C) \to \operatorname{Ext}_R^2(B,C) = 0$ where the first Ext vanishes as A is torsion-free. Hence $\operatorname{Ext}_R^2(M,C) = 0$ for all R-modules M, proving (ii).

Conversely, suppose C satisfies (i) and (ii). Let F be a torsion-free R-module and E its injective hull. Thus $\operatorname{Ext}_R^1(E,C) = 0$, and the exact sequence $0 \to F \to E \to E/F \to 0$ induces the exact sequence $0 = \operatorname{Ext}_R^1(E,C) \to \operatorname{Ext}_R^1(F,C) \to \operatorname{Ext}_R^2(E/F, C) = 0$ where the last group vanishes in view of (ii). We infer that C is cotorsion. □

By (VI.5.2), $\operatorname{i.d.}_R C \leq 1$ if and only if $\operatorname{Ext}_R^1(J,C) = 0$ for all ideals J of R. Hence,

COROLLARY 3.2. [D] An R-module C is cotorsion exactly if $\operatorname{Ext}_R^1(A,C) = 0$ for A equal to Q and to ideals J of R. □

It is worthwhile mentioning the following two consequences of (3.1).

COROLLARY 3.3. [D] A bounded R-module $T \neq 0$ is cotorsion if and only if $i.d._R T = 1$.

Proof. If T is bounded, then $Ext^1_R(Q,T) = 0$ as a bounded divisible module. The rest follows from (3.1). □

COROLLARY 3.4. [D] Let $0 \to A \to C \to B \to 0$ be an exact sequence of R-modules with C cotorsion. B is cotorsion whenever $i.d._R A \leq 1$.

Proof. If C is cotorsion, then by (3.1) the end terms in the induced exact sequence

$$Ext^1_R(F,C) \to Ext^1_R(F,B) \to Ext^2_R(F,A) \to Ext^2_R(F,C)$$

vanish for every torsion-free F. Consequently, B is cotorsion if and only if $Ext^2_R(F,A) = 0$ for all torsion-free F. □

Before proceeding to the discussion of cotorsion hulls, it should be pointed out:

PROPOSITION 3.5. [D] An R-module is RD-injective if and only if it is both torsion-ultracomplete and cotorsion.

Proof. Sufficiency follows from the exact sequence $RDext^1_R(M/tM,C) = Ext^1_R(M/tM,C) \to RDext^1_R(M,C) \to RDext^1_R(tM,C)$ induced by the RD-exact sequence $0 \to tM \to M \to M/tM \to 0$. □

Since torsion-free R-modules are necessarily torsion-ultracomplete, from (3.5) we get at once:

PROPOSITION 3.6. [D] A torsion-free R-module is RD-injective exactly if it is cotorsion. □

EXERCISES

1. [D] For an h-reduced cotorsion module C, the following isomorphism holds: $C \cong Ext^1_R(Q/R, C)$.

2. [D] Every cotorsion module is the epimorphic image of a torsion-free RD-injective module.

4. THE COTORSION HULL   245

3. Let $I \neq 0$ be an ideal of the valuation domain R. Show that R is cotorsion if and only if I is.

§4. THE COTORSION HULL

It is a somewhat more challenging task to establish cotorsion hulls than torsion-ultracompletions. Here again, we rely on RD-injective hulls, but in a different way.

Let N be a submodule of M such that M/N is torsion-free. We shall say that N is <u>torsion-free-essential</u> in M and M is a <u>torsion-free-essential extension</u> of N if there exists no non-zero submodule H in M such that

(a) $H \cap N = 0$, and

(b) $M/(H + N)$ is torsion-free.

Again, suppose N is a submodule of M with M/N torsion-free. It is straightforward to verify that the set of all submodules H of M satisfying (a) and (b) is inductive, hence it contains a maximal member H_0, and $(N \oplus H_0)/H_0 \cong N$ is torsion-free-essential in M/H_0.

By the <u>cotorsion hull</u> of an R-module M is meant an R-module $M^\bullet$ such that

(i) M is torsion-free-essential in $M^\bullet$;

(ii) $M^\bullet/M$ is torsion-free;

(iii) $M^\bullet$ is cotorsion.

THEOREM 4.1. [D] (Warfield [5]) Every R-module M has a cotorsion hull; it is unique up to isomorphism over M.

<u>Proof.</u> First assume M torsion-free. Then its RD-injective hull $\hat{M}$ satisfies (i)-(iii), so $M^\bullet = \hat{M}$ can be chosen.

In the general case, let M be any R-module and $0 \to G \to F \to M \to 0$ a free resolution of M. Then G is torsion-free, and so is $\hat{G}$. Using the injection map $G \to \hat{G}$, we form a commutative diagram with exact rows and left pushout square:

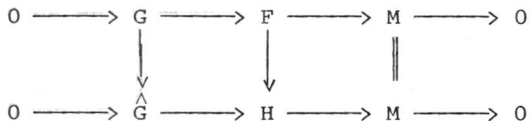

The middle vertical map is monic, and $H/F \cong \hat{G}/G$. The latter module is torsion-free, thus H is torsion-free. We now argue that M is isomorphic to the submodule $H/\hat{G}$ of the module $\hat{H}/\hat{G}$ such that the corresponding factor module $\cong \hat{H}/H$ is torsion-free. From (3.1) and (3.4) it follows at once that $\hat{H}/\hat{G}$ is a cotorsion module. In this way, we can embed $M \cong H/\hat{G}$ in a cotorsion module $C \cong \hat{H}/\hat{G}$ with C/M torsion-free.

Moreover, we claim that this embedding is torsion-free-essential. To prove this, choose N such that $\hat{G} \leq N \leq \hat{H}$, $H/\hat{G} \cap N/\hat{G} = 0$ and $\hat{H}/(H + N)$ is torsion-free. In view of the torsion-freeness of $N/\hat{G}$, $N = \hat{G} \oplus N_0$ for a suitable submodule N_0 of N. We thus have $H + N = H \oplus N_0$. As H is RD-essential in $\hat{H}$, $N_0 = 0$ follows. In other words, $N/\hat{G} = 0$, establishing our claim. Consequently, M has a cotorsion hull $M^{\bullet} \cong C$.

It remains to show that $M^{\bullet}$ is unique up to isomorphism over M. This is readily verified by standard arguments. □

The cotorsion hull $M^{\bullet}$ is actually a minimal cotorsion module containing M with $M^{\bullet}/M$ torsion-free in the following strong sense.

PROPOSITION 4.2. [D] Let C be a cotorsion module containing M such that C/M is torsion-free. Then the identity map of M extends to an embedding $M^{\bullet} \to C$.

Proof. Using the injective property of cotorsion modules relative to exact sequences with torsion-free third terms, we infer that the identity map of M extends to a homomorphism $\eta : M^{\bullet} \to C$. Clearly, $\operatorname{Ker} \eta \cap M = 0$ and $\operatorname{Im} \eta/M$, as a submodule of C/M, is torsion-free. Since $\operatorname{Im} \eta/M \cong M^{\bullet}/(M + \operatorname{Ker} \eta)$, from the torsion-free-essential character of M in $M^{\bullet}$ we deduce $\operatorname{Ker} \eta = 0$. □

Actually, we have proved more:

COROLLARY 4.3. [D] If M and C are as stated in (4.2) and if N

4. THE COTORSION HULL

is any torsion-free-essential extension of M, then the identity map of M can be extended to an embedding $N \to C$. □

We infer that a torsion-free-essential extension of M can always be embedded in $M^\bullet$. It follows that $M^\bullet$ is a <u>maximal torsion-free-essential extension</u> of M in the sense that if $M^\bullet < E$, then E is no longer a torsion-free-essential extension of M. Moreover, all maximal torsion-free-essential extensions of M are isomorphic to $M^\bullet$.

Unfortunately, it is not possible to say much about the structure of cotorsion modules in general, not even if R is a discrete rank one valuation domain. The reason for this is that the torsion submodules of cotorsion modules can be arbitrary torsion modules of whose structures we hardly know anything. The best we can do is to reduce the classification problem of cotorsion modules to that of their torsion submodules. This is done via the next theorem.

THEOREM 4.4. [D] (Warfield [5]) Every cotorsion module C can be decomposed, uniquely up to isomorphism, in the following way:

$$C = E \oplus F$$

where

(i) E is a cotorsion hull $T^\bullet$ of the torsion submodule T of C;

(ii) F is a torsion-free RD-injective R-module.

<u>Proof</u>. By making use of (4.3), we can find a cotorsion hull E of T in C. To show that C/E is torsion-free, we argue as follows. Write $t(C/E) = A/E$ and assume this is $\neq 0$. As A/T is torsion-free and $A > E$, A is no loger a torsion-free-essential extension of T; consequently, there is a submodule $H \neq 0$ of A such that $H \cap T = 0$ and $A/(H + T)$ is torsion-free. Observe that $E/[(H + T) \cap E] \cong (E + H + T)/(H + T)$ is torsion-free where $(H + T) \cap E = (H \cap E) + T$. Hence $H \cap E = 0$, because T is torsion-free-essential in E. This is a contradiction: A/E is torsion and $H \neq 0$ is torsion-free. Consequently, C/E is torsion-free.

We can now finish the proof quickly. As E is cotorsion and C/E torsion-free, we obtain $C = E \oplus F$ for a suitable submodule $F \cong C/E$ of E. As a summand, F is cotorsion, so (3.6) implies (ii). The uniqueness statement is obvious. □

In view of (XI.4.9), the torsion-free RD-injective modules over valuation domains can be characterized in terms of numerical invariants. These invariants, along with the torsion submodule T of C, provide a complete system of invariants for C.

EXERCISES

1. [D] Let E_i ($i \in I$) be a chain of torsion-free-essential extensions of the R-module M. Then their union $\cup E_i$ is likewise a torsion-free-essential extension of M.

2. [D] The cotorsion hull of a divisible module is again divisible. This can fail if 'divisibility' is replaced by 'h-divisibility'.

3. [D] The cotorsion hull of an infinite direct sum of injective torsion modules does not contain, in general, any maximal injective submodule.

4. [D] For a torsion-ultracomplete module, the RD-injective hull coincides with the cotorsion hull.

NOTES

This chapter demonstrates the power and versatility of several techniques developed so far for modules over domains. Surprisingly, once a theory for RD-injectives is developed, it becomes a relatively easy task to study torsion-ultracomplete and cotorsion modules.

The theory of cotorsion abelian groups was generalized to modules over domains by Matlis [3]. His definition was weaker: he called a module C cotorsion if $\text{Ext}^1_R(Q,C) = 0$. He also assumed that C was h-reduced, i.e. $\text{Hom}_R(Q,C) = 0$. He studied cotorsionness in relation to R-completeness and established several remarkable results. We follow a different pattern: our definition places emphasis on RD-injectivity rather than R-completeness. The material on cotorsion modules comes, to a great extent, from an unpublished manuscript of Warfield [5].

If R is not a domain, cotorsion modules can be defined in terms of flatness, cf. Enochs [1].

Torsion-ultracompleteness was introduced in Fuchs [3], while the more general concept of torsion-completeness was developed by Schoeman [1]. Although we have not treated the material in its fullest form, we went far enough to get acquainted with their main features, as they are expected to play a role in the theory of torsion modules.

PROBLEM 24. Investigate the torsion submodules of R-ultracomplete R-modules, for any domain R.

XIII. Torsion Modules

We have accumulated a fair amount of information about various classes of modules over valuation domains, including classes of torsion modules (divisible, RD-injective, etc.), and now we have come to the point of testing our arsenal on torsion modules in general.

If one begins his quest of a theory modeled on the theory of torsion abelian groups, very soon he discovers that the most fundamental lemmas which are the cornerstones of the theory are out of reach even in a watered down version. It is tempting to suppose that the underlying domain is almost maximal, but even this hypothesis would not eliminate the inherent difficulties. As a matter of fact, the lack of the maximum condition in the ring poses a very stubborn barrier which is impossible to come by.

Our results on torsion modules are quite meagre. The central topic is the study of pure uniserial submodules. We believe this is ideally suited to the role of a prelude to a more coherent theory of torsion modules. Unfortunately, no such theory is in sight at this moment.

1. PURE POLYSERIAL SUBMODULES

Our results are centered around two major classes which are at opposite ends of the spectrum. The first class consists of modules in which pure uniserial submodules are abundant; these are the separable modules studied in §§2-3. The other class is made up of modules which contain no pure uniserial submodules at all. §6 is devoted to investigating these modules. □

§1. EMBEDDING IN PURE POLYSERIAL SUBMODULES

In this section we investigate conditions to be imposed on the indicator of an element a of a module M over a valuation domain R, in order that a be contained in a pure standard polyserial submodule of M.

We start with an analysis of the indicators of elements in direct sums of standard uniserial modules.

By (VIII.3.4), if U is a standard uniserial module and $0 \neq a \in U$, then the indicator $i_U(a)$ assumes a constant non-limit height value up to Ann a.

Let now $M = \bigoplus_1^n U_i$, where the U_i are standard uniserial modules. Every $a \in M$ can be uniquely written in the form $a = \sum_1^n a_i$ ($a_i \in U_i$), and clearly

$$h_M(a) = \inf(h_{U_i}(a_i) \mid 1 \leq i \leq n).$$

Hence it is obvious that

(a) $i_M(a)$ assumes only non-limit height values.

Manifestly, $rh_M(ra) = h_M(a)$ for all $r \notin \bigcup_i$ Ann a_i. Furthermore $i_M(a)$ can have a gap at an ideal L of R only if $L =$ Ann a_i for some i. Hence the two following conditions are obvious:

(b) $i_M(a)$ has no irregularities other than gaps.

(c) $i_M(a)$ has but a finite number of gaps.

Finally, recall that uniserial modules are smooth, and smoothness is preserved under the formation of direct sums. Thus we have

(d) $i_M(a)$ is smooth.

We shall show that conditions (a)-(d) for $i_M(a)$ ensure that $a \in M$ can be embedded in a pure standard polyserial submodule of M.

THEOREM 1.1. [VD] (Fuchs-Salce [3]) Let M be an R-module and $a \in M$. Suppose $i_M(a)$ satisfies (a)-(d). Then a can be embedded in a pure standard polyserial submodule N of M such that the length of a pure-composition series of N equals the number n of gaps in $i_M(a)$, or $n+1$, according as Ann $a \neq 0$ or Ann $a = 0$.

Proof. We induct on n. If $n = 0$, then Ann $a = 0$. If $n = 1$ and $K =$ Ann $a \neq 0$, then the only gap in $i_M(a)$ occurs at K. In both cases (a) and (b) guarantee that the hypothesis of (VIII.3.4) is satisfied. Hence a is contained in a pure standard uniserial submodule of M.

Suppose now that $i_M(a)$ has n gaps, where $n \geq 1$ if Ann $a = 0$, and $n \geq 2$ if Ann $a \neq 0$. Let the last proper gap in $i_M(a)$ occur at the ideal L. Because of (d), we can write $a = b + c$ with Ann $b \geq L$ and $h_M(c) \geq i_M(a)^L = \sigma$. We claim that

$$rh_M(rc) = \sigma \text{ for all } r \notin \text{Ann } c.$$

This is obvious if $r \in L$, since then $rc = ra$ and $i_M(ra)$ has no irregularities except possibly at Ann $ra = r^{-1}$Ann a. If $r \notin L$, the claim follows by (VIII.5.1). As $h_M(c) = \sigma$ has to be non-limit, from (VIII.3.4) we infer that $c \in M$ is contained in a pure standard uniserial submodule U of M.

Our next step is to show that

$$h_{M/U}(r(a + U)) = h_M(ra) \text{ for all } r \notin L.$$

Assume, by way of contradiction, that there exist $x \in M$, $s \in R$, $r \in R \setminus L$ such that $sx - ra \in U$, but $ra \notin sM$. In this case, $h_M(sx - ra) = h_M(ra) < h_M(c)$. Let $t \in R$ be such that $tr \in L$ but $tra \neq 0$. Then $h_M(tra) = h_M(trc) > t^{-1}h_M(ra)$ and we have

$$t^{-1}h_M(ra) = h_M(tsx-tra) \geq \inf(h_M(tsx), h_M(tra)) > t^{-1}h_M(ra),$$

a contradiction. We conclude that the indicator of $a + U = b + U$ is the same as the indicator of a up to Ann$(b + U)$. Thus it has exactly $n-1$ gaps and it satisfies (a) and (b). To verify (d) for $b + U$, observe that $\tau = i_M(a)^J = i_{M/U}(b + U)^J$ for all

$J > \text{Ann}(b + U)$. If $a \in M[J] + M^\tau$, i.e. $a = x + y$ with $x \in M[J]$ and $y \in M^\tau$, then $b + U = (x + U) + (y + U)$, where $x + U \in (M/U)[J]$ and $y + U \in (M/U)^\tau$. An obvious induction completes the proof. □

If R is an almost maximal or a maximal valuation domain, then we are led to the following result.

COROLLARY 1.2. Let M be a (torsion) module over an (almost) maximal valuation domain. An element $a \in M$ can be embedded in a summand of M which is a direct sum of uniserial modules if and only if $i_M(a)$ satisfies conditions (a)-(d).

Proof. The necessary part is clear. By (IX.5.1) and (XI.4.2), polyserial (torsion) modules over an (almost) maximal valuation domain are direct sums of uniserial modules and are pure-injective. □

EXERCISE

1. Show that, in (1.2), if a summand of M containing a is chosen minimal, then its uniserial summands are uniquely determined up to isomorphism.

§2. SEPARABLE MODULES

In this section we shall study separable modules over valuation rings, i.e. modules such that every finite set of elements can be embedded in a direct summand which is a direct sum of uniserial modules. If these uniserial modules are standard, we will talk about standard separable modules.

We start with a couple of lemmas. The first is obvious.

LEMMA 2.1. [VR] (i) Direct sums of (standard) separable modules are likewise (standard) separable.

(ii) Indecomposable (standard) separable modules are (standard) uniserial. □

LEMMA 2.2. [VR] Fully invariant submodules of separable modules are again separable.

Proof. Let N be a fully invariant submodule of the separable module M, and let $a_1,\ldots,a_k \in N$. Then each $a_j \in U_1 \oplus \ldots \oplus U_m$, where each U_i is a uniserial module, and $M = U_1 \oplus \ldots \oplus U_m \oplus M'$. As N is fully invariant, it decomposes accordingly as $N = (U_1 \cap N) \oplus \ldots \oplus (U_m \cap N) \oplus (M' \cap N)$. Here the $U_i \cap N$ are uniserial and the a_j belong to their direct sum. □

LEMMA 2.3. [VR] Suppose that $a \in M$ is contained in a summand of M which is a finite direct sum of uniserial modules. If A is any summand of M containing a, then a is contained in a summand of A which is a direct sum of uniserial modules.

Proof. Let $M = A \oplus B = U_1 \oplus \ldots \oplus U_n \oplus M'$ with $a \in (U_1 \oplus \ldots \oplus U_n) \cap A$. By the exchange property of uniserial modules (see (VII.2.7)), we have $M = U_1 \oplus \ldots \oplus U_n \oplus A' \oplus B'$ with suitable submodules $A' \leq A$ and $B' \leq B$. Thus $A = A' \oplus A''$ with $a \in A'' = (U_1 \oplus \ldots \oplus U_n \oplus B') \cap A$. Note that A'' is isomorphic to a summand of $U_1 \oplus \ldots \oplus U_n$. Hence, by (VII.4.2), it is itself a direct sum of uniserial modules. □

The preceding result has remarkable consequences.

PROPOSITION 2.4. [VR] A module is separable if any single element is embeddable in a direct summand which is a direct sum of uniserial modules.

Proof. Suppose that the elements of M have the stated property. Let $a_1,\ldots,a_n \in M$. We induct on n to show that these elements are embeddable in a direct summand which is a direct sum of uniserial modules. The hypothesis gives the claim for $n = 1$. Assume now $n > 1$, and that $a_1,\ldots,a_{n-1} \in U_1 \oplus \ldots \oplus U_k$, where the U_i are uniserial and $M = U_1 \oplus \ldots \oplus U_k \oplus N$. Let a be the projection of a_n in N. Then (2.3) ensures that a is contained in a summand of N which is a direct sum of uniserial modules. □

PROPOSITION 2.5. [VR] Summands of separable modules are again separable.

2. SEPARABLE MODULES

Proof. Let $M = A \oplus B$ be a separable module. By (2.4) it is enough to show that, if $a \in A$, then a belongs to a summand of A which is a direct sum of uniserial modules. This is guaranteed by (2.3). □

The following is an easy consequence of the definitions and the smoothness of direct sums of uniserials.

PROPOSITION 2.6. [VR] Separable modules are smooth. □

In the rest of this section we concentrate our attention on the domain case. It is easy to give examples for separable modules which are not direct sums of uniserial modules.

EXAMPLE 2.7. Let M be an injective torsion R-module, R an almost maximal valuation domain. As every element $a \in M$ can be embedded in a uniserial summand of M, M is separable.

EXAMPLE 2.8. Let I be a non-zero proper ideal of an almost maximal valuation domain R such that R/I is not Artinian. Let M be a product of infinitely many copies of R/I. Every element $a \in M$ has constant indicator equal to a non-limit height, therefore it can be embedded in a standard uniserial summand of M. Consequently, M is separable. But it is not a direct sum of cyclic modules, since it is pure-injective, while an infinite direct sum of copies of R/I is not R-complete.

We now characterize separable torsion modules over almost maximal valuation domains in terms of indicators of their elements.

THEOREM 2.9. [VD] (Fuchs-Salce [3]) A torsion module M over an almost maximal valuation domain R is separable if and only if, for every $a \in M$, the indicator $i_M(a)$ satisfies conditions (a)-(d) of §1.

Proof. Necessity follows from the discussion at the beginning of §1 (so far almost maximality is not required). For the sufficiency, observe that (1.1) guarantees that a belongs to a pure polyserial submodule N of M. But in the present case polyserial torsion modules are direct sums of uniserial modules and are pure-injective. Therefore N is a summand of M. □

From the definition and from §1(a)-(d) we deduce that a separable torsion module over a valuation domain satisfies the hypothesis of (X.3.3). Thus, we have

COROLLARY 2.10. [VD] Every α-basic submodule of a standard separable torsion module over a valuation domain is basic. □

EXERCISES

1. Show that an abelian p-group is separable if and only if its reduced part has no elements of infinite height.

2. Prove that a divisible torsion module over an almost maximal valuation domain is separable if and only if it is h-divisible.

3. [VD] Prove or disprove: the product of separable torsion R-modules is again separable.

4. [VD] Let M be a separable torsion module and $a,b \in M$. There is an endomorphism [automorphism] of M carrying a into b if and only if $i_M(a) \geq i_M(b)$ [$i_M(a) = i_M(b)$].

5. [VD] Find an example of a separable torsion module all of whose elements $\neq 0$ have limit heights.

§3. SUBMODULES OF SEPARABLE MODULES

We gave in (VII.4.5) an example of an equiheight submodule of a direct sum of cyclic modules which was not smooth. This example shows, in particular, that the class of separable modules is not closed under taking equiheight submodules.

Here we give a slight modification of (VII.4.5) that will be useful at the end of this section.

EXAMPLE 3.1. Let R be a valuation domain whose maximal ideal $P = R_P$ is principal. Suppose there is an immediate non-zero prime successor ideal P', say, generated by the elements r_α ($\alpha < \lambda$) with $v(r_\alpha) > v(r_{\alpha+1})$ for all $\alpha < \lambda$. Consider the following

3. SUBMODULES OF SEPARABLE MODULES

direct sum of cyclic modules:

$$M = Rx \oplus Ry \oplus \bigoplus_{\alpha} Rx_\alpha$$

where $\text{Ann } x = P'$, $\text{Ann } y = prR$ with $0 \neq r \in P'$, and $\text{Ann } x_\alpha = pr_\alpha R$. Let $a = x - py$ and

$$N = \langle a; x_\alpha + y \ (\alpha < \lambda) \rangle.$$

It is straightforward to verify that N is equiheight in M, $N[P'] = 0$ and $i_N(a)^{P'} = p^{-1}R/R$. Obviously $h_N(a) = 0$. This N is not smooth, because $a \notin N[P'] + pN = pN$.

Separability is inherited by equiheight submodules if an additional condition on the annihilators of the elements is imposed. We need a preliminary result which is of independent interest.

PROPOSITION 3.2. [VR] Uniserial pure submodules of separable modules are summands.

Proof. Let U be uniserial and pure in the separable module M. Choose $0 \neq u \in U$. There exist uniserial modules $U_1, \ldots, U_n$ such that $U_1 \oplus \ldots \oplus U_n$ is a summand of M containing u. Let ϕ be the restriction to U of the projection of M onto $U_1 \oplus \ldots \oplus U_n$. Then ϕ is monic, keeping u fixed, and ϕU is pure in $U_1 \oplus \ldots \oplus U_n$. By (IX.5.6), ϕU is a summand of $U_1 \oplus \ldots \oplus U_n$, and so of M, $M = \phi U \oplus M'$. As $\phi U \cap U \geq Ru$ and U is pure in M, $M = U \oplus M'$ follows by a straightforward argument. □

We can now prove the announced result.

THEOREM 3.3. [VD] Equiheight submodules of standard separable modules, whose elements have principal ideal annihilators, are likewise separable.

Proof. Let N be an equiheight submodule of the standard separable module M; then $i_N(a) = i_M(a)$ for all $a \in N$, and N is smooth by (VIII.5.5). As the heights of the elements are non-limit, every element of N can be embedded in a pure polyserial submodule S of N. By making use of (3.2), we can induct on the length of S to conclude that it is a direct sum of uniserial modules and a summand of N. □

An immediate consequence of (3.3) is:

COROLLARY 3.4. [VD] Pure submodules of pure-projective modules are separable.

Proof. A pure-projective module is a direct sum of cyclic modules with principal annihilator ideals. In the present case, pure submodules are equiheight. Thus a simple appeal to (3.3) completes the proof. □

We conclude this section with a characterization of discrete rank one valuation domains, in terms of their separable modules.

THEOREM 3.5. For a valuation domain R, the following are equivalent:
1) R is a discrete rank one valuation domain.
2) Every cohesive torsion R-module is separable.

Proof. 1) $\implies$ 2) is a well-known result for abelian p-groups and for modules over discrete rank one valuation domains.

2) $\implies$ 1) A cohesive torsion R-module $M \neq 0$ contains a non-zero pure standard uniserial submodule. By (6.4) infra, P is then principal and has an immediate prime successor P'. If $P' \neq 0$, (3.1) shows how to construct a torsion R-module with cyclic heights which is not separable. Hence $P' = 0$ and R is discrete rank one. □

EXERCISES

1. [VD] Prove that a smooth, equiheight submodule of a standard separable module is likewise separable.

2. Over a maximal valuation domain, equiheight submodules of separable modules are separable.

§4. DIRECT SUMS OF CYCLIC MODULES

This section centers around issues related to direct sums of cyclic torsion modules. So far, the only approach to the study of these modules is homological, viz. by considering them as projective objects relative to cyclically-pure-exact sequences (Simmons [1]). We confine ourselves to valuation rings.

Unfortunately, no reasonably general criterion has as yet been found for a torsion module to be a direct sum of cyclics. In this respect, we can offer only a rather weak result.

THEOREM 4.1. (Fuchs-Salce [1]) Let R be an almost maximal valuation ring and M a countably generated torsion R-module. M is a direct sum of cyclic modules if and only if it is the union of an ascending chain of finitely generated pure submodules.

Proof. The necessity being obvious, assume M is the union of the ascending chain $0 = M_0 < M_1 < \ldots < M_n < \ldots$ where each M_n is finitely generated and pure in M. As R is almost maximal, by (XI.4.2), each M_n is both a direct sum of cyclics and pure-injective. Consequently, $M_{n+1} = M_n \oplus A_n$ for some submodule A_n of M_{n+1}, for every n, where A_n is necessarily a direct sum of cyclics. It follows that $M = \oplus A_n$, as desired. □

One of the major problems on direct sums of cyclic modules is the submodule problem: what can be said about submodules of direct sums of cyclics? And in particular, when is such a submodule likewise a direct sum of cyclic modules? The balance of this section is devoted to partial answers to these questions.

First, an easy but relevant observation for valuation rings: a summand of a direct sum of cyclics is again a direct sum of cyclics. In fact, this is a special case of (VII.4.2).

Recalling what we said at the beginning of this section, it is natural to concentrate on the cyclically pure submodules. We require a technical lemma.

LEMMA 4.2. [VR] (Simmons [1]) Suppose that M is a smooth R-module in which all the heights are principal. Then the cyclically pure submodules of M are likewise smooth.

Proof. Given an element $a \neq 0$ in a cyclically pure submodule N of M and an ideal $L \neq 0$ of M, by smoothness we can write

$$a = b + c \text{ with Ann } b \geq L \text{ and } h(c) \geq i(a)^L.$$

The hypothesis on the heights in M ensures that $i(a)$ is the same whether computed in M or in N. Set $h(c) = r^{-1}R/R$ $(r \in R)$ and let $x \in M$ be such that $rx = c$. We appeal to cyclic purity to obtain a $y \in N$ such that $\text{Ann}(x-y) = \text{Ann}(x+N)$. Obviously, $\text{Ann}(c+N) \geq L$ whence $\text{Ann}(rx-ry) \geq L$ follows. Now write $b' = b + (c-ry) = a - ry \in N$ and $c' = ry \in N$. Then $a = b' + c'$ with $\text{Ann } b' \geq L$ and $h(c') \geq r^{-1}R/R \geq i(a)^L$. □

The following is the most general result we can establish for cyclically pure submodules of direct sums of cyclics.

THEOREM 4.3. [VR] (Simmons [1]) Cyclically pure submodules of direct sums of cyclic R-modules are separable.

Proof. The assertion follows at once from §3, Ex.1, taking into account the preceding lemma. □

In particular, we conclude that over valuation rings the countably generated cyclically pure submodules of direct sums of cyclic modules are themselves direct sums of cyclics.

At this stage, it seems to be convenient to introduce the concept of cyclically-pure-projective dimension. We refer to (II.6.3) to guarantee the existence of enough cyclically-pure-projective modules (over any ring). Using the obvious resolutions, it should be clear what is meant by the cyclically-pure-projective dimension of a module; as in the projective case (IV.§1), this dimension is independent of the selected resolution. Hence, the following is obvious.

THEOREM 4.4. [VR] (Simmons [1]) Let M be a direct sum of cyclic R-modules and N a cyclically pure submodule of M. Then N itself

4. DIRECT SUMS OF CYCLIC MODULES

is a direct sum of cyclics if and only if the cyclically-pure-projective dimension of M/N is at most 1. □

In view of this result, it seems desirable to obtain conditions on a module to have cyclically-pure-projective dimension ≤ 1. Such conditions are given by the next two lemmas.

LEMMA 4.5. (Simmons [1]) A countably generated module over an almost maximal valuation ring has cyclically-pure-projective dimension ≤ 1.

Proof. Let the module M be generated by the elements a_n ($n < \omega$). For every finite subset α of ω, let C_α be isomorphic to the submodule $\langle a_n \mid n \in \alpha \rangle$ and ϕ_α a fixed isomorphism between them. By (IX.1.1), C_α is a direct sum of cyclic modules. Thus $C = \oplus C_\alpha$ (with α running over all finite subsets of ω) is a direct sum of countably many cyclic modules. The ϕ_α's induce an epimorphism $\phi : C \to M$. The kernel of ϕ is cyclically pure in C, since every $x \in M$ is contained in some $\langle a_n \mid n \in \alpha \rangle$ for a suitable α, so x has a preimage in C_α with annihilator equal to Ann x. It is readily checked that Ker ϕ is countably generated whence the remark made after (4.3) completes the proof. □

If the module M in this proof is uniserial, then the ring need not be almost maximal to ensure that finitely generated submodules are direct sums of cyclic modules, and we obtain:

LEMMA 4.6. [VR] (Simmons [1]) Every countably generated uniserial module is of cyclically-pure-projective dimension ≤ 1. □

EXERCISES

1. [VR] If C is a countable direct sum of torsion cyclic modules, then every cyclically pure submodule of C is again a direct sum of cyclics.

2. [VR] (Simmons [1]) A uniserial module which can be generated by $\aleph_n$ elements has cyclically-pure-projective dimension $\leq$ n+1.

262   XIII.   TORSION MODULES

3. (Simmons [1]) Let R be an almost maximal valuation domain. A module which can be generated by $\aleph_n$ elements is of cyclically-pure-projective dimension $\leq n+1$.

§5. TORSION MODULES OF PROJECTIVE DIMENSION ONE

There is an interesting class of torsion modules that deserves special attention as it looks more tractable. Here we wish to initiate its study with a couple of results (some of which were needed in VI.§3). Throughout we assume that R is a valuation domain.

We start with an easy lemma.

LEMMA 5.1. [VD] Let T be a countably generated R-module and $0 \neq r \in R$. Suppose that T is the union of an ascending chain $T_0 \leq \ldots \leq T_n \leq \ldots$ of submodules ($n < \omega$) where each T_n is a finite direct sum of copies of R/Rr. Then T is the direct sum of countably many copies of R/Rr.

Proof. It is readily checked that, for each n, T_n is pure in T_{n+1}. As T_{n+1}/T_n is finitely presented, and so pure-projective (cf. (II.3.4)), we obtain $T_{n+1} = T_n \oplus C_{n+1}$ for some $C_{n+1} \leq T_{n+1}$. Set $C_0 = T_0$. Obviously, each C_n is a direct sum of copies of R/Rr. It is routine to verify that $T = \oplus C_n$. □

Call an R-module T R/Rr-<u>homogeneous</u> if every element of T is contained in a submodule of T that is isomorphic to R/Rr. Evidently, $rT = 0$ holds for an R/Rr-homogeneous module T. It is clear that an R/Rr-homogeneous submodule is pure in every R-module annihilated by r.

LEMMA 5.2. [VD] Let M have projective dimension 1, and suppose $rM = 0$ for some $0 \neq r \in R$. An R/Rr-homogeneous tight submodule T of M is a summand of M.

Proof. As $p.d.M/T = 1$, by (IV.4.7) there is a well-ordered continuous ascending chain $T = M_0 < M_1 < \ldots < M_\alpha < \ldots < M_\tau = M$ of submodules such that each $M_{\alpha+1}/M_\alpha$ is cyclically presented. We

5. PROJECTIVE DIMENSION ONE

establish the existence of a continuous well-ordered ascending chain $0 = X_0 < X_1 < \ldots < X_\alpha < \ldots < X_\tau = X$ such that $M_\alpha = T \oplus X_\alpha$ for each $\alpha \leq \tau$. As the definition of X_α at a limit ordinal α is evident, assume that the X_α's have been constructed as desired up to and including β. We can view $M_{\beta+1}/X_\beta$ as an extension of $M_\beta/X_\beta \cong T$ by the pure-projective R-module $M_{\beta+1}/M_\beta$. Since T is necessarily pure in $M_{\beta+1}/X_\beta$, this module splits:

$$M_{\beta+1}/X_\beta = M_\beta/X_\beta \oplus X_{\beta+1}/X_\beta$$

for some $X_{\beta+1} \leq M_{\beta+1}$. Hence $M_{\beta+1} = T \oplus X_{\beta+1}$ follows at once. We conclude $M = T \oplus X$. □

Our main result on R/Rr-homogeneous R-modules is as follows.

THEOREM 5.3. [VD] An R/Rr-homogeneous R-module of projective dimension 1 is the direct sum of submodules, each $\cong$ R/Rr.

Proof. Let T be R/Rr-homogeneous of projective dimension 1, and $\mathcal{T}$ a tight system in T. A countably generated submodule $S_1 \in \mathcal{T}$ is by (IV.4.4) the union of an ascending chain $F_1 \leq F_2 \leq \ldots$ of finitely presented submodules. Hypothesis implies that each summand of F_n is embeddable in a submodule $\cong$ R/Rr of T, thus F_n is contained in a finite direct sum F'_n of copies of R/Rr. There is a countably generated $S_2 \in \mathcal{T}$ that contains $\cup F'_n$. Repeat this process with S_2 playing the role of S_1, and keep doing so to obtain an ascending chain $S_1 \leq S_2 \leq \ldots$ in $\mathcal{T}$. Evidently, $T_1 = \cup S_n \in \mathcal{T}$ is countably generated and R/Rr-homogeneous. (5.2) implies that T_1 is a summand of T; thus T/T_1 is likewise R/Rr-homogeneous.

It is now easy to construct a continuous well-ordered ascending chain

$$0 = T_0 < T_1 < \ldots < T_\alpha < \ldots \qquad (\alpha < \tau)$$

of submodules of T such that each $T_\alpha \in \mathcal{T}$ is a summand of T, $T_{\alpha+1}/T_\alpha$ is countably generated for each $\alpha < \tau$ and $T = \cup\{T_\alpha \mid \alpha < \tau\}$. It follows that $T \cong \oplus_\alpha (T_{\alpha+1}/T_\alpha)$ where the summands are, because of (5.1), direct sums of copies of R/Rr. □

COROLLARY 5.4. [VD] If D is a divisible R-module of projective dimension 1, then D[r] is a direct sum of copies of R/Rr for each $0 \neq r \in R$.

Proof. The exact sequence $0 \longrightarrow D[r] \longrightarrow D \xrightarrow{r} D \longrightarrow 0$ shows that D[r] is tight in D. The R/Rr-homogeneity of D[r] is a consequence of the divisibility of D. An appeal to (5.3) completes the proof. □

Clearly, the number of summands R/Rr in D[r] is independent of $r \neq 0$.

EXERCISES

1. [VR] Use (IX.5.6) rather than (II.3.4) to prove (5.1).

2. [VD] Let M be an R-module, $0 \neq r \in R$ such that p.d.M/rM = 1. If T is an R/Rr-homogeneous submodule of M, $T \cap rM = 0$, and if T is tight in M/rM, then T is a summand of M.

§6. MODULES WITH ZERO α-INVARIANTS

We exhibited in X. §4 an example of a torsion module over a valuation domain which failed to contain any non-zero pure standard uniserial submodules. It is this property that we focus our attention on in this section. The underlying rings will be valuation domains.

Given a module M, we say that a non-zero element $a \in M$ has ultimately constant non-limit indicator if there exists $0 \neq sa \in M$ ($s \in R$) such that $i_M(sa)$ is, up to Ann sa, constantly equal to a non-limit height. (We have already considered this property in the study of α-basic submodules; see (X.3.3)).

PROPOSITION 6.1. [VD] For a torsion R-module M, the following are equivalent:

1) M contains no non-zero pure standard uniserial submodules.

2) No element $\neq 0$ of M has ultimately constant non-limit indicator.

3) The α-invariants of M, corresponding to non-limit heights, are all zero.

Proof. This is an immediate consequence of (VIII.3.4) and (X.1.1). □

We can derive easily:

COROLLARY 6.2. [VD] A cohesive module which does not contain any non-zero pure standard uniserial modules has trivial socle.

Proof. Every non-zero element of the socle has ultimately constant non-limit indicator. □

The following lemma gives relevant information about indicators.

LEMMA 6.3. [VD] Let M be a smooth cohesive module with zero α-invariants. If $0 \neq a \in M$ and $i_M(a)$ has a gap at the ideal L, then $i_M(a)$ increases on the left at L.

Proof. Let $a = b + c$ with $b \in M[L]$ and $c \in M^\sigma$, $\sigma = i(a)^L$. By (VIII.5.1), $i(a) = i(b)$ up to L. If $i(a)$ does not increase on the left at L, then b has ultimately constant non-limit indicator, which is absurd, since $b \neq 0$. Thus $i(a)$ increases on the left at L. □

Our next goal is to characterize the class of valuation domains R such that every cohesive torsion R-module $\neq 0$ contains non-zero pure uniserial submodules. Manifestly, this class includes all the discrete rank one valuation domains.

THEOREM 6.4. (Zanardo [1]) For a valuation domain R, the following are equivalent.

1) Every cohesive torsion R-module $\neq 0$ contains a non-zero pure uniserial submodule.

2) For every prime ideal J of R, JR_J is a principal ideal of R_J.

3) R is discrete and the set of prime ideals of R is well-ordered by the opposite inclusion.

Proof. 1) $\Longrightarrow$ 2) Assume that JR_J is not a principal ideal of R_J for some prime ideal J of R. By X.§4, there exists an R_J-module A, all of whose elements have cyclic heights, which fails to contain pure uniserial submodules $\neq 0$. A is naturally an R-module, that we denote by A_R to avoid confusion. By (VIII.1.6), A_R is cohesive. We show that A_R likewise fails to contain non-zero pure uniserial submodules. Let $0 \neq a \in A_R$; there exist $r,s \in R_J$ such that $a \notin rA$ and $0 \neq sa \in srA$. Write $r = qp^{-1}$, $s = tp^{-1}$ ($q,t \in R$; $p \in R\setminus J$). Then $0 \neq ta$, because p is a unit in R_J, and $v(p) < v(q)$, otherwise $a \in rA$. It follows that $r \in R$, thus $0 \neq ta \in trA_R$ and $a \notin rA_R$ show that a has no ultimately constant non-limit indicator in A_R. In view of (6.1), 2) follows.

2) $\Longrightarrow$ 3) Let J be a non-zero prime ideal of R. By hypothesis, $JR_J = qR_J$ for some $q \in J$. By (I.1.6), $J' = \bigcap_{n<\omega} J^n$ is a prime ideal of R contained in J. Notice that $J > J'$, because $q \notin J'$. Clearly there are no prime ideals between J and J'. It follows that every non-zero prime ideal has an immediate successor in the ordering of the opposite inclusion. Moreover, every non-empty set of prime ideals has a supremum in this ordering. Therefore, the set of all prime ideals is well-ordered, with P as minimum element and $\{0\}$ as maximum element. Recall now that R discrete means that for any two consecutive prime ideals $P_1 > P_2$ of R, the factor ring $R_{P_1}/P_2 R_{P_1}$ is a discrete rank one valuation domain. This fact is an obvious consequence of the hypothesis.

3) $\Longrightarrow$ 1) Suppose there exists a cohesive R-module M without non-zero pure uniserial submodules. For an ordinal σ, let $\{J_\alpha \mid \alpha \leq \sigma\}$ be the well-ordered set of primes of R; thus $J_0 = P$, $J_\sigma = 0$, and $\alpha < \beta$ means $J_\alpha > J_\beta$. For every $\alpha < \sigma$, $R_{J_\alpha}/J_{\alpha+1}R_{J_\alpha}$ is by hypothesis a discrete rank one valuation domain, thus $J_\alpha R_{J_\alpha} = q_\alpha R_{J_\alpha}$ for some $q_\alpha \in J_\alpha$ (in particular, 2) follows). Choose β to be minimal with $M[J_\beta] \neq 0$. Then $\beta > 0$, because $M[P] = 0$ by (6.2). Pick $0 \neq a \in M[J_\beta]$; in view of (6.1), there exist $r,s \in R$, $b \in M$ such that $0 \neq ra = rsb$ and $a \notin sM$. If γ is the minimal

6. MODULES WITH ZERO α-INVARIANTS

ordinal such that $q_\gamma^n \in rR$ for some $n \in \omega$, then $J_{\gamma+1} < rR \leq J_\gamma$ and $\gamma < \beta$ because of $r \notin J_\beta$. Choose $m \in \omega$ to be minimal such that $q_\gamma^m a = q_\gamma^m sb$; evidently, $1 \leq m \leq n$. Consider the element

$$0 \neq x = q_\gamma^{m-1}(a - sb) \in M[q_\gamma].$$

Observe that J_γ is generated by elements of the form $t^{-1}q_\gamma$ for certain $t \in R \setminus J_\gamma$, and for such a t, $M[t] = 0$ as $M[t] \leq M[J_\gamma] = 0$. Therefore, $t(t^{-1}q_\gamma x) = q_\gamma x = 0$ implies $t^{-1}q_\gamma x = 0$, i.e. J_γ annihilates x. We reached a contradiction, and 1) follows. □

It is worth giving an example to illustrate (6.4).

EXAMPLE 6.5. [VD] Let σ be an arbitrary ordinal, and consider the free abelian group $\Gamma = \bigoplus_{\alpha < \sigma} \mathbb{Z}x_\alpha$. Define an ordering on Γ by setting

$$\sum_{1}^{n} m_{\alpha_i} x_{\alpha_i} > 0 \quad \text{if} \quad m_{\alpha_n} > 0$$

where $0 \neq m_{\alpha_i} \in \mathbb{Z}$ for all i and $\alpha_1 < \alpha_2 < \ldots < \alpha_n$. It is easy to check that the convex subgroups of Γ are exactly the subgroups $\Gamma_\beta = \bigoplus_{\alpha < \beta} \mathbb{Z}x_\alpha$ ($\beta \leq \sigma$). Obviously, $\Gamma_{\beta+1}/\Gamma_\beta$ is order-isomorphic to $\mathbb{Z}$ for all $\beta+1 \leq \sigma$. If R is a valuation domain such that $\Gamma(R) \cong \Gamma$, then R satisfies condition 3) of (6.4).

It is worthwhile pointing out that (6.4) may fail for modules which are not cohesive. In fact, if p.d.Q ≥ 2, then the module ∂ of VI.§3 is h-reduced and contains no uniserial divisible submodules. In this case, all elements $\neq 0$ have limit heights.

EXERCISES

1. [VD] Let M be a smooth cohesive R-module with zero α-invariants. If $0 \neq a \in M$ and $L = \{r \in R \mid rh(ra) > h(a)\} = Ps$ for some $s \in R$, then $i(a)^L = h(sa)$.

2. Let R be a valuation domain with value group $\mathbb{Z}^\omega$ (with the lexicographic ordering). Show that there are cohesive R-modules without pure, standard uniserial modules $\neq 0$.

3. (Zanardo [1]) Show that the following is equivalent to the conditions in (6.4): every ideal of R is isomorphic to a prime ideal of R.

NOTES

Very little of the properties of torsion modules over valuation domains was known until recently, even particular examples were not easy to come by. Abelian group theory was much too special to provide a clear guidance in laying a sound groundwork for a general theory. It came as a pleasant surprise that the study of separability opened the way to the introduction of great many concepts and techniques, and to the discovery of new phenomena.

We can not help but note that the mere fact that the theory of torsion modules over valuation domains could be developed this far, primarily motivated by abelian groups (and subjects like simple presentation have not even been touched upon), is certainly a tribute to the advanced state of art in abelian groups. The reader is, however, cautioned that such a development has the shortcoming of not being based on its own foundations and the theory has to undergo vast changes before it becomes an independent branch of module theory.

Most of the results in this chapter are due to the authors. Some of them are formulated here in a more general setting than published originally. For the results in §4 and §6, see Simmons [1] and Zanardo [1].

PROBLEM 25. Do there exist indecomposable torsion modules whose Goldie dimensions are large infinite cardinals?

XIV. Torsion-Free Modules

For numerous reasons, the study of torsion-free modules has a flavor different from the torsion case. Though the results that follow are only of a fragmentary nature so far, the torsion-free case is more developed. Notably, several aspects of torsion-free abelian groups admit generalizations to torsion-free modules over valuation domains. Some suggestive results have been developed, but at this moment it is not clear how this approach will further elucidate the structure of torsion-free modules over valuation domains.

§1. PRELIMINARIES

Our point of departure for a study of torsion-free modules over domains, and in particular, over valuation domains, is the discussion of the rank one case. The rank one modules are the building blocks of torsion-free modules as every element is contained in a unique rank one RD-submodule. (Recall that by the rank of an

R-module M is meant the Q-dimension of $M \otimes_R Q$.)

For any domain R, the torsion-free R-modules of rank 1 are precisely the R-submodules $\neq 0$ of Q. Every homomorphism $\alpha : J \to L$ between two non-zero submodules is simply a multiplication by a suitable $q \in Q$. Therefore,

$$\mathrm{Hom}_R(J,L) \cong L : J.$$

Manifestly, $J \cong L$ if and only if $J = qL$ for some $0 \neq q \in Q$.

If R is a valuation domain, then each rank one torsion-free R-module, which is not isomorphic to Q, is isomorphic to an ideal of R. Consequently,

THEOREM 1.1. *The rank one torsion-free modules over a valuation domain R are characterized by the isomorphism classes of ideals of R, plus the isomorphism class of Q.* □

A more explicit information can be obtained by using the notion of height, introduced in Chapter VIII. Let $\sigma = J/R$ be a height (torsion-free modules are always cohesive), and M a torsion-free module over the valuation domain R. The σ-<u>invariant</u> of M is defined as

$$\alpha_M(\sigma) = M^\sigma/M^{\sigma+}$$

which is an S/P_S-vectorspace where $S = \mathrm{End}_R J$ and P_S is its maximal ideal.

LEMMA 1.2. [VD] *If L is a rank one torsion-free module, and $\sigma = J/R$ is a height, then*

$$\dim \alpha_L(\sigma) = \begin{cases} 1 & \text{if } L \cong J, \\ 0 & \text{otherwise}. \end{cases}$$

<u>Proof</u>. This follows from the simple observation that L contains an element of height σ exactly if $L \cong J$. □

From what has been said above it is clear that the endomorphism ring of a rank one torsion-free module is isomorphic to a subring of Q containing R. Hence it is a valuation domain if R is a valuation domain, and from (II.7.2) we conclude:

1. PRELIMINARIES

LEMMA 1.3. [VD] The rank one torsion-free modules have the exchange property. □

Recall that in Chapter X we have established the existence of basic submodules B in torsion-free modules M over valuation domains. As they satisfy $\alpha_B(\sigma) = \alpha_M(\sigma)$ for each σ, basic submodules are unique up to isomorphism. We reiterate that in a torsion-free module over a valuation domain every maximal RD-independent system of rank one submodules generates a basic submodule.

We emphasize that M/B need not be divisible, i.e. B need not be dense in M in the R-topology. In fact, R is a basic submodule in its RD-injective hull $\hat{R}$, but $\hat{R}/R$ is divisible only if $\tilde{R} = \hat{R}$ (which is the case exactly if R is almost maximal). However, we can prove:

LEMMA 1.4. Let R be an almost maximal valuation domain. Basic submodules B of finite rank torsion-free R-modules M are dense in M.

Proof. If M/B were not divisible, then it would contain an RD-submodule C/B isomorphic to an ideal of R. Owing to (VI.5.4), B is a summand of C. Here C is RD in M, so B could not have been RD-essential in M. □

Next we establish an inequality between $|M|$ and $|B|$. With no extra effort we can prove a more general result, valid over arbitrary domains.

THEOREM 1.5. [D] Let M be a torsion-free R-module and B a submodule of M, generated by a maximal RD-independent family of rank one submodules of M. Then

$$|M| \leq |B|^m \quad \text{where} \quad m = |R| \aleph_0.$$

Proof. B is RD-essential in M as is readily seen from its definition. Hence M is contained in the RD-injective hull $\hat{B}$ of B. Consequently, an upper bound for $|M|$ is given by the cardinality of $\hat{B} \cong \text{Hom}_R(K, E_R(K \otimes B))$ where $K = Q/R$; cf. XI.§1, Ex.2. It is well known that the injective hull of any R-module N is contained in $\text{Hom}_{\mathbb{Z}}(R,D)$ where D is the $\mathbb{Z}$-injective hull of N. Hence the inequality $|E(N)| \leq |N|^m$ should be clear. We conclude

and
$$|E_R(K \otimes B)| \leq |K \otimes_R B|^m \leq |B|^m$$
$$|\text{Hom}_R(K, E_R(K \otimes B))| \leq (|B|^m)^m = |B|^m. \quad \square$$

We now return to valuation domains R. If J is an ideal of R, then by a J-<u>homogeneous</u> R-module is meant a torsion-free R-module M such that all rank one RD-submodules of M are isomorphic to J. <u>Homogeneous</u> R-module will mean a module which is J-homogeneous for some J. The next lemma is trivial.

LEMMA 1.6. [VD] A torsion-free R-module is J-homogeneous exactly if one (and hence each) of its basic submodules is J-homogeneous. $\quad\square$

If M is J-homogeneous and if S denotes the R-endomorphism ring $\text{End}_R J$, then M becomes an S-module in the natural way. S is a localization of R and is again a valuation domain, thus switching from R to S, the study of J-homogeneous torsion-free R-modules can be reduced to the case $\text{End } J \cong R$. A further reduction is possible to the R-homogeneous case, by making use of the following straightforward result.

PROPOSITION 1.7. [VD] (Fuchs-Viljoen [1]) Let J be an ideal of R such that $\text{End}_R J \cong R$. There is a bijection between the class C_J of non-isomorphic J-homogeneous torsion-free R-modules and the class C_R of non-isomorphic R-homogeneous torsion-free R-modules, viz.

$$M \longmapsto \text{Hom}_R(J,M) \quad (M \in C_J)$$

and

$$N \longmapsto J \otimes_R N \quad (N \in C_R).$$

These correspondences are inverse to each other.

<u>Proof</u>. By hypothesis, $\text{Hom}_R(J,J) = R$. As Hom preserves purity, $\text{Hom}_R(J,M) \in C_R$ for all $M \in C_J$. Tensoring also preserves purity, thus $N \in C_R$ implies $J \otimes_R N \in C_J$. Observe that every non-zero homomorphism $R \to N$ is monic and induces a monomorphism

2. COMPLETELY DECOMPOSABLE MODULES

$J \to J \otimes_R N$. Every homomorphism $\phi : J \to J \otimes_R N$ is completely determined by the image of any non-zero element $x \in J$, and since $\text{Hom}(J,J) = R$, ϕx is of the form $x \otimes y$ for some $y \in N$. Hence the isomorphism $\text{Hom}_R(J, J \otimes N) \cong N$ follows readily. Every element of $M \in C_J$ is clearly the image of some element $a \in J$ under a suitable map $\alpha : J \to M$, thus the map $a \otimes \alpha \longrightarrow \alpha a \in M$ is an isomorphism between $J \otimes \text{Hom}_R(J,M)$ and M. □

EXERCISES

1. [VD] (Franzen [2]) A valuation domain R is
   (a) almost maximal exactly if basic submodules of finite rank torsion-free R-modules are dense;
   (b) a discrete rank one valuation domain if basic submodules in all torsion-free R-modules are necessarily dense.

2. [D] Define basic submodules for arbitrary domains as B is defined in (1.5). Show that such basic submodules need not be isomorphic (not even for $R = \mathbb{Z}$).

3. [VD] Direct products of J-homogeneous R-modules need not be J-homogeneous.

§2. COMPLETELY DECOMPOSABLE MODULES

In this section, R will denote a valuation domain. We call a torsion-free R-module <u>completely decomposable</u> if it is a direct sum of rank one R-modules. For instance, basic submodules are completely decomposable.

We have the fundamental result:

THEOREM 2.1. Over a valuation domain R, any two direct decompositions of a completely decomposable torsion-free R-module M into direct sums of rank one modules are isomorphic. Every summand of M is likewise completely decomposable.

<u>Proof</u>. Both assertions follow from (II.7.3) as the endomorphism rings of rank one torsion-free modules are valuation domains

(see (I.4.6)), while the modules themselves are small in the category of torsion-free R-modules.

It should be observed that the first part of the theorem is a corollary to (X.3.2). In fact, a completely decomposable module is a basic submodule of itself. □

A noteworthy property is the following.

THEOREM 2.2. [VD] Every rank one pure submodule of a completely decomposable torsion-free module is a summand.

Proof. It suffices to prove that a rank one pure submodule L of $M = J_1 \oplus \ldots \oplus J_k$ with J_i of rank 1 is a summand of M. Pick any $0 \neq a \in L$ and write $a = a_1 + \ldots + a_k$ ($a_i \in J_i$). For the heights we clearly have $h_M(a) = h_{J_1}(a_1) \cap \ldots \cap h_{J_k}(a_k)$. As the set of heights is totally ordered, it follows that for some i, say for $i = 1$, $h_M(a) = h_{J_1}(a_1)$ holds. This implies that $L \cong J_1$ and the projection of L on J_1 is surjective. We conclude that J_1 can be replaced by L, $M = L \oplus J_2 \oplus \ldots \oplus J_k$. □

The completely decomposable modules form a projective class relative to certain exact sequences.

A submodule C of a torsion-free R-module M will be called a balanced submodule if C is pure in M and for every $a \in M$, there is a $c_0 \in C$ such that

$$h_M(a + c_0) = h_{M/C}(a + C),$$

i.e. every coset mod C can be represented by an element of the same height as the coset. In case C is a balanced submodule of M, the exact sequence $0 \to C \to M \to M/C \to 0$ will be called balanced-exact.

LEMMA 2.3. [VD] Completely decomposable torsion-free modules have the projective property relative to all balanced-exact sequences of torsion-free modules.

Proof. Evidently, it is enough to check the projective property for rank one torsion-free modules. Let $J \neq 0$ be a submodule of Q; without loss of generality, $R \leq J$ may be assumed.

2. COMPLETELY DECOMPOSABLE MODULES

If $0 \longrightarrow C \longrightarrow A \overset{\alpha}{\longrightarrow} B \longrightarrow 0$ is a balanced-exact sequence of torsion-free R-modules and $\phi : J \to B$ is a homomorphism, then by balancedness, there exists an $a \in A$ such that

$$\alpha a = \phi(1) \quad \text{and} \quad h_A(a) = h_B(\phi(1)).$$

As $h_B(\phi(1)) \geq J/R$, there is a (unique) homomorphism $\psi : J \to A$ such that $\psi(1) = a$. Therefore, $\alpha\psi(1) = \phi(1)$, whence torsion-freeness makes it possible to conclude that $\alpha\psi = \phi$. ☐

The question of 'enough' balanced-projectives can be settled at once:

LEMMA 2.4. [VD] For every torsion-free R-module M there is a balanced-exact sequence

$$0 \to B \to C \to M \to 0$$

of torsion-free modules with C completely decomposable.

Proof. Let $\{M_i \mid i \in I\}$ be the set of all rank one pure submodules of M. For every $i \in I$, select a copy C_i of M_i along with a fixed isomorphism $\phi_i : C_i \to M_i$. Set $C = \bigoplus_{i \in I} C_i$, and define $\phi : C \to M$ as the map induced by the ϕ_i ($i \in I$). By construction, ϕ is surjective. Moreover, Ker ϕ is balanced in C, since every $a \in M$ is contained in some M_i, so $\phi_i^{-1}a \in C_i \leq C$ will be a preimage of a of the same height as a. ☐

This leads to the following characterization.

THEOREM 2.5. A torsion-free module over a valuation domain is balanced-projective if and only if it is completely decomposable.

Proof. If M is balanced-projective, then the balanced-exact sequence in (2.4) splits. Thus M is isomorphic to a summand of a completely decomposable torsion-free module, and hence itself completely decomposable [cf. (2.1)]. The converse is covered by (2.3). ☐

There is an interesting generalization of complete decomposability, called separability. This notion is analogous to separability discussed in XIII.§2 for torsion modules.

A torsion-free module M is said to be <u>separable</u> if every finite set of its elements is contained in a completely decomposable summand (of finite rank) of M. This "local" complete decomposability is easily recognizable:

LEMMA 2.6. [VD] A torsion-free R-module M is separable if and only if, for each a $\in$ M, the pure submodule of M, generated by a, is a summand of M.

<u>Proof.</u> Combining the definition of separability with (2.2), the necessity of the stated condition is evident. On the other hand, sufficiency follows by a straightforward induction on the number of elements in the finite set to be embedded in a completely decomposable summand. □

An immediate consequence of (2.6) is:

COROLLARY 2.7. [VD] Summands of separable torsion-free modules are likewise separable. □

One can derive easily:

THEOREM 2.8. [VD] A separable torsion-free module of countable rank is completely decomposable. □

The easiest examples for separable modules which are not necessarily completely decomposable are the pure submodules of completely decomposable modules. In fact,

THEOREM 2.9. [VD] Pure submodules of separable torsion-free modules are separable.

<u>Proof.</u> Let A be a rank one pure submodule of the R-module N which is pure in a separable module M. Then A is pure in M, hence a summand of M, cf. (2.2). We conclude that A is a summand of N. □

EXERCISES

1. [VD] Show that the balanced-exact sequences form a proper class.

2. [D] Define balanced-exactness for arbitrary domains by requiring the rank one torsion-free modules to have the projective property. Show that there are enough balanced-projectives and characterize them as summands of completely decomposable modules.

3. [VD] Fully invariant submodules of separable R-modules are separable.

4. [VD] Direct sums of separable R-modules are again separable.

5. (Franzen [1]) Let R be an almost maximal valuation domain and Q countably generated. Then the product of countably many copies of R is separable if and only if either R is discrete rank one or maximal.

§3. FINITE RANK MODULES OVER ALMOST MAXIMAL VALUATION DOMAINS

Having studied the rank one modules and their direct sums, we proceed to the finite rank case. Unfortunately, no satisfactory classification is known, not even for discrete valuation domains. Pleasant exceptions are the maximal valuation domains in which case all finite rank torsion-free modules are completely decomposable. The general case being unmanageable, we shall be content with the study of almost maximal valuation domains.

Let R denote an almost maximal valuation domain and M a finite rank torsion-free R-module. In view of (1.4), the basic submodules B of M are dense in M, i.e. for every $r \neq 0$ in R, $M = rM + B$.

LEMMA 3.1. Submodules of a reduced, completely decomposable torsion-free module M of finite rank over an almost maximal valuation domain R are again completely decomposable.

Proof. M is isomorphic to a finite direct sum of ideals of R, hence Q can not be an epic image of M. The same holds for submodules N of M, thus N modulo its basic submodule B is 0. We conclude that N = B is completely decomposable. ☐

This lemma leads to the following characterization of almost maximality.

THEOREM 3.2. (Franzen [1]) A valuation domain R is almost maximal if and only if all submodules of finite rank free R-modules are completely decomposable.

Proof. The necessity follows from (3.1) immediately. In order to verify sufficiency, assume R not almost maximal, i.e. i.d.$R \geq 2$, or equivalently, $\text{Ext}_R^1(J,R) \neq 0$ for some ideal J of R. Let the exact sequence $0 \longrightarrow R \longrightarrow M \xrightarrow{\alpha} J \longrightarrow 0$ represent a non-split extension of R by J. If M were decomposable, say, $M = M_1 \oplus M_2$ with M_i of rank 1, then either $\alpha M_1 = J$ or $M_2 = J$; e.g. in the first case $M = R \oplus M_1$ would follow, a contradiction. Thus M is indecomposable. Choose any $0 \neq x \in J$; then $rJ \leq Rx$ for some $0 \neq r \in R$. The submodule $M^* = \alpha^{-1}(Rx)$ is a free R-module, such that $rM \leq M^*$. If F is a free R-module with $rF = M^*$, then it contains (a copy of) M. □

We can now proceed to a similar characterization of maximal valuation domains.

THEOREM 3.3. (Kaplansky [1]) A valuation domain R is maximal if and only if all torsion-free R-modules of finite rank are completely decomposable if and only if all torsion-free R-modules of finite rank are cotorsion [pure-injective].

Proof. As the pure-injective hull of the R-module R is its maximal immediate extension (cf. (XI.5.9)), the stated condition is necessary. On the other hand, all rank one modules over a maximal valuation domain are by (XI.4.2) pure-injective whence the assertion is immediate. □

It is useful to have some information about the R-completions of finite rank modules.

LEMMA 3.4. [VD] Let M be a reduced torsion-free R-module with finite rank dense basic submodule B. Then for the R-completions $\widetilde{M}, \widetilde{B}, \widetilde{R}$, we have canonical isomorphisms

3. FINITE RANK MODULES

$$\tilde{M} \cong \tilde{B} \cong B \otimes_R \tilde{R}.$$

Proof. B being dense in M, it is evident that the completion of B in the induced topology is just $\tilde{M}$. But this topology is nothing else than the R-topology of $\tilde{M}$, whence $\tilde{B} \cong \tilde{M}$ is clear. Completions commute with finite direct sums, so it only remains to show that $\tilde{L} \cong L \otimes \tilde{R}$ for an ideal L of R.

Note that $L \cong L \otimes R$ is a basic submodule of $L \otimes \tilde{R}$, thus $L \otimes \tilde{R}$ embeds as a pure and dense submodule in $\tilde{L}$. The isomorphism $L \otimes \tilde{R} \cong L\tilde{R}$ shows that $L \otimes \tilde{R}$ is isomorphic to an ideal of $\tilde{R}$, and therefore, in view of (V.1.3), it is R-complete. Consequently, $\tilde{L} \cong L \otimes \tilde{R}$. □

We conclude this section with the discussion of the so-called splitting fields for modules.

The following lemma will be required; we state it in a more general form than needed here.

LEMMA 3.5. (Menini [1]) Let R be a valuation domain, $\hat{R}$ a maximal immediate extension of R, and Q, $\hat{Q}$ their quotient fields. If K is any field between Q and $\hat{Q}$, and $S = K \cap \hat{R}$, then for every torsion-free R-module M,

$$S \otimes_R M = (K \otimes_R M) \cap (\hat{R} \otimes_R M).$$

Proof. The inclusion $\leq$ being obvious, let x be an element of $(K \otimes M) \cap (\hat{R} \otimes M)$. The coset $x + (S \otimes_R M)$ belongs to $(\hat{R} \otimes_R M)/(S \otimes_R M) \cong (\hat{R}/S) \otimes_R M$; the last isomorphism is evident in view of the exact sequence $0 \to S \otimes M \to \hat{R} \otimes M \to \hat{R}/S \otimes M \to 0$ induced by the RD-exact sequence $0 \to S \to \hat{R} \to \hat{R}/S \to 0$. As $(\hat{R}/S) \otimes M$ is torsion-free and as $rx \in S \otimes M$ for some $0 \neq r \in R$, we obtain $x \in S \otimes M$, as desired. □

Keeping the notations of the preceding lemma, the field K is said to be a <u>splitting field</u> for a finite rank torsion-free R-module M if the S-module $S \otimes_R M$ is completely decomposable. From (3.3) it follows that $K = \hat{Q}$ is a splitting field for every torsion-free R-module M of finite rank. In general, there exist much smaller splitting fields.

THEOREM 3.6. (Menini [1]) Let R be an almost maximal valuation domain and M a reduced torsion-free R-module of finite rank. In $\hat{Q}$ there exists a unique splitting field K_M for M which is contained in all other splittng fields for M. This K_M is a finitely generated field extension of Q.

Proof. Let K be a splitting field for M, and $S = K \cap \tilde{R}$ (in the present case, $\hat{R} = \tilde{R}$). Let $B = \oplus_{i=1}^{k} U_i$ (U_i uniserial) be a basic submodule of M. By (1.4), M/B is divisible, thus it has a Q-basis $\{a_j + B\}_{j=1}^{m}$. Clearly, $S \otimes_R M$ will be the direct sum of $S \otimes_R U_i$ ($i = 1,\ldots,k$) and an m-dimensional K-vectorspace D. We have a commutative diagram with exact rows:

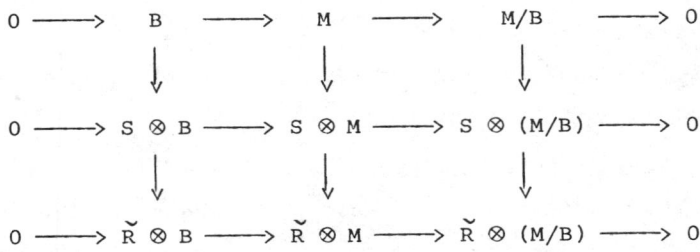

where the vertical arrows are the obvious embeddings induced by the ring embeddings $R \to S \to \tilde{R}$. The middle row is splitting, so we can write

$$1 \otimes a_j = \sum_{i=1}^{k} \alpha_{ij} (1 \otimes u_{ij}) + d_j \quad (j = 1,\ldots,m) \qquad (1)$$

where $\alpha_{ij} \in S$, $u_{ij} \in U_i$, $d_j \in D$. Evidently, (1) holds true in $\tilde{R} \otimes M$ as well, and hence (1) is the same for every splitting field K provided that both a basic submodule with its decomposition $B = \oplus U_i$ and representatives a_j of a Q-basis of M/B are kept fixed. Consequently, $\alpha_{ij} \in K$ for every splitting field K for M. In other words, the field extension

$$K_M = Q(\alpha_{ij}; i = 1,\ldots,k; j = 1,\ldots m)$$

is contained in every splitting field for M.

It remains to show that K_M itself is a splitting field. The equations (1) make sense in K_M where they are interpreted as definitions of d_j ($j = 1,\ldots,m$). These d_j span the divisible part of $\tilde{R} \otimes M$, whence (3.5) implies that they belong to the divisible part of $S \otimes_R M$. We conclude that $S \otimes_R M$ is splitting. □

EXERCISES

1. A finite rank, reduced, torsion-free module M over an almost maximal valuation domain has a rank 1 summand if and only if $\text{Hom}_R(M,R) \neq 0$.

2. [D] Let M be torsion-free of finite rank and N a submodule of M. If $N \cong M$, then $rM \leq N$ for some $0 \neq r \in R$.

3. [VD] Give an estimate for the number of generators of K_M in terms of the rank of M, in (3.6).

§4. RANK ONE DENSE BASIC SUBMODULES

As a first step towards the classification of finite rank torsion-free modules, we discuss a special class which includes the rank one modules and, over almost maximal valuation domains, all indecomposable rank two modules. This study was initiated by Fuchs-Viljoen [1]; cf. Franzen [2] for a more direct approach.

Throughout this section, R will denote a valuation domain.

Let M be a reduced torsion-free R-module satisfying the following two properties:

(i) the basic submodules B of M are of rank 1;
(ii) M/B is divisible for each basic submodule B.

A module M with (i) and (ii) can be referred to as one <u>with rank one dense basic submodules</u>. Such an M is evidently indecomposable. In connection with (ii), it should be noted:

LEMMA 4.1. [VD] If M is a finite rank torsion-free R-module and M/B_0 is divisible for some basic submodule B_0 of M, then M/B

is divisible for all basic submodules B of M.

   Proof. Let S be a maximal immediate extension of R. From the pure-exact sequence $0 \to B \to M \to M/B \to 0$ we derive the pure-exact sequence

$$0 \to S \otimes B \to S \otimes M \to S \otimes (M/B) \to 0.$$

Here $S \otimes B$ is a finite rank torsion-free module over a maximal valuation domain, so the sequence is splitting: $S \otimes M \cong (S \otimes B) \oplus (S \otimes M/B)$. The second summand is divisible if and only if M/B is divisible. Comparing the direct decompositions of $S \otimes M$ for B and B_0, the assertion follows at once. □

   In view of (3.4), the following is clear.

PROPOSITION 4.2. [VD] A torsion-free R-module has a dense basic submodule isomorphic to the ideal J of R if and only if it is isomorphic to a pure and dense R-submodule of $\tilde{J}$. □

   Observe that if $\tilde{M} \cong \tilde{J}$, then every R-endomorphism of M extends uniquely to $\tilde{J}$. Furthermore, the endomorphism ring of $\tilde{J}$ is isomorphic to $\tilde{S}$ where $S = \mathrm{End}_R J$. Therefore, $\mathrm{End}_R M$ is a subring of $\tilde{S}$ containing S, and we conclude:

LEMMA 4.3. [VD] The endomorphism ring of a torsion-free R-module M whose basic submodules are dense and isomorphic to an ideal J of R is a valuation domain between $\mathrm{End}_R J \cong S$ and $\tilde{S}$. □

   Now, a simple appeal to (II.7.3) leads us to the following result.

THEOREM 4.4. [VD] (Fuchs-Viljoen [1]) Let $M = \oplus_{i \in I} M_i$ where each M_i is a torsion-free R-module with a rank 1 dense basic submodule. Then every direct summand N of M is isomorphic to a direct sum $\oplus_{i \in K} M_i$ for a subset K of I, and any two such decompositions of N are isomorphic. □

   We now turn to the construction of R-modules with rank 1 dense basic submodules. In view of (1.7), it will suffice to deal with the case where the basic submodules are cyclic.

4. RANK ONE DENSE BASIC SUBMODULES

Let M be a torsion-free R-module with cyclic dense basic submodules. By (4.2), there exists an embedding $f : M \to \tilde{R}$ such that Im f is pure in $\tilde{R}$. With this f, we associate the Q-subspace V_f of $\tilde{Q}$ spanned by Im f; thus $V_f \cong \text{Im } f \otimes Q$. We claim:

LEMMA 4.5. [VD] For an embedding $f : M \to \tilde{R}$ with Im f pure in $\tilde{R}$, we have
$$\text{Im } f = V_f \cap \tilde{R}.$$
Conversely, if V is any subspace of $\tilde{Q}$, then $V \cap \tilde{R}$ is a torsion-free R-module with cyclic dense basic submodules.

Proof. As Im $f \leq V_f \cap \tilde{R}$ is obvious, we need to prove the converse containment only. This is an immediate consequence of the definition of V_f along with the purity of Im f in $\tilde{R}$.

To verify the second claim, notice that $V \cap \tilde{R}$ is always a pure R-submodule of $\tilde{R}$, hence its basic submodules have to be isomorphic to R. □

In view of this lemma, all torsion-free R-modules of rank n with cyclic dense basic submodules can be constructed first by selecting n elements $\pi_1, \ldots, \pi_n \in \tilde{Q}$ which are independent over Q, and then forming $V \cap \tilde{R}$ where V is the Q-subspace of $\tilde{Q}$ spanned by $\pi_1, \ldots, \pi_n$. It is clear that there is no loss of generality in assuming that $\pi_i \in \tilde{R}$ (moreover, that they are units of $\tilde{R}$); then $V \cap \tilde{R}$ will coincide with the pure R-submodule of $\tilde{R}$ generated by $\pi_1, \ldots, \pi_n$.

The main question that still remains to be answered is the isomorphism problem: under what conditions on the Q-subspaces V_1 and V_2 are the R-modules $V_1 \cap \tilde{R}$ and $V_2 \cap \tilde{R}$ isomorphic? An answer to this question will also provide information about the various embeddings of M in $\tilde{R}$ as pure submodules.

We start with a crucial lemma which is stated in a more general form so that we can derive from it additional information (viz. (4.7)).

LEMMA 4.6. [VD] Let M and N be R-modules with cyclic dense basic submodules, and let $f : M \to \tilde{R}$, $g : N \to \tilde{R}$ be embeddings as pure submodules. Every homomorphism $\eta : M \to N$ is of the form

$$\eta = g^{-1} \rho f \quad \text{for some } \rho \in \tilde{R}.$$

Proof. If $B = Ra$ is a basic submodule of M, then fa is a unit in $\tilde{R}$; thus there is a $\rho \in \tilde{R}$ such that $g\eta a = \rho f a$. Hence $g\eta$ and ρf coincide on B, that is, the map $g\eta - \rho f : M \to \tilde{R}$ induces a homomorphism from M/B into $\tilde{R}$. Since M/B is divisible, this last map is 0. Hence $g\eta = \rho f$, as claimed. □

The following consequence is readily verified.

COROLLARY 4.7. [VD] Under the hypotheses of (4.6), there is a non-zero homomorphism $\eta : M \to N$ if and only if there is a non-zero $\rho \in \tilde{R}$ such that

$$\rho V_f \leq V_g.$$

Such an η is necessarily monic. □

It is convenient to introduce the following definition. Let K be a subfield of a field L. Two K-subspaces, V and W, of L will be called L-_equivalent_ if there exists a non-zero $\rho \in L$ such that $\rho V = W$. This is obviously an equivalence relation (preserving K-isomorphism).

Now the condition (4.7) can be rephrased as follows. V_g contains a Q-subspace of $\tilde{Q}$ which is $\tilde{Q}$-equivalent to V_f. Hence we derive at once:

COROLLARY 4.8. [VD] Under the hypotheses of (4.6), the modules M and N are isomorphic exactly if the Q-subspaces V_f and V_g are $\tilde{Q}$-equivalent. □

A consequence of this corollary is that the embedding of M in $\tilde{R}$ is unique up to $\tilde{Q}$-equivalence.

We are now ready to verify our main result.

THEOREM 4.9. [VD] (Fuchs-Viljoen [1]) For reduced torsion-free modules M with rank 1 dense basic submodules, the following are complete and independent systems of invariants:

4. RANK ONE DENSE BASIC SUBMODULES

(i) the isomorphy class of the ideal of R isomorphic to basic submodules of M;

(ii) the $\tilde{Q}$-equivalence class of finite-dimensional Q-subspaces V_f of $\tilde{Q}$, associated with the embeddings $f : M \to \tilde{Q}$.

Proof. Given M as stated, it is clear that the isomorphy class of basic submodules of M is an invariant for M. That (ii) too is an invariant is shown by (4.8).

To see that (i) and (ii) form a complete system of invariants, note that by (4.8), $V_f \cap \tilde{R}$ is unique up to $\tilde{Q}$-equivalence no matter what embedding f has been selected. This intersection has by (4.6) dense cyclic basic submodules. It remains only to refer to (1.7) to conclude that $M \cong J \otimes (V_f \cap \tilde{R})$ where J is an ideal of R isomorphic to basic submodules of M.

Using an arbitrary finite-dimensional vector space V (rather than V_f), the argument in the preceding paragraph proves that these invariants are indeed independent. □

It is easy to decide when two ideals of R are isomorphic or when two Q-subspaces of $\tilde{Q}$ are $\tilde{Q}$-equivalent, so the classification in (4.9) seems satisfactory.

EXERCISES

1. [VD] Let M and N be torsion-free of the same finite rank and with rank one dense basic submodules. If their basic submodules are isomorphic and $N \leq M$, then $N = rM$ for some $r \in R$.

2. [VD] Let M, N be reduced torsion-free R-modules with isomorphic rank one dense basic submodules. If there are non-zero maps $M \to N$ and $N \to M$, then $M \cong N$.

3. [VD] A torsion-free R-module with rank one dense basic submodules has the exchange property.

4. Every indecomposable rank two torsion-free module over an almost maximal valuation domain has rank one dense basic submodules.

5. [VD] (Franzen [2]) For an ideal J of R and a Q-subspace V of $\widetilde{Q}$, $J\widetilde{R} \cap V$ has rank one dense basic submodules isomorphic to J.

6. [VD] Let $\pi_1 = 1, \pi_2, \ldots, \pi_n$ be units of $\widetilde{R}$ which are independent over R. Suppose $\pi_{ir} \in R$ are chosen such that $\pi_{ir} - \pi_i \in r\widetilde{R}$ ($i = 1, \ldots, n$) and for all $0 \neq r \in R$. If V is a Q-vectorspace with a basis $a_1, \ldots, a_n$, then setting
$$x_{ir} = r^{-1}(a_i - \pi_{ir}a_1) \qquad (0 \neq r \in R),$$
the elements $a_1, \ldots, a_n$ and all these x_{ir} ($2 \leq i \leq n$, $0 \neq r \in R$) generate an R-submodule M of V. Show that M has cyclic dense basic submodules.

7. [VD] Show that the $\widetilde{Q}$-equivalence of two Q-subspaces V_1 and V_2 of $\widetilde{Q}$ can be decided in finite extensions of Q.

8. Let R be an almost maximal valuation domain. The indecomposable R-homogeneous torsion-free R-modules of rank 2 form a rigid system (i.e. there is no non-trivial homomorphism between non-isomorphic members and all endomorphism rings are subrings of $\widetilde{Q}$).

§5. CHAINS OF PURE SUBMODULES

This section is devoted to results, which are, in a sense, generalizations of Auslander's Lemma in the special case where R is a valuation domain and the chain consists of pure submodules of a torsion-free R-module. These results are motivated by the study of pure submodules in free modules; as a matter of fact, applications may be found in the next section.

THEOREM 5.1. [VD] (Dimitric-Fuchs [1]) Let M be a torsion-free R-module and m a non-negative integer. Assume
$$0 = M_0 \leq M_1 \leq \cdots \leq M_\alpha \leq \cdots \leq M_{\omega_m} = M \qquad (1)$$
is a well-ordered continuous chain of submodules such that

5. CHAINS OF PURE SUBMODULES

(a) each M_α is pure in M;
(b) $p.d.M_\alpha \leq m$ for each $\alpha < \omega_m$.

Then $p.d.M \leq m$.

Proof. Because of (IV.5.1), there is a tight system T_α in each M_α ($\alpha < \omega_m$) such that the members are pure submodules of M_α. Using these T_α, we first verify:

LEMMA 5.2. [VD] Under the hypotheses of (5.1), given any $\aleph_m$-generated submodule H of M, there is a submodule $\overline{H}$ of M satisfying

(α) H is contained in $\overline{H}$;
(β) the rank of $\overline{H}$ is $\leq \aleph_m$;
(γ) $\overline{H} \cap M_\alpha \in T_\alpha$ for each $\alpha < \omega_m$;
(δ) $\overline{H} + M_\alpha$ is pure in M for each α.

Proof of (5.2). We describe two processes whose combination will yield a submodule with the required properties.

First, for each $\alpha < \omega_m$ choose an $\aleph_m$-generated submodule $T_{1\alpha} \in T_\alpha$ such that $H \cap M_\alpha \leq T_{1\alpha}$. Clearly, the submodule $H_2 = \langle H, T_{1\alpha} \ (\alpha < \omega_m)\rangle$ is $\aleph_m$-generated. Repeat this process with H_2 in place of H to obtain a submodule $H_3 = \langle H_2, T_{2\alpha} \ (\alpha < \omega_m)\rangle$ with $\aleph_m$-generated $T_{2\alpha} \in T_\alpha$ such that $H_2 \cap M_\alpha \leq T_{2\alpha}$, etc. The union H_* of the chain $H < H_2 \leq \ldots \leq H_n \leq \ldots$ of these $\aleph_m$-generated submodules evidently satisfies

$$H_* \cap M_\alpha \in T_\alpha \quad \text{for all } \alpha < \omega_m$$

(just recall that the T_α's are closed under taking unions of chains).

Next, consider the submodule $(H + M_\alpha)/M_\alpha$ of M/M_α. View M/M_α as the union of its submodules M_β/M_α ($\alpha < \beta < \omega_m$). Since $p.d.M_{\beta+1}/M_\beta \leq m+1$ because of (b), from Auslander's Lemma we infer that $p.d.M/M_\alpha \leq m+1$ as well. From (IV.5.1) it follows that all pure submodules of rank $\leq \aleph_m$ in M/M_α are $\aleph_m$-generated. Consequently, there exists an $\aleph_m$-generated pure submodule $(H_\alpha^2 + M_\alpha)/M_\alpha$ of M/M_α that contains $(H + M_\alpha)/M_\alpha$; here $H_\alpha^2 (\geq H)$

can be chosen so as to be $\aleph_m$-generated. The submodule $H^2 = \langle H, H_\alpha^2 \ (\alpha < \omega_m)\rangle$ is again of rank $\leq \aleph_m$, so the same process can be repeated with H^2 playing the role of H to obtain a larger $\aleph_m$-generated submodule H^3 such that $(H^3 + M_\alpha)/M_\alpha$ contains the purification of $(H^2 + M_\alpha)/M_\alpha$ in M/M_α, etc. The union H^* of the chain $H \leq H^2 \leq \ldots \leq H^n \leq \ldots$ will be of rank $\leq \aleph_m$ and for each $\alpha < \omega_m$ it will satisfy:

$$H^* + M_\alpha \text{ is pure in } M.$$

To conclude the proof of (5.2), we alternate the two processes and define $\overline{H}$ as the union of the ascending chain $H \leq H_* \leq (H_*)^* \leq ((H_*)^*)_* \leq \ldots$ Obviously $\overline{H}$ will satisfy $(\alpha)-(\delta)$. □

Resuming the proof of (5.1), we proceed to establish the existence of a continuous chain

$$0 = H_0 \leq H_1 \leq \ldots \leq H_\nu \leq \ldots \leq H_\lambda = M \quad (2)$$

of submodules in M with the following properties:
  (i) $H_{\nu+1}/H_\nu$ is $\aleph_m$-generated for $\nu < \lambda$;
  (ii) $H_\nu \cap M_\alpha \in T_\alpha$ for $\alpha < \omega_m$ and $\nu < \lambda$;
  (iii) $H_\nu + M_\alpha$ is pure in M for $\alpha < \omega_m$, $\nu < \lambda$.
Here λ denotes a suitable ordinal.

Define H_ν by transfinite induction. Setting $H_0 = 0$, assume that the H_μ ($\mu < \nu$) have been selected so as to have properties (i)-(iii).

If ν is a limit ordinal, then we have no choice other than setting $H_\nu = \cup_{\mu < \nu} H_\mu$. (ii) and (iii) will obviously hold for H_ν.

If ν is a successor ordinal, say $\nu = \mu + 1$, and if $H_\mu < M$, then in (1) we pass mod H_μ and consider the chain

$$0 = H_\mu/H_\mu \leq \ldots \leq (H_\mu + M_\alpha)/H_\mu \leq \ldots \quad (\alpha < \omega_m). \quad (3)$$

By $(H_\mu + M_\alpha)/H_\mu \cong M_\alpha/(H_\mu \cap M_\alpha)$ and $H_\mu \cap M_\alpha \in T_\alpha$, we see that in (3) all modules have projective dimensions $\leq m$. Thus (3) is a chain like (1), so that (5.2) can be applied to a non-zero cyclic submodule of M/H_μ to obtain a submodule $\overline{H}/H_\mu$ of M/H_μ satisfying $(\beta)-(\delta)$. It only remains to put $H_\nu = \overline{H}$ and to check that (i)-(iii) hold (which is routine), completing the proof of (3).

In order to verify (5.1), by Auslander's Lemma it will be enough to show that in (2) $p.d. H_{\nu+1}/H_\nu \leq m$ for each $\nu < \lambda$. Note that $H_{\nu+1}/H_\nu$ is the union of the following continuous well-ordered ascending chain:

$$0 \leq [(H_{\nu+1} \cap M_1) + H_\nu]/H_\nu \leq \cdots \leq [(H_{\nu+1} \cap M_\alpha) + H_\nu]/H_\nu \leq \cdots \quad (4)$$

with $\alpha < \omega_m$. Because of (iii), here $(H_{\nu+1} \cap M_\alpha) + H_\nu = H_{\nu+1} \cap (H_\nu + M_\alpha)$ is pure in M; thus the chain (4) consists of pure submodules of $H_{\nu+1}/H_\nu$. Therefore, (i) implies that the modules in (4) are $\aleph_m$-generated; cf. (II.5.4). Furthermore, in view of (ii),

$$[(H_{\nu+1} \cap M_\alpha) + H_\nu]/H_\nu \cong (H_{\nu+1} \cap M_\alpha)/(H_\nu \cap M_\alpha)$$

has projective dimension $\leq m$. From (IV.3.8) we now conclude that $p.d. H_{\nu+1}/H_\nu \leq m$. □

The special case $m = 0$ is most interesting:

COROLLARY 5.3. [VD] The union of a countable ascending chain $0 = F_0 \leq F_1 \leq \cdots \leq F_n \leq \cdots$ of free R-modules is again free whenever each F_n is pure in F_{n+1}. □

Another corollary is as follows.

COROLLARY 5.4. [VD] (Dimitrić-Fuchs [1]) Let M be a torsion-free R-module and m, k non-negative integers. If (1) is a well-ordered continuous chain of submodules of M such that (a) of (5.1) holds and $p.d. M_\alpha \leq m+k$ for each $\alpha < \omega_m$, then $p.d. M \leq m+k$.

Proof. We induct on k, the case $k = 0$ being covered by (5.1). Suppose $k \geq 1$. For each α, consider the canonical projective resolution of M_α:

$$0 \longrightarrow H_\alpha \longrightarrow F_\alpha = \bigoplus_{a \in M_\alpha} Rx_a \xrightarrow{\phi_\alpha} M_\alpha \longrightarrow C \quad (5)$$

where $\phi_\alpha(x_a) = a$. Here H_α is pure in F_α and $p.d. H_\alpha \leq m+k-1$. The embeddings $M_\alpha \to M_\beta$ for $\alpha < \beta$ give rise to a direct system of exact sequences (5) whose direct limit is the resolution (5) for

$\alpha = \omega_m$. As $H_{\omega_m} = \cup \{H_\alpha \mid \alpha < \omega_m\}$, the induction hypothesis applies to H_{ω_m}, and we conclude that $p.d.H_{\omega_m} \leq m+k-1$. Hence $p.d.M \leq m+k$, indeed. □

If we wish to consider longer chains in (1), and to retain the same conclusion on $p.d.M$, then we need a cardinality restriction on the M_α's as well as an additional hypothesis on the $M_{\alpha+1}/M_\alpha$'s.

THEOREM 5.5. [VD] (Dimitrić-Fuchs [1]) Assume that m is a non-negative integer and

$$0 = M_0 \leq M_1 \leq \cdots \leq M_\alpha \leq \cdots \leq M_{\omega_{m+1}} = M \qquad (6)$$

is a continuous well-ordered chain of submodules of the torsion-free R-module M, satisfying the following conditions:

(i) each M_α is pure in M;
(ii) each M_α is $\aleph_{m+1}$-generated;
(iii) $p.d.M_\alpha \leq m$ for each $\alpha < \omega_{m+1}$;
(iv) the pure submodules of rank $\leq \aleph_{m-1}$ in $M_{\alpha+1}/M_\alpha$ are $\aleph_{m-1}$-generated for each $\alpha < \omega_{m+1}$.

Then $p.d.M \leq m$.

Proof. We start off with the construction of another chain (7) replacing the given (6). For each $\alpha < \omega_{m+1}$, we fix a tight system T_α in M_α, consisting of pure submodules. We want a continuous chain

$$0 = A_0 \leq A_1 \leq \cdots \leq A_\alpha \leq \cdots \leq A_{\omega_{m+1}} = M \qquad (7)$$

of submodules, subject to the conditions:

1) each A_α is $\aleph_m$-generated ($\alpha < \omega_{m+1}$);
2) $A_\alpha \in T_\alpha$ whenever $\alpha < \omega_{m+1}$ is a non-limit ordinal;
3) $A_\alpha \cap M_\beta \in T_\beta$ for all $\beta < \alpha < \omega_{m+1}$;
4) $A_\alpha + M_\beta$ is pure in M_α for $\beta < \alpha < \omega_{m+1}$.

Observe that $\alpha < \omega_{m+1}$ implies $\text{cof } \alpha \leq \omega_m$, thus (5.2) can be applied in the same way as in the proof of (2) to establish a chain (7) with the desired properties. In order to ascertain that $A_{\omega_{m+1}} = M$, all what we have to do is to well-order a generating

5. CHAINS OF PURE SUBMODULES

set of M: $\{a_\alpha \mid \alpha < \omega_{m+1}\}$, and to demand that $a_\beta \in A_\alpha$ for all $\beta < \alpha$. This will hold whenever $A_{\alpha+1}$ is constructed so as to include a_α ($\alpha < \omega_{m+1}$).

Once (7) has been established, it is sufficient to verify that

$$p.d. A_{\alpha+1}/A_\alpha \leq m \quad \text{for} \quad \alpha < \omega_{m+1}.$$

In the exact sequence

$$0 \to (A_{\alpha+1} \cap M_\alpha)/A_\alpha \to A_{\alpha+1}/A_\alpha \to A_{\alpha+1}/(A_{\alpha+1} \cap M_\alpha) \to 0,$$

the last non-zero module is $\cong (A_{\alpha+1} + M_\alpha)/M_\alpha$. This is $\aleph_m$-generated and 4) implies that it has property (iv). A simple reference to (IV.3.5) shows that its projective dimension is at most m. Hence it remains only to show that

$$p.d. (A_{\alpha+1} \cap M_\alpha)/A_\alpha \leq m. \tag{8}$$

We distinguish two cases according as α is a successor or a limit ordinal.

In the first alternative, 2) ensures $A_\alpha \in \mathcal{T}_\alpha$. Furthermore, by 3), $A_{\alpha+1} \cap M_\alpha \in \mathcal{T}_\alpha$ likewise, whence (8) follows at once.

If α is a limit ordinal, then we view $(A_{\alpha+1} \cap M_\alpha)/A_\alpha$ as the union of its submodules $[(A_{\alpha+1} \cap M_\beta) + A_\alpha]/A_\alpha$ for $\beta < \alpha$. These are isomorphic to $(A_{\alpha+1} \cap M_\beta)/(A_\alpha \cap M_\beta)$; here both intersections belong to $\mathcal{T}_\beta$, so the projective dimension of their quotient is $\leq m$. Furthermore, $(A_{\alpha+1} \cap M_\beta) + A_\alpha = A_{\alpha+1} \cap (M_\beta + A_\alpha)$ are pure submodules. Since cof $\alpha \leq \omega_m$, (5.1) can be applied to conclude that (8) holds true. □

Again the case $m = 0$ deserves special attention:

COROLLARY 5.6. [VD] If

$$0 = F_0 \leq F_1 \leq \cdots \leq F_\alpha \leq \cdots \quad (\alpha < \omega_1)$$

is a continuous chain of $\aleph_1$-generated free R-modules F_α such that F_α is pure in $F_{\alpha+1}$ for each α and in $F_{\alpha+1}/F_\alpha$ the finite rank submodules are finitely generated, then the union $\cup F_\alpha$ is again a free R-module. □

EXERCISE

1. [VD] If the lengths of the chains (1) and (6) are limit ordinals λ (not necessarily initial ordinals), then (5.1) and (5.5) continue to hold if $\operatorname{cof} \lambda \leq \omega_m$ and $\operatorname{cof} \lambda \leq \omega_{m+1}$, respectively, are assumed.

§6. PURE SUBMODULES OF FREE MODULES

In this section, we consider exclusively modules over valuation domains R.

Submodules of free R-modules need not be free, not even when they are pure, so the problem of studying pure submodules of free R-modules arises. So far they have not been investigated systematically and here we can only establish a few relevant properties.

THEOREM 6.1. [VD] Pure submodules of free R-modules are separable, and their pure submodules of countable rank are free.

Proof. This is an immediate consequence of (2.9) and (2.8), respectively. ☐

The second assertion in (6.1) generalizes easily:

PROPOSITION 6.2. [VD] In a free R-module, pure submodules of rank $\aleph_m$ have projective dimension $\leq m$.

Proof. A pure submodule of a free R-module of rank $\aleph_m$ can have projective dimension $d \geq m+1$ only if it contains a finite rank pure submodule of projective dimension d (see (IV.3.7)). However, this would contradict (6.1). ☐

A submodule H of a free R-module F is pure exactly if F/H is torsion-free; hence the equivalence of conditions (i) and (ii) in the following theorem is obvious.

THEOREM 6.3. [VD] (Dimitrić-Fuchs [1]) For a valuation domain R, the following are equivalent:

(i) pure submodules of free R-modules are free;

(ii) $\text{p.d.} M \leq 1$ for all torsion-free R-modules M;

6. PURE SUBMODULES OF FREE MODULES

(iii) $gl.d.R \leq 2$ and $p.d.Q = 1$.

Proof. It remains to verify the equivalence of (ii) and (iii). Observe that (ii) implies $p.d.I \leq 1$ for all the ideals I of R. Therefore $gl.d.R = \sup p.d.I + 1 \leq 2$, and (iii) follows.

Assume now that (iii) holds. Then $p.d.I \leq 1$ for all the ideals I of R. Given any torsion-free R-module M, we can find a chain (like in the proof of (II.5.1)) of submodules of M whose factors are rank one torsion-free modules. By hypothesis, these factors have projective dimensions ≤ 1, so by Auslander's Lemma, $p.d.M \leq 1$. □

With the aid of (6.3) one can show that if R admits non-free pure submodules in free R-modules, then this already occurs at the cardinality $\aleph_1$.

LEMMA 6.4. [VD] If R is such that not all pure submodules of free R-modules are free, then the free R-module of rank $\aleph_1$ contains a non-free pure submodule.

Proof. By (6.3), either $gl.d.R \geq 3$ or $p.d.Q \geq 2$. Thus either an ideal I of R or Q has projective dimension ≥ 2, i.e. by (IV.2.2) or (IV.2.4), requires at least $\aleph_1$ generators. In any case, there is a rank one torsion-free R-module M with exactly $\aleph_1$ generators (which can not be countably generated). If $0 \to H \to F \to M \to 0$ is a free resolution of this M with F free of rank $\aleph_1$, then $p.d.M = 2$ implies $p.d.H = 1$. Thus H is a non-free pure submodule of F. □

As far as the converse of (6.1) is concerned, we offer a counterexample; cf. Dimitric-Fuchs [1].

EXAMPLE 6.5. [VD] There is a valuation domain R which has a separable torsion-free R-module M of rank $\aleph_1$ whose $\aleph_0$-generated pure submodules are free, but M is not embeddable as a pure submodule in a free R-module. Let R be such that $gl.d.R = 2$ and $p.d.Q = 1$. Choose a free resolution $0 \to H \to F \to Q \to 0$ with H, F countably generated free. Using a fixed isomorphism $\phi : F \to H$, it is easy to construct a chain

$$0 = F_0 < F_1 = H < F_2 = F < F_3 < \ldots$$

of countably generated free R-modules such that $F_{n+1}/F_n \cong Q$ for $n \geq 0$. By (5.3), $\cup_{n<\omega} F_n$ is likewise free, so we can proceed transfinitely and get a well-ordered continuous ascending chain of countably generated free R-modules F_α for every $\alpha < \omega_1$ such that $F_{\alpha+1}/F_\alpha \cong Q$ for each $\alpha < \omega_1$. Let M be $\cup F_\alpha$ for $\alpha < \omega_1$. Then the F_α define an $\aleph_1$-filtration of M in the sense of IV.§2. Eklof's theorem shows p.d.$M = 1$. By (6.3), M is not isomorphic to any pure submodule of a free R-module. As every countable rank submodule of M is contained in some F_α, it is readily checked that M is separable and its $\aleph_0$-generated pure submodules are free.

EXERCISES

1. [VD] Generalize (6.3) by adding m to all projective dimensions occurring in (i)-(iii). (E.g. (i) becomes: pure submodules of torsion-free R-modules of projective dimension $\leq m$ have projective dimension $\leq m$.)

2. [VD] Show that in (6.5), Q can be replaced by any countably generated torsion-free module of projective dimension 1, and M will still have the same properties.

3. There exist valuation domains R with gl.d.$R = 2$ and p.d.$Q = 1$.

§7. SLENDER MODULES

We follow the pattern of abelian group theory, and investigate a remarkable class of torsion-free modules. In developing the theory of slender modules, we shall attempt to deduce as much as possible from the sole hypothesis that our modules are torsion-free over an arbitrary domain. Deeper results, however, can be obtained only under certain restrictions of which the most useful is that Q is countably generated. In our presentation, we follow Dimitrić [1].

The following notation will be used. The direct sum and the direct product of countably many copies of R will be denoted by

7. SLENDER MODULES

$$F = \bigoplus_{n<\omega} Re_n \quad \text{and} \quad F^* = \prod_{n<\omega} Re_n,$$

respectively. F will be viewed as a submodule of F^* consisting of the finite vectors. The elements of F^* can be written either as formal sums $\sum_{n<\omega} r_n e_n$ ($r_n \in R$) or as infinite vectors $(r_0, \ldots, r_n, \ldots)$.

A torsion-free R-module M is called <u>slender</u> if, for every homomorphism $\eta : F^* \to M$, almost all of ηe_n are 0.

The following observations are useful.

1. Submodules of slender modules are slender. Hence if there exists a slender R-module $M \neq 0$ at all, R is slender as an R-module.

2. Injective modules are not slender. Hence slender modules are reduced.

3. RD-injective modules are not slender. In fact, F is an RD-submodule of F^*, thus every homomorphism of F into an RD-injective module extends to a homomorphism of F^*.

Before stating further results, we furnish F^* with the product topology where R is given the discrete topology. We may think of the submodules

$$W_k = \prod_{n \geq k} Re_n \qquad (k = 0, 1, \ldots)$$

as a base of neighborhoods of 0 in F^*. In this topology, F is a dense submodule of F^*.

This topology of F^* is a metrizable linear topology; in general, such a topology of an R-module M is defined in terms of a countable set of submodules of M, $\{V_n\}_{n<\omega}$ with $\cap V_n = 0$, constituting a subbase of neighborhoods of 0. Without loss of generality, $V_0 \geq V_1 \geq \ldots \geq V_n \geq \ldots$ may be and will be assumed. This topology is non-discrete if and only if V_n is never 0.

LEMMA 7.1. [D] An R-module which is complete in a non-discrete metrizable linear topology is not slender.

<u>Proof.</u> Suppose that $\{V_n\}$ defines such a topology on M. Choose any $a_n \neq 0$ in V_n, and define an R-homomorphism $\eta_0 : F \to M$

by setting $\eta_0(e_n) = a_n$. Then $\eta_0(W_n \cap F) \leq V_n$ shows that η_0 is a continuous homomorphism. By the completeness of M and the density of F in F^*, there exists a unique continuous extension $\eta : F^* \to M$ of η_0. This is a forbidden homomorphism for slender modules. □

The next results exhibit algebraic properties of slender modules.

LEMMA 7.2. [D] For a slender module M, we have

$$\text{Hom}_R(F^*/F, M) = 0.$$

Proof. Suppose there is a non-zero $\eta : F^* \to M$ such that $\eta F = 0$, and let $x = \Sigma r_n e_n$ have a non-zero image under η. Define an endomorphism χ of F^* via

$$\chi : \sum_n s_n e_n \longmapsto \sum_n r_n(s_0 + \ldots + s_n)e_n$$

($s_n \in R$). Then $\eta\chi : F^* \to M$ satisfies, for every n, $\eta\chi e_n = \eta(0,\ldots,0,r_n,r_{n+1},\ldots) = \eta x \neq 0$. This contradicts the definition of slenderness. □

As $W_k \cong F^*$, we see that in view of (7.2), the slenderness of M is equivalent to the property that every homomorphism $\eta : F^* \to M$ is continuous; here M carries the discrete topology. It also follows that ηF^* is finitely generated if M is slender.

THEOREM 7.3. [D] (Dimitrić [1]) Direct sums of slender modules are again slender.

Proof. For finite direct sums the assertion is obvious. Let $\{M_i \mid i \in I\}$ be an infinite set of slender modules and $\eta : F^* \to \oplus M_i = M$. From the definition of slenderness it is immediate that for at most countably many indices i can $\pi_i \eta F \neq 0$ hold where π_i is the ith coordinate projection $M \to M_i$. The preceding lemma implies that $\pi_i \eta F^* \neq 0$ only if $\pi_i \eta F \neq 0$. What we have to show is that $\pi_i \eta F^* \neq 0$ cannot hold for infinitely many indices i.

Since F and F^* can be constructed by using any infinite subset of the e_n's, $\eta e_n \neq 0$ may be assumed for all n. Let $\eta e_0 \in M_0 \oplus \ldots \oplus M_k$; then this direct sum being slender, only finitely many ηe_n's can have non-zero coordinates in $M_0 \oplus \ldots \oplus M_k$.

7. SLENDER MODULES

We ignore all these, except for e_0. Let $\eta e_1 \in M_{k+1} \oplus \ldots \oplus M_\ell$, and repeat what has been done with ηe_0, etc. In this way, we argue that - after changing the notation and replacing $M_0 \oplus \ldots \oplus M_k$, $M_{k+1} \oplus \ldots \oplus M_\ell, \ldots$ by single slender modules, without loss of generality $0 \neq \eta e_n \in M_n$ may be assumed.

Slender modules are reduced, so for each n, there is an $r_n \in R$ such that $\eta e_n \notin r_n M_n$. With these r_n, form the element

$$x = \sum_{n=0}^{\infty} r_0 \ldots r_{n-1} e_n \in F^*.$$

There is a k such that $\eta x \in M_0 \oplus \ldots \oplus M_{k-1}$. Visibly, $\pi_i \eta F^* = \eta R e_i$, and we infer that $\eta x = \eta(s_0 e_0 + \ldots + s_{k-1} e_{k-1})$ for some $s_0, \ldots, s_{k-1} \in M$. Therefore,

$$\eta(s_0 e_0 + \ldots + s_{k-1} e_{k-1} - \sum_{n=0}^{k} r_0 \ldots r_{n-1} e_n) = \eta(\sum_{n \geq k+1} r_0 \ldots r_{n-1} e_n).$$

This element belongs to $M_0 \oplus \ldots \oplus M_k$ and is divisible by $r_0 \ldots r_k$, so $\eta(r_0 \ldots r_{k-1} e_k) \in r_0 \ldots r_k M_k$. Hence torsion-freeness implies $\eta e_k \in r_k M_k$, a contradiction. □

The most important result on slender modules is the analogue of J. Łoś' theorem.

THEOREM 7.4. [D] Let A_i ($i \in I$) be a set of torsion-free R-modules and $\eta : \Pi A_i \to M$ where M is a slender module. Then
  (i)   for almost all $i \in I$, $\eta A_i = 0$;
  (ii) if I is a non-measurable set and $\eta A_i = 0$ for all $i \in I$, then $\eta = 0$.

Proof. The proof given for abelian groups e.g. in Fuchs [2] carries over verbatim. □

We now attempt to imitate R. Nunke's characterization of slender abelian groups in terms of subgroups. A preliminary lemma is required.

LEMMA 7.5. [D] Let M be a torsion-free R-module with a non-discrete metrizable linear topology. Then M contains either a copy of R with a non-discrete linear topology or a copy of F

with the topology induced by the product topology of F^*.

If, moreover, M is complete, then it contains the completion of R in a non-discrete metrizable linear topology or a copy of F^* in the product topology.

Proof. Let M be as stated and $\{V_n\}$ a descending chain of submodules forming a base of neighborhoods about 0. If M contains an a such that $Ra \cap V_n \neq 0$ for each n, then $Ra \cong R$ is a submodule as desired. If there is no such a, then for every $0 \neq a \in M$, there is a V_n such that $Ra \cap V_n = 0$. In this case we can select a sequence $a_1, \ldots, a_k, \ldots$ of non-zero elements of M along with a sequence of integers, $n_1 < \ldots < n_k < \ldots$, satisfying

$$a_i \in V_{n_i} \text{ and } Ra_i \cap V_{n_{i+1}} = 0 \quad (i = 1, 2, \ldots).$$

Then $\oplus Ra_i$ will be algebraically and topologically isomorphic to F.

The second part of our lemma follows at once from the first if we just note that complete modules contain the completions of submodules in their induced topologies. □

We are now ready to prove:

THEOREM 7.6. [D] (Dimitrić [1]) A torsion-free R-module M is slender if and only if it satisfies the following conditions:

(i) $\text{Hom}_R(F^*/F, M) = 0$;

(ii) M contains no completion of R in any non-discrete, metrizable linear topology;

(iii) M contains no copy of F^*.

Proof. The necessity of these conditions follows at once from (7.2) and (7.1).

To verify their sufficiency, suppose that M satisfies (i)-(iii), and let $\eta : F^* \to M$. By way of contradiction, assume $\eta e_n \neq 0$ for infinitely many n. Then $\eta W_n \neq 0$ for each n, thus $\{\eta W_n\}$ is a base of neighborhoods of a non-discrete topology of M. Suppose $a \in \cap_n \eta W_n$, and let $b_n = (0, \ldots, 0, r_{nn}, r_{n,n+1}, \ldots) \in W_n$ satisfy $\eta b_n = a$, for every n. Define an endomorphism α of F^* via

7. SLENDER MODULES

$$\alpha(s_0, \ldots, s_n, \ldots) =$$

$$(r_{00}s_0, (r_{01} - r_{11})s_0 + r_{11}s_1, \ldots, \sum_{i=0}^{n}(r_{in} - r_{i+1,n})s_i, \ldots).$$

Manifestly, $\alpha(1,\ldots,1,\ldots) = b_0$ and $\alpha e_n = b_n - b_{n+1}$. Therefore $a \in \eta\alpha F^*$ and $\eta\alpha F = 0$, so condition (i) implies $a = 0$. This proves that $\{\eta W_n\}$ gives rise to a non-discrete metrizable topology of M. Since F^* was complete in the product topology, ηF^* is complete in the $\{\eta W_n\}$-topology. The same holds for M, too.

To finish the proof, we appeal to (7.5). This contradicts either condition (ii) or (iii). □

Significant simplification takes place in the last theorem if the quotient field Q of R is assumed to be countably generated. In this case, the rather obscure condition (i) can be replaced by the simple hypothesis that M be reduced, and as a consequence, we can characterize slender modules by their submodules.

If Q is countably generated, then the R-topology on the reduced part A of F^*/F will be non-discrete and metrizable. Moreover, because of V.§2.Ex.4, it will be complete in this topology. This observation is used in the proof of the next result.

THEOREM 7.7. [D] (Dimitrić [1]) Suppose that the quotient field Q of R is countably generated. Then a torsion-free R-module M is slender if and only if

(i) it is reduced;

(ii) it contains no completion of R in any non-discrete metrizable linear topology;

(iii) it contains no submodule isomorphic to F^*.

Proof. A comparison with (7.6) shows that all what we have to show is that from our conditions (i)-(iii), $\text{Hom}_R(F^*/F, M) = 0$ follows. By virtue of (i), here F^*/F can be replaced by its reduced part A which is R-complete (see our remark above). Every non-zero homomorphism $\phi : A \to M$ induces a non-discrete topology on M if $\{\phi(rA)\}_{0 \neq r \in R}$ are declared to be a base of neighborhoods about 0.

As $\phi(rA) \leq rM$ and M is reduced, this is a metrizable topology on M. Moreover, M has to be complete in this topology, since A was complete in its metrizable R-topology. By (7.5), we are again in conflict with (ii) or (iii). □

The following corollaries are worth mentioning.

COROLLARY 7.8. [D] Let the quotient field Q of the domain R be countably generated. A proper submodule of Q is slender if and only if it is not complete in any non-discrete metrizable linear topology.

Proof. In fact, such a module always satisfies (i) and (iii) of the preceding theorem. □

If R is a valuation domain, then its only non-discrete Hausdorff linear topology is its R-topology. Consequently,

COROLLARY 7.9. [VD] Let R be a valuation domain whose quotient field is countably generated. R is slender if and only if it is not R-complete. □

If Q is uncountably generated, then the criterion which we obtain from (7.6) is less explicit:

COROLLARY 7.10. A valuation domain R with $p.d.Q \geq 2$ is slender if and only if $\operatorname{Hom}_R(F^*/F, R) = 0$. □

EXERCISES

1. [D] R is a slender module if and only if all ideals of R are slender if and only if R contains a slender ideal $\neq 0$.

2. [D] Let R be a slender module and I a non-measurable index set. There exist natural isomorphisms
$$\operatorname{Hom}_R(\bigoplus_{i \in I} R, R) \cong \prod_{i \in I} R, \quad \operatorname{Hom}_R(\prod_{i \in I} R, R) \cong \bigoplus_{i \in I} R.$$

3. [D] Let A_i ($i \in I$) be slender R-modules of rank 1 and I a non-measurable set. Then every rank 1 summand of $\prod A_i$ is isomorphic to a summand of a finite direct sum of the A_i's.

NOTES

In this chapter, we have concentrated on torsion-free modules over valuation domains, but ignored those over arbitrary domains. The reason is that we do not know much about the general theory. The interested reader will find some information in the excellent monograph by Matlis [5].

Torsion-free modules even over valuation domains are quite complex. Since no satisfactory classification theory is available for torsion-free abelian groups, not even in the local case, we can not hope for any essential progress for valuation domains in general. The case of dense rank one basic submodules, discussed in Fuchs-Viljoen [1], is just a modest beginning. Additional classes of interest have been dealt with by Franzen [1], generalizing results by Arnold [1] from discrete rank one valuation domains to arbitrary almost maximal valuation domains.

The Grothendieck group of finite rank torsion-free modules has been discussed by Fakhruddin [1].

The following question is open:

PROBLEM 26. [VD] Are the direct decompositions of finite rank torsion-free modules into indecomposable summands unique up to isomorphism?

In this connection, note that the endomorphism rings of finite rank torsion-free modules over valuation domains R need not be local. In fact, their endomorphism rings can be quite complex (as we were informed by B. Franzen, A. L. S. Corner's theorem on the endomorphism rings of torsion-free abelian groups carries over with obvious modifications).

The results on completely decomposable modules are straightforward generalizations of the abelian group case. Separability looks more challenging, but so far not much is known about it except for a recent study by Franzen [2]. He also studied reflexivity.

The results in §§5-6 are the groundwork for an extensive study of pure submodules in free modules over valuation domains. Some results in §5 were influenced by P. Hill's work on chains of free and κ-free abelian groups.

Slender modules over arbitrary domains were discussed by Dimitrić [1], generalizing earlier results by various authors (E. L. Lady, B. de Marco–A. Orsatti, G. Heinlein). The results will definitely be useful in studying products of modules.

References

Anderson, F. W. - Fuller, K. R. [1] Rings and Categories of Modules, Springer, 1974.

Arnold, D. [1] A duality for torsion-free modules of finite rank over a discrete valuation ring, Proc. London Math. Soc. 24 (1972), 204-216.

Auslander, L. [1] On the dimension of modules and algebras III, Nagoya Math. J. 9 (1955), 67-77.

Azumaya, G. [1] Corrections and supplementaries to my paper concerning Remak-Krull-Schmidt's theorem, Nagoya Math. J. 1 (1950), 117-124.

Bazzoni, S. - Fuchs, L. [1] On modules of finite projective dimension over valuation domains, Proceedings of Conference on Abelian Groups and Modules in Udine, 1984.

Becker, T. [1] Real closed rings and ordered valuation rings, Zeitschr. f. Math. Logik 29 (1983), 417-425.

Brandal, W. [1] Almost maximal integral domains and finitely generated modules, Trans. Amer. Math. Soc. 183 (1973), 203-222.

Brandal, W. [2] Commutative rings whose finitely generated modules decompose, Lecture Notes in Math. 723, Springer (1979).

Cartan, H. - Eilenberg, S. [1] Homological Algebra, Princeton University Press, 1956.

Cherlin, G. L. - Dickmann, M. A. [1] Anneaux réals clos et anneaux des fonctions continues, C. R. Acad. Sci. Paris 290 (1980), 1-4.

Cherlin, G. L. - Dickmann, M. A. [2] Real closed rings and rings of continuous functions (to appear).

Crawley, P. - Jónsson, B. [1] Refinements for infinite direct decompositions of algebraic systems, Pac. J. Math. 14 (1964), 797-855.

de la Rosa, B. - Fuchs, L. [1] On h-divisible torsion modules over domains (to appear).

Dimitrić, R. [1] Slender modules over domains, Comm. in Alg. 11 (1983), 1685-1700.

Dimitrić, R. - Fuchs, L. [1] On torsion-free modules over valuation domains (to appear).

Eklof, P. C. [1] Homological dimension and stationary sets, Math. Z. 180 (1982), 1-9.

Enochs, E. [1] Flat covers and flat cotorsion modules, Proc. Amer. Math. Soc. 92 (1984), 179-184.

Facchini, A. [1] Decompositions of algebraically compact modules (to appear).

Fakhruddin, S. M. [1] The Grothendieck group of torsion-free modules of finite rank over a valuation ring, Journ. Alg. 17 (1971), 25-33.

Fakhruddin, S. M. [2] Modules over Prüfer domains, Trans. Amer. Math. Soc. 159 (1971), 469-487.

Fleischer, I. [1] Modules of finite rank over Prüfer rings, Annals of Math. 65 (1957), 250-254.

Fleischer, I. [2] Maximality and ultracompleteness in normed modules, Proc. Amer. Math. Soc. 9 (1958), 151-157.

Franzen, B. [1] Exterior powers and torsion-free modules over almost maximal valuation domains, Abelian Group Theory, Lecture Notes in Math. 1006, Springer (1983), 599-606.

REFERENCES

Franzen, B. [2] On the separability of a direct product of free modules over a valuation domain, Archiv d. Math. 42 (1984), 131-135.

Fuchs, L. [1] Partially Ordered Algebraic Systems, Pergamon Press, 1963.

Fuchs, L. [2] Infinite Abelian Groups, Vol. I-II, Academic Press, 1970 and 1973.

Fuchs, L. [3] Torsion-complete modules over Prüfer domains, Proc. Conference at Univ. of Southwestern Louisiana.

Fuchs, L. [4] On projective dimensions of modules over valuation domains, Abelian Group Theory, Lecture Notes in Math. 1006, Springer (1983), 589-598.

Fuchs, L. [5] Modules over Valuation Domains, Vorlesungen Fachb. Math. Univ. Essen (1983).

Fuchs, L. [6] On divisible modules over domains, Proceedings of Conference on Abelian Groups and Modules in Udine, 1984.

Fuchs, L. - Salce, L. [1] Uniserial modules over valuation rings, J. Algebra 85 (1983), 14-31.

Fuchs, L. - Salce, L. [2] Prebasic submodules over valuation rings, Annali Mat. Pura Appl. 32 (1982), 257-274.

Fuchs, L. - Salce, L. [3] Separable torsion modules over valuation domains, Archiv d. Math. 41 (1983), 17-24.

Fuchs, L. - Viljoen, G. [1] On finite rank torsion-free modules over almost maximal valuation domains, Comm. in Alg. 12 (1984), 245-258.

Gill, D. T. [1] Almost maximal valuation rings, J. London Math. Soc. 4 (1971), 140-146.

Gruson, L.-Jensen, C. U. [1] Modules algébriquement compacts et foncteurs $\varprojlim^{(i)}$, C. R. Acad. Sci. Paris 276 (1973), 1651-1653.

Hamsher, R. M. [1] On the structure of a one-dimensional quotient field, Journ. of Alg. 19 (1971), 416-425.

Hattori, A. [1] On Prüfer rings, J. Math. Soc. Japan 9 (1957), 381-385.

Herrmann, P. [1] Self-projective modules over valuation rings (to appear).

Kaplansky, I. [1] Modules over Dedekind rings and valuation rings, Trans. Amer. Math. Soc. 72 (1952), 327-340.

Kaplansky, I. [2] Projective modules, Ann. of Math. 68 (1958), 372-377.

Kaplansky, I. [3] The homological dimension of a quotient field, Nagoya Math. J. 27 (1966), 139-142.

Kaplansky, I. [4] Fields and Rings, Univ. of Chicago Press, Chicago 1969.

Klatt, G. B. - Levy, L. S. [1] Pre-self-injective rings, Trans. Amer. Math. Soc. 137 (1969), 407-419.

Lady, E. L. [1] Splitting fields for torsion-free modules over discrete valuation rings I, Journ. Alg. 49 (1977), 261-275.

Lady, E. L. [2] On classifying torsion-free modules over discrete valuation rings, Lecture Notes in Math. 616, Springer (1977), 168-172.

Lafon, J. P. [1] Sur un problème d'Irving Kaplansky, C. R. Acad. Sci. Paris, Ser. A. 268 (1969), 1309-1311.

Lafon, J. P. [2] Anneaux locaux commutatifs sur lesquels tout module de type fini est somme directe de modules monogènes, Journ. Alg. 17 (1971), 575-591.

Matlis, E. [1] Injective modules over Prüfer rings, Nagoya Math. J. 15 (1959), 57-69.

Matlis, E. [2] Divisible modules, Proc. Amer. Math. Soc. 11 (1960), 385-391.

Matlis, E. [3] Cotorsion Modules, Mem. Amer. Math. Soc. 49 (1964).

Matlis, E. [4] Decomposable modules, Tran. Amer. Math. Soc. 125 (1966), 147-179.

Matlis, E. [5] Torsion-free Modules, Univ. of Chicago Press, 1972.

Megibben, C. [1] Absolutely pure modules, Proc. Amer. Math. Soc. 26 (1970), 561-566.

Menini, C. [1] The minimal splitting field for a finite rank torsion-free module over an almost maximal valuation domain, Comm. in Alg. 11 (1983), 1803-1815.

Monari-Martinez, E. [1] On pure-injective modules, Proceedings of Conference on Abelian Groups and Modules in Udine, 1984.

Nishi, M. [1] On the ring of endomorphisms of an indecomposable injective module over a Prüfer ring, Hiroshima Math. J. 2 (1972), 271-283.

Ohm, J. - Vicknair, P. [1] Monoid rings which are valuation rings, Comm. in Alg. 11 (1983), 1355-1368.

Osofsky, B. [1] Global dimension of valuation rings, Trans. Amer. Math. Soc. 127 (1967), 136-149.

Osofsky, B. [2] Upper bounds on homological dimensions, Nagoya Math. J. 32 (1968), 315-322.

Osofsky, B. [3] Global dimension of commutative rings with linearly ordered ideals, J. London Math. Soc. 44 (1969), 183-185.

Osofsky, B. [4] Projective dimension of "nice" directed unions, J. Pure Applied Alg. 13 (1978), 179-219.

Rangaswamy, K. M. - Vanaja, N. [1] Quasi- projectives in abelian and module categories, Pac. J. Math. 43 (1972), 221-238.

Salce, L. [1] Moduli di torsione su domini di valutazione, Notes, Trento, 1980.

Salce, L. - Zanardo, P. [1] On a paper of I. Fleischer, Abelian Group Theory, Proceedings Oberwolfach Conference 1981, Lecture Notes in Mathematics 874, Springer (1981), 76-86.

Salce, L. - Zanardo, P. [2] Finitely generated modules over valuation rings, Comm. in Alg. 12 (1984), 1795-1812.

Schilling, O.F.G. [1] The Theory of Valuations, Math. Surveys, Vol. 4, Amer. Math. Soc., New York, 1950.

Schoeman, M. [1] Torsion-complete modules over a domain (to appear).

Shores, T. S. [1] On generalized valuation rings, Mich. J. Math. 21 (1974), 405-409.

Shores, T. S. - Lewis, W. J. [1] Serial modules and endomorphism rings, Duke Math. J. 41 (1974), 889-909.

Simmons, J. H. [1] Cyclic purity: a generalization of purity to modules, Thesis, Tulane University 1983.

Small, L. W. [1] Some remarks on the homological dimension of a quotient field, Mimeographed notes, Univ. of Cal., Berkeley 1966.

Soileau, P. L. [1] The tensor and torsion products of modules over valuation domains, Thesis, Tulane University 1984.

Stenström, B. [1] Pure submodules, Arkiv für Math. 7 (1967), 159-171.

Vámos, P. [1] Classical rings, J. Algebra 34 (1975), 114-129.

Vasconcelos, W. V. [1] Injective endomorphisms of finitely generated modules, Proc. Amer. Math. Soc. 25 (1970), 900-901.

Warfield, R. B., Jr. [1] Purity and algebraic compactness for modules, Pac. J. Math. 28 (1969), 699-719.

Warfield, R. B., Jr. [2] Decompositions of injective modules, J. Math. 31 (1969), 263-276.

Warfield, R. B., Jr. [3] Decomposability of finitely presented modules, Proc. Amer. Soc. 25 (1970), 167-172.

Warfield, R. B., Jr. [4] Relatively injective modules (unpublished manuscript, 1969).

Warfield, R. B., Jr. [5] A theory of cotorsion modules (unpublished manuscript, 1970).

Warfield, R. B., Jr. [6] Exchange rings and decompositions of modules, Math. Ann. 199 (1972), 31-36.

Warfield, R. B., Jr. [7] Serial rings and finitely presented modules, J. Algebra 37 (1975), 187-222.

Zanardo, P. [1] Valuation domains without pathological modules, (to appear).

Zanardo, P. [2] Indecomposable finitely generated modules over valuation domains (to appear).

Ziegler, M. [1] Model theory of modules, Ann. Pure Appl. Logic 26 (1984), 149-213.

Zimmermann-Huisgen, B. - Zimmermann, W. [1] Algebraically compact rings and modules, Math. Z. 161 (1978), 81-93.

Notation

Notation	Definition	Page
R	commutative ring with 1	
P	maximal ideal of valuation ring R	2
Q, Q(R)	ring (field) of quotients of R	3
U(R)	group of units of R	
Γ, $\Gamma(R)$	value group of R	7, 11
Γ^+	positivity domain of Γ	7
R_L	localization of R at prime ideal L	4
v	valuation of a field	10
N(R)	nilradical of R	2
Z(R)	set of zero-divisors of R	2
$H\Gamma_\lambda$	Hahn product of groups Γ_λ	8
$J : I = \{x \in Q \mid xI \subseteq J\}$		14
$M^\# = \{r \in R \mid rM < M\}$		15, 34, 144

$M_{\#} = \{r \in R \mid ra = 0 \text{ for some } 0 \neq a \in M\}$		15, 34, 144
$\text{End}_R M$	endomorphism ring of M	16
$\text{Aut}_R L$	automorphism group of L	16
I_F	$\cup\{$ideals I of R with R/I not maximal$\}$	22
$N \leq M$ $(N < M)$	N is (proper) submodule of M	31
$\langle \ldots \rangle$, $\langle \ldots \rangle_*$	submodule, pure submodule generated by $\ldots$	31, 48
$rM = \{ra \mid a \in M\}$, $IM = \{ra \mid r \in I, a \in M\}$		32
tM	torsion submodule of M	33
$\text{Ann } a$, $\text{Ann } M$	annihilator of a, M	32
$M[I] = \{a \in M \mid \text{Ann } a \geq I\}$, $M[I^+] = \{a \in M \mid \text{Ann } a > I\}$		32
$A(M) = \{\text{Ann } a \mid 0 \neq a \in M\}$		34, 141
dM, hM	divisible, h-divisible part of M	36, 37
∂	a particular divisible R-module	123
$M^1 = \cap\{rM \mid 0 \neq r \in R\}$	the first Ulm submodule of M	35
RDext	the module of RD-extensions	59
$p.d.M$, $i.d.M$	projective, injective dimension of M	73
$gl.d.R$	global dimension of R	73
$\tilde{M}$	R-completion of M	95
$\hat{M}$	RD-injective hull of M	212
$\overline{M}$	torsion-completion of M	235
$\overline{\overline{M}}$	torsion-ultracompletion of M	241
$M^\bullet$	cotorsion hull of M	245
$E(M)$	injective hull of M	130
$PE(M)$	pure-injective hull of M	219
$H_M(a)$	height ideal of a in M	157

$h_M(a)$	height of a in M	157
Σ	set of heights	157
$M^\sigma = \{a \in M \mid h_M(a) \geq \sigma\},\ M^{\sigma+} = \{a \in M \mid h_M(a) > \sigma\}$		159
$i_M(a)$	indicator of a in M	162
$i_M(a)_L = \sup\{rh(ra) \mid r \in RL\},\ i_M(a)^L = \inf\{rh(ra) \mid r \in L\}$		165
$\ell(M)$	length of finitely generated M	175
$g(M)$	Goldie dimension of M	178
$I(u) = \{a \in R \mid u \notin aS + R\}$		182
$\alpha_M(\sigma, I)$	(σ, I)-invariant of M	196
$\lvert \ldots \rvert$	cardinality of $\ldots$	
$\subseteq,\ \subset$	containment (proper containment)	
$\aleph_\alpha$	the αth infinite cardinal	
ω_α	the initial ordinal of cardinality $\aleph_\alpha$	
$\mathrm{cof}\ \alpha$	cofinality of α	

Author Index

Anderson, F. W.   31
Arnold, D.   301
Auslander, L.   73
Azumaya, G.   54

Bazzoni, S.   87, 88, 90, 93
Becker, T.   30
Brandal, W.   21, 23, 173

Cartan, H.   26, 57, 60, 66, 71, 130
Cherlin, G. L.   30
Corner, A. L. S.   301
Crawley, P.   53, 54

de la Rosa, B.   119, 120, 122, 139
de Marco, B.   302
Dickman, M. A.   30

Dimitrić, R.   286, 289, 290, 292-294, 296, 298, 299, 302

Eilenberg, S.   26, 57, 60, 66, 71, 130
Eklof, P. C.   75
Enochs, E.   249

Facchini, A.   231
Fakhruddin, S. M.   301
Fleischer, I.   55, 115, 131, 181, 194
Franzen, B.   273, 277, 278, 281, 286, 301
Fuller, K. R.   31

Gill, D. T.   6, 22, 132, 174, 193
Gruson, L.   231

AUTHOR INDEX

Hahn, H. 9
Hamsher, R. M. 117, 118
Harrison, D. K. 99
Hattori, A. 56, 72
Heinlein, G. 302
Herrmann, P. 90-93
Hill, P. 302
Hölder, O. 8

Jensen, C. U. 231
Jensen, R. B. 148
Jónsson, B. 53, 54

Kaplansky, I. 30, 55, 56, 71, 73, 76, 93, 174, 193, 278
Klatt, G. B. 26
Krull, W. 10-12, 30

Lady, E. L. 302
Lafon, J. P. 174, 193
Levy, L. S. 26
Lewis, W. J. 19, 141, 144-147, 155
Łoś, J. 297

Matlis, E. 15, 28, 37, 55, 60, 94, 96-99, 101, 102, 106, 107, 114-117, 121, 123, 131, 132, 134, 174, 193, 242, 248, 301
Megibben, C. 48
Menini, C. 279, 280
Monari-Martinez, E. 224, 225, 227, 232

Nishi, M. 14, 19, 131, 133
Nunke, N. 297

Ohm, J. 30
Orsatti, A. 302
Osofsky, B. 75, 76, 93
Ostrowski, A. 115

Rangaswamy, K. M. 92

Schilling, O. F. G. 5, 25
Schoeman, M. 234-238, 249
Shelah, S. 147, 155
Shores, T. S. 19, 30, 141, 144-147, 155
Simmons, J. H. 51-53, 56, 111, 193, 259-262, 268
Small, L. W. 76, 93
Soileau, P. L. 17, 67-69
Stenström, B. 212, 214, 231

Vámos, P. 132
Vanaja, N. 92
Vasconcelos, W. V. 188
Vicknair, P. 30
Viljoen, G. 139, 272, 281, 282, 284, 301

Warfield, R. B. Jr. 39, 43, 46, 47, 53-56, 100, 131, 210, 212-214, 219, 220, 222, 223, 226, 230-232, 242, 243, 245, 247, 249

Zanardo, P. 175-177, 179-181, 183, 185-188, 193, 194, 265, 268
Ziegler, M. 225, 231
Zimmermann, W. 219, 231
Zimmermann-Huisgen, B. 219, 231

Subject Index

α-invariants, 198
abelian group, totally ordered, 7
   archimedean, 8
   discrete, 8, 9
annihilator, filtration, 108, 110-112
   of element, 32
   of module, 32
   sequence, 174
Auslander's Lemma, 73

basic submodule, 203, 271
Bézout domain, 27

Cauchy net, 95
Cauchy sequence, 237
chain, continuous, 49
club, 75
composition series, pure-, 174

◇, 149

Dedekind domain, 27
defect, 229
dimension, global 73, 76
   Goldie, 178
   hd-, 121
   injective, 73, 134-136
   projective, 73, 75-87
divide, 35
divisibility, 35-38
   condition, 113
   relative, 39-43
domain, Bézout 27
   completely integrally closed, 6
   Dedekind, 27
   h-local, 28
   integrally closed, 6
   Prüfer, 27
   real-closed, 30
   valuation, 1

Eklof's Theorem, 75
element, of infinite height, 157
   positive, 7
   regular, 2

SUBJECT INDEX

strictly positive, 7
exchange property, 53
extension, immediate, 4, 12
  pure-essential, 219
  RD-essential, 211
  torsion-free essential,
    245, 247

field, residue-class, 2, 12
  splitting, 279-281
filter, 8
  prime, 8
  principal, 8
filtration, 107
  annihilator, 108, 110-112
  discrete, 108
  functorial, 108
  R-, 108
  topological, 108
  trivial, 108
finite intersection property, 108
Fleischer rank, 181

gap, 165
  proper, 165
Goldie dimension, 178

Hahn product, 8
height, 157-160
  -filtration, 160
  ideal, 157
  limit, 157
  non-limit, 157
homomorphism, locally
    extendable, 110
  partial, 215
hull, cotorsion, 245-248
  pure-injective, 219
  RD-injective, 211, 212

ideal, 13
  archimedean, 15
  equivalence of, 14
  fractional, 26

invertible, 26
topology, 101
increase on the left (right),
  165
indicator, 162-164
  irregular, 165
  irregularities of, 165-166
  smooth, 168
  ultimately constant
    non-limit, 264
invariants, 198

κ-filtration, 75
Kaplansky's Lemma, 73

length, of finitely generated
    module, 175
  of polyserial module, 189
limit, 95
  height, 157

Matlis duality, 99
module, 31
  A-projective, 90
  absolutely pure, 48
  algebraically compact, 215
  almost maximal, 112
  balanced-projective, 275
  bounded, 33
  cocyclic, 210
  coherent, 75
  cohesive, 162
  compact, 216
  completely decomposable,
    273-277
  cotorsion, 243-248
  cyclic, 33
  cyclically presented, 33
  cyclically-pure-projective
    260
  divisible, 36, 116-118
  F-complete, F-ultracomplete,
    109
  finitely cogenerated, 218
  finitely generated, 33, 42,
    174-178

finitely presented, 33, 43
  special, 46
flat, 71
fully transitive, 171
h-divisible, 37, 119-123
h-reduced, 37
homogeneous, 134, 262, 272
injective, 111, 130-134
κ-divisible, 118
κ-generated, 48
linearly compact, 216
maximal, 112
mixed, 33
polyserial, 189-194, 228-230, 251-253
  standard, 189
projective, 70-72
pure-injective, 214-220
pure-projective, 258
quasi-injective, 136-138
quasi-projective, 90-92
R-complete, 95-107
R-ultracomplete, 112-114
RD-injective, 100, 210-214
RD-projective, 41
reduced, 36
separable, 253-256, 276
  standard, 253
slender, 295-300
smooth, 168
splitting, 33
superdecomposable, 226
topological, 94
torsion, 33
torsion-complete, 233-238
torsion-free, 33, 77-82
  of finite rank, 277-281
torsion-ultracomplete, 238-242
transitive, 171
ultracomplete, 109, 220, 240
uniserial, 33, 140-147, 152, 223
  non-standard, 147-152
  standard, 141, 199

net, Cauchy, 95
nilradical, 2

object, countably small, 54
  small, 54
order-homomorphism, 7
order-isomorphism, 7

positivity domain, 7
power series, formal, 24
product, Hahn, 8
  lexicographic, 9
Prüfer domain, 27
pseudo-convergent, 109
pseudo-limit, 108
pure-composition series, 174
purification, 48

R-completion, 95
R-filtration, 108
R-topology, 95
rank, 269
  Fleischer, 181
residual, 14
residue class field, 2, 12
resolution, projective, 72
ring, formal power series, 24
  of quotients, 3
  valuation, 1

σ-invariant, 270
sequence, balanced-exact, 274
  bounded Cauchy, 237
  Cauchy, 237
  cyclically-pure-exact, 52
  ∂-exact, 125
  hd-exact, 119
  pure-exact, 45
  RD-exact, 40
set, u-independent, 182
  totally ordered, 7
socle, 34
stacked basis theorem, 44
subgroup, convex, 7
  dense, 8
submodule, 31
  α-basic, 201-205
  balanced, 274, 275

    basic   203, 271
    cyclically pure, 51, 111, 260
    dense,   95
    equiheight,   161
    f-closed,   109
    pure,   44-48
       generated by..., 48
    pure-essential,   219
    pure-independent family of, 202
    RD,   39
    RD-essential,   211
    relatively divisible,   39
    tight,   83
    torsion,   33
    torsion-free-essential,   245
    Ulm,   35, 65, 105
subset, stationary,   75
subspace, equivalence of,   284
support,   23
system, tight,   87-89

topology, linear,   95
torsion-completion,   235

torsion-ultracompletion,   241

Ulm submodule,   35, 65, 105

valuation,   10
   domain,   1, 11, 27
      discrete rank one,   3
   ring,   1, 151
      almost maximal,   20, 22,
         62, 92, 113, 114, 132,
         174, 278
      Artinian,   3
      maximal,   20, 22, 23, 92,
         278
      maximally complete,   5, 23
      Noetherian,   3
   value group,   11, 12

zero-divisor,   2

JUN 0 5 1989